Springer Series in Materials Science

Volume 290

The Springer Series in Materials Science covers the complete spectrum of materials research and technology, including fundamental principles, physical properties, materials theory and design. Recognizing the increasing importance of materials science in future device technologies, the book titles in this series reflect the state-of-the-art in understanding and controlling the structure and properties of all important classes of materials.

More information about this series at http://www.springer.com/series/856

Ihsan Boustani

Molecular Modelling and Synthesis of Nanomaterials

Applications in Carbon- and Boron-based Nanotechnology

Springer

Ihsan Boustani
Faculty of Mathematics and Natural Science
University of Wuppertal
Wuppertal, Nordrhein-Westfalen, Germany

ISSN 0933-033X ISSN 2196-2812 (electronic)
Springer Series in Materials Science
ISBN 978-3-030-32725-5 ISBN 978-3-030-32726-2 (eBook)
https://doi.org/10.1007/978-3-030-32726-2

This Springer imprint is published by the registered company Springer Nature Switzerland AG
The registered company address is: Gewerbestrasse 11, 6330 Cham, Switzerland

Preface

This book presents carbon- and boron-based nanomaterials predicted by molecular modelling and numerical simulation tools, then confirmed by modern experimental methods. It begins by summarizing basic theoretical methods, then showing the influence of the magic numbers of the atomic alkali metal clusters on the development of the clusters into nanostructures, theoretically and experimentally. It continues into a discussion of atomic clusters and nanostructures, focusing primarily on boron and carbon. It thoroughly explores the one-, two-, and three-dimensional structures of boron and carbon, closing with their potential applications in nanotechnology. These include myriad possibilities, from nano-coating and nano-sensing to nano-battery with high borophene capacity. Materials scientists will find the broad discussion of computational modelling as well as its specific application to boron and carbon an essential reference.

The evolution of small to nanosized clusters (nanosize $\approx$ the width of ten small abreast atoms) in form of nanotubes, nanowires, nano-sheets, nanocages induced the epitome nanomaterials and led consequently to nanostructured Materials-by-Design. That enables us to construct nanostructures with desired properties, like definite electron conductivity or the range of band gaps for applications in semiconductors and nanoelectronics, or neutron protection, absorption, and radiation shielding for nuclear reactors and aero-space vehicles and astronauts, or coating, hardening, resistance, anti-abrasion and anti-corrosion for surfaces and interfaces, or drug delivery and cancer therapy in medicine, or energetic materials which have a high amount of chemical energy that can be stored and released, or can be used as propulsion for carrier rockets, or finally molecular sensor or wheels to attract dirty molecules.

Conventionally and traditionally, the experiment observed, measured and announced always novel materials and related properties as Quasicrystals and Buckminster-Fullerenes discovered in 1984 and 1985 by Shechtman and Kroto, respectively. Furthermore, the Femtochemistry was developed by Zeweil in 1999, and the superconductivity of MgB_2 was observed in 2001 by Nagamatsu. However, the theory was always lagging behind to explain and reproduce the experimental data. However, the recent development of the theory and the rapid evolution of the computational tools reverse the research process and enabled the molecular modelling to establish new research fields to predict and simulate novel structures and properties, so far unknown in nature. Nowadays, there are a number of new materials, which are predicted and discovered by theory, more precisely by molecular modelling, and are already confirmed by experiments.

The success of the molecular modelling can be documented with many examples. For instance, Boustani, the author of this Book, predicted in the 80s and 90s the structures of lithium and sodium clusters, and the existence of planar boron clusters and planar sheets, known as borophene, as well as the existence of boron nanotubes. The predicted nanostructures were unknown in nature at that time. Around ten years later, these species were detected and confirmed experimentally. New crystalline form of boron was proposed and predicted by Oganov via theory, named γ-boron. This new boron solid was later experimentally observed.

All these novel and potential applications disembogue into the term nanotechnology. The interplay between the chemical modelling and the experimental verification was in the recent years successfully performed. It also demonstrates the near-term capability and viability of controlled synthesis for specific engineering of definite material properties.

Wuppertal, Germany Ihsan Boustani

Contents

Part I
Molecular Modelling

Chapter 1
Molecular Modelling

Abstract The basics of molecular modelling in molecular and quantum mechanics are presented. Starting from the classical molecular mechanics and molecular dynamics and related force field equations and algorithms, over the minimization and optimization procedures as well as the gradient and Newton–Raphson Methods, the relationship between the trajectory and the equation of motion are explained. The quantum mechanics is also the fundament of the modern quantum chemistry. The quantum chemistry methods, like the Hartree–Fock and post Hartree–Fock theory, the density functional theory and related most popular hybrid functionals, are introduced.

1.1 Concepts of Molecular Modelling

Molecular Modelling (Concept): Molecular modelling is simply a general term for computer simulation of atomic, molecular, cluster or solid systems to reproduce their geometrical and electronic structures on the one side, and chemical, physical, as well as mechanical properties, on the other side. Molecular modelling can depict one-, two-, or three-dimensional geometrical structures of in nature existing molecules or solids, or can predict and simulate novel structures. The procedure proceeds through five steps: (i) the identification, purpose and definition of the problem, (ii) constructing and building the model, (iii) carrying out the calculations, (iv) analyzing the evaluating the results, and finally, (v) significance or validity of results in comparison with theoretical or experimental data.

1.1.1 Areas of Molecular Modelling

Understanding and solving chemical problems, reactions, and geometrical structures of molecular systems is based on three areas. The first one is the determination of the molecular objects, which will be investigated, like molecules, clusters, DNA. The second area includes the mathematical algorithm and numerical methods to solve

I. Boustani, *Molecular Modelling and Synthesis of Nanomaterials*, Springer Series in Materials Science 290, https://doi.org/10.1007/978-3-030-32726-2_1

the physical laws like atomic and molecular interactions. The third area consists of the goals that to be achieved like chemical and physical properties, geometrical and electronic structures, ground and excited states and corresponding spectra, thermodynamical and activation energies and finally comparison between theoretical models and experimental evidence of real systems.

Molecular Modelling and Computational Chemistry: Molecular models can help to understand their geometric and electronic structures, their behavior and shape and to visualize their electron distribution. With the help of molecular methods one can also manipulate and simulate the structures and properties of molecules and can give insights into their dynamics and reactions. Nowadays, the concept of **Molecular Modelling** is purely tightly bonded to "**Computational Chemistry**", which uses mathematical approximations and computer software to solve chemical problems and simulate chemical reactions [1].

1.1.2 The Basics of Molecular Modelling

The Principle of Molecular Modelling: Calculating the total energy of the molecular systems, is the benchmark for any investigation. The total energy is the sum of the kinetic and potential energies. In order to find the ground state energy of certain geometrical configuration of a system there are of course different procedures, like Newton–Raphson or conjugate gradient methods, which carry out the energy minimization. This procedure occurs through energy lowering via displacements of the atomic positions as long as the energy minimum is achieved.

Molecular Mechanics and Quantum Mechanics: There are two well-known mathematical models in computer simulation: the classical or molecular mechanics and the quantum mechanics. The molecular mechanics is a numerical method for determining the geometrical structure of molecules or atomic clusters, while the quantum mechanics is a numerical method for calculating the motion in molecules or clusters. The inter-atomic interaction in both simulations can be calculated in two ways: (i) using classical potential function (this can be pairwise or multi center interaction) (ii) using quantum mechanical methods (this can be ab initio Hartree–Fock, density functional theory, semi-empirical or tight-binding methods, or any other quantum mechanical methods).

1.2 Molecular Mechanics

The term molecular mechanics refers to the use of Newtonian mechanics to model molecular systems. The potential energy of molecules in molecular mechanics, which can be called as positional energy, is calculated by using force field concept.

1.2.1 Force Field Equation

A force field is a set of functions and constants used to describe the potential energy of the molecule. This potential is composed of bonded contributions arising from bond vibrations, bond angle bending and dihedral angle or torsion around bonds, and of non-bonding contributions arising from van der Waals and Coulomb electrostatic interactions [2]:

$$E_{\text{pot}} = E_{\text{bond}} + E_{\text{angle}} + E_{\text{dihedral}} + E_{\text{vdW}} + E_{\text{el.static}} \tag{1.1}$$

This is the force field equation describing the components of the potential energy. The function of the potential energy is usually called the potential energy surface. The potential energy terms of the force field are respectively corresponding to:

$$E_{\text{bond}} = \sum_{r} K_b (r - r_0)^2 \text{.....} \; describes\ bond\ stretching\ energy \tag{1.2}$$

$$E_{\text{angle}} = \sum_{\theta} K_\theta (\theta - \theta_0)^2 \quad \text{.........} \quad bond\ angle\ deformation \tag{1.3}$$

$$E_{\text{dihedral}} = \sum_{\phi} K_\phi \left[1 + \cos(n\phi - \gamma)\right] \; \text{...} \; dihedral\ deformation \tag{1.4}$$

$$E_{\text{vdW}} = \sum_{ij} \left(\frac{A_{ij}}{r_{ij}^{12}} - \frac{B_{ij}}{r_{ij}^{6}} \right) \quad \text{........} \quad van\ der\ Waals\ interaction \tag{1.5}$$

$$E_{\text{electrostatic}} = \sum_{ij} \left(\frac{q_i q_j}{\varepsilon r_{ij}} \right) \quad \text{........} \quad Coulomb\ interaction \tag{1.6}$$

The stretching, bond angle, and torsion energies are those which deviate from the ideal bond distance, bond angle and dihedral angle of molecule. The non-bonding interaction energy contains dipole-dipole and van-der-Waals interaction energies. The force field molecular mechanics is described by a set of equations of motion, which are derived from the classical mechanics. The force field gives how the constituents of molecule relate to each other and how the forces determine the structure of the molecule.

The potential energy is a function of the nuclear positions of the conformation atoms: $E_{\text{pot}} = f(r)$. The nuclear positions can be expressed either in Cartesian representations, as independent variables, or however, in terms of internal coordinates expressed as a Z-Matrix containing bond stretching or lengths, bond angles and torsion or dihedral angles. The function E_{pot} of the potential energy does not include contributions of kinetic energy. The latter is made by the motion of the involved atoms and can be calculated using molecular dynamics based on the second Newton's law of motion $F = m\ \ddot{x}(t)$, which will be discussed in Sect. 1.3. The main objective or the prototypical application of molecular mechanics is to find the lowest energy conformation of a molecule.

1.2.2 Minimization Procedure

The search for the lowest energy is termed as energy minimization. The minimization of energy can be obtained through the determination the forces acting on the atoms. This can be achieved with the gradient of the potentials:

$$F = -\nabla\ E_{\text{pot}} = -grad\ E_{\text{pot}} \tag{1.7}$$

This equation gives the link between the force and mutual potential energy and presents the force as the first derivative of the energy. The energy minimization procedure begins with the first derivative, ∂E/∂r, for all coordinates (r) of the initial structure of the system determining the energy and the slope, direction, and magnitude of the function at a definite point, telling us the steepness of the slope. If the slope is positive, it is an indication that the coordinate is too large. And if the slope is negative, then the coordinate is too small. By a positive slope, the numerical minimization technique then adjusts or rather reduces the coordinate. The calculation of the energy and the slope will be repeated until the slope is near zero. Then the minimum has converged and the derivatives are close to zero and the forces are disappeared. The search for the energy minimum on the potential energy surface uses an iterative formula to work in a step-wise fashion. This formula is:

$$r_{\text{new}} = r_{\text{old}} + correction \tag{1.8}$$

The second derivative of the system ∂^2E/∂^2r, known as the Hessian matrix, describes the curvature of the potential energy surface at the coordinate r. Thus, the goal of the geometry optimization algorithm is to find the coordinates which correspond to the minimum value of energy as soon as the values of both derivatives are performed:

$$\forall\, i\ ,\ grad\ F = \frac{\partial E}{\partial r_i} = 0\ ,\ and\ \frac{\partial^2 E}{\partial^2 r_i} > 0 \tag{1.9}$$

That means, the force or the first derivative of energy is zero and the value of the second derivative of the energy is positive.

1.2.3 Optimization Methods

There are a few numbers of optimization methods, which are broadly categorized as local and global optimization methods using diverse optimization algorithms. The local optimization methods use the information from the neighborhood of the current approximation and always converge to the nearest local extremum close to the starting point. Thus, the local optimization methods are further categorized in gradient-independent algorithms, such as Downhill simplex methods, and in gradient-based algorithms such as least-squares fitting, steepest descent and Newton–Raphson methods. The global optimization methods can be classified as deterministic, stochastic and heuristic. Stochastic optimization algorithms incorporate some probabilistic elements in the objective function or the algorithm implementation. Some of popular stochastic optimization algorithms are simulated annealing, stochastic tunnelling and parallel tempering. The most popular global optimization algorithms are heuristic and meta-heuristic algorithms, which include evolutionary algorithms (genetic algorithms and evolutionary strategies), swarm-based optimization algorithms (particle swarm optimization and ant colony optimization) and Tabu[1] search [3].

Local Minimization Methods: As mentioned above, there are two local optimization categories to perform energy minimization or geometry optimization. The first category contains non-gradient methods, like Downhill and Powell simplex methods. The Downhill simplex method starts from an initial simplex. Each step of the method consists in an update of the current simplex. These updates are carried out using four operations: reflection, expansion, contraction, and multiple contraction. The second category contains the gradient-based minimization methods, which are related to the first-order and the second-order in the Taylor expansion of the potential energy surface corresponding to the first and second derivative of the energy [4].

1.2.4 Gradient Methods

The well-known gradient methods are the (i) steepest descent, (ii) conjugate gradient, and (iii) Newton–Raphson methods. The steepest descent method is a first-order minimization method which uses the information of the first derivative, the gradient $\vec{g}_k$, to find local minima on the potential energy surface. This method moves along the negative gradient $-\vec{g}_k$ downhill the energy landscape. The conjugate gradient requires also the first derivative and makes use of the gradient history to decide a better direction for the next optimization step. It does not exhibit a zigzag behavior during convergence, but still tends to be quite slow in very non-linear cases. The Newton–Raphson method is a second-order minimization procedure which needs the first and the second derivatives to find the minima [5].

[1]The word Tabu comes from the Tongan word to indicate things that cannot be touched because they are sacred.

Newton–Raphson for Dimer: The Newton–Raphson method is the most effective optimization technique. In order to take an impression of this method and to understand the tasks of the first and second derivatives, which are indicated respectively as gradient and Hessian matrix, one can introduce a simple system of a diatomic molecule of a bond length **r**, which is the only coordinate to be minimized [6]. In this case it is only to consider the potential energy $\mathbf{E}_{\text{bond}}$, which can be expressed in a harmonic function term of bond stretching:

$$E_{\text{bond}} = \frac{1}{2} K_b (r - r_0)^2 \tag{1.10}$$

where K_b is the force constant for the bond, and r_0 is the equilibrium bond length. By calculating the first derivative of E_{bond} one obtains:

$$\frac{\mathrm{d}E_{\text{bond}}}{\mathrm{d}r} = K_b (r - r_0) \tag{1.11}$$

and the second derivative of E_{bond}:

$$\frac{\mathrm{d}^2 E_{\text{bond}}}{\mathrm{d}r^2} = K_b \tag{1.12}$$

Solving the equation of the first derivative (1.11) for the change in bond length after $(r_0 - r)$ one gets:

$$(r_0 - r) = -\frac{1}{K_b} \frac{\mathrm{d}E_{\text{bond}}}{\mathrm{d}r} \tag{1.13}$$

After substitution K_b of (1.12) in the latter (1.13), a new coordinate for the harmonic potential is obtained:

$$(r_0 - r) = -\frac{1}{\frac{\mathrm{d}^2 E_{\text{bond}}}{\mathrm{d}r^2}} \frac{\mathrm{d}E_{\text{bond}}}{\mathrm{d}r} \tag{1.14}$$

1.2.5 General Newton–Raphson Method

Since the most force field terms are non-harmonic potentials, the derivatives of the potentials are not linear. However, many optimization steps are necessary to find minimum for the potential energy function E_{pot} of (1.1). Therefore, in order to generalize the minimization of this potential E_{pot}, one can consider instead an energy function ***f(x)*** and carry out the Taylor expansion around x_n, where $(x = x_n + \Delta x)$:

$$f_{\text{Taylor}}(x_n + \Delta x) = f(x_n) + f'(x_n)\Delta x + \frac{1}{2} f''(x_n)\Delta^2 x + \cdots \tag{1.15}$$

The Taylor expansion can be approximated to the second order polynomial around Δx. It attains its extremum when its derivative with respect to Δx is equal zero:

$$\Big[f(x_n + \Delta x) - f(x_n)\Big]/\Delta x = f'(x_n) + f''(x_n)\Delta x = 0 \tag{1.16}$$

solving the right-hand side of the this equation for Δx:

$$\Delta x = -f'(x_n)/f''(x_n) \tag{1.17}$$

Since $\Delta x = x - x_n$, then the final formula for new coordinates $\mathbf{x_{n+1}}$ is now:

$$x_{n+1} = x_n - f'(x_n)/f''(x_n), \quad n = 0, 1, \ldots \tag{1.18}$$

After comparison (1.8) and (1.18) and replacing **r** by **x**, we find the term $correction = -f'(x_n)/f''(x_n)$. Since the first derivative $f'(x_n)$ is the gradient $\nabla f(x_n)$, and the second derivative $f''(x_n)$ is the Hessian matrix $Hf(x_n)$, we introduce the inverse of the Hessian matrix as $[Hf(x_n)]^{-1}$ and rewrite the last equation as:

$$x_{n+1} = x_n - [Hf(x_n)]^{-1}\nabla f(x_n), \quad n \geq 0. \tag{1.19}$$

The use of second derivatives in Newton–Raphson minimization is responsible for the excellent convergence properties. However, the inversion of the Hessian is time consuming. An approach, named truncated Newton–Raphson, has been developed that uses conjugate gradients to determine the directions for the minimization and then the Hessian to determine the minimum in that direction. The "direction" of the minimization determines the particular bond lengths and angles that will be changed. The minimum in that "direction" determines how much to change those bond lengths and angles. Truncated Newton–Raphson method has similar and often better convergence characteristics to the non-truncated Newton–Raphson one without a significant difference in time [6, 7].

1.3 Molecular Dynamics

Equation of Motion for Dimer: The classical molecular dynamics is mainly based on the second Newton's law of motion. This kinetic energy is called also the motional energy. We have already seen that the forces are derived from the gradients of the interaction potentials, as given in (1.7). The Newton equation has now the following form:

$$m.a = F = -\nabla \ E(r) \tag{1.20}$$

For instance, a simple system for motion can be taken after considering $E(r) = k/2(r - r_0)^2$, which is the equation of motion of a dimer, as given in (1.10):

$$m.\frac{\mathrm{d}r^2}{\mathrm{d}t^2} = F = -\frac{\mathrm{d}E(r)}{\mathrm{d}r} = -kr \tag{1.21}$$

Thus, from the first and the fourth terms of (1.21) arises the differential equation:

$$m.\frac{\mathrm{d}r^2}{\mathrm{d}t^2} = -k.r \tag{1.22}$$

for which the solution after setting $w = \sqrt{k/m}$, is known:

$$r(t) = A\sin(wt) \tag{1.23}$$

1.3.1 *Trajectory and Equation of Motion*

In general, from the knowledge of the atomic forces and masses, it is possible to determine the acceleration of each atom in the system. Integration the equations of motion yields a solution that describes the positions, velocities and accelerations of the atoms along a series of extremely small time steps (on the order of femtoseconds $= 10^{-15}$ s). The resulting series of snapshots of structural changes over time is called a trajectory. From this trajectory, the average values of properties can be determined. The method is deterministic; once the positions and velocities of each atom are known, the state of the system can be predicted at any time in the future or the past. Thus, by the molecular dynamics simulation all atoms are described by the Newton's equation of motion:

$$F = m.a = m.\frac{\mathrm{d}^2r}{\mathrm{d}t^2} \tag{1.24}$$

where the acceleration **a** is the derivatives of the position **r** or of the velocity **v**:

$$a = \frac{\mathrm{d}v}{\mathrm{d}t} = \frac{\mathrm{d}^2r}{\mathrm{d}t^2} \tag{1.25}$$

By a constant acceleration **a** and after integration we obtain the following expression for the velocity:

$$v = a.t + v_0 \tag{1.26}$$

Since the velocity **v** is the first derivative of the position $v = \mathrm{d}r/\mathrm{d}t$, then by a constant velocity the position **r** can be obtained by an integration obtaining the following expression for the position:

$$r = v.t + r_0 \tag{1.27}$$

After insertion (1.26) in (1.27) we receive:

$$r = a.t^2 + v_0 t + r_0 \tag{1.28}$$

where v_0 and r_0 are the initial velocity, and position, respectively. The acceleration is given as the derivative of the potential energy with respect to the position **r**, as expressed in (1.21):

$$a = -\frac{1}{m}\frac{\mathrm{d}E}{\mathrm{d}r} \tag{1.29}$$

1.3.2 Initialization of Positions and Velocities

In order to calculate a trajectory one only needs the initial positions of the atoms, an initial distribution of velocities and the acceleration, which is determined by the gradient of the potential energy function. The equations of motion are deterministic, e.g., the positions and the velocities at the time zero (t_0) determine the positions and velocities at all other times (**t**). The initial positions simply correspond to the Cartesian coordinates of the initial geometry of the system, which can be obtained from experimental structures, such as the x-ray crystal structure [8]. The initial velocities, however, are generally obtained by selecting speeds from the Maxwell–Boltzmann distribution for a specified temperature, like 300 K or 1000 K, and then selecting random directions for the velocities. Thus, the Maxwell–Boltzmann distribution for the velocities, $F(v)$, is given as:

$$F(v) = 4\pi v^2 \left(\frac{m}{2\pi k_B T}\right)^{3/2} exp\left(\frac{-mv^2}{2k_B T}\right) \tag{1.30}$$

The maximum or the **m**ost **p**robable velocity $\mathbf{v}_{mp}$ corresponds to the maximum of the distribution, which can be determined by the derivative of $\mathbf{F}(\mathbf{v})$ after ($\mathbf{v}$) has the form:

$$v_{mp} = \left(\frac{2k_B T}{m}\right)^{1/2} \tag{1.31}$$

However, since the distribution of the velocities is not symmetric, the average velocity $\mathbf{v}_{\mathrm{ave}} = \langle \mathbf{v} \rangle$ of the distribution is in general relative larger than the most probable value $\mathbf{v}_{\mathbf{mp}}$:

$$\langle v \rangle = \left(\frac{8\ k_B\ T}{\pi\ m}\right)^{1/2} = \frac{2}{\sqrt{\pi}} v_{mp} \tag{1.32}$$

Temperature and Kinetic Energy: In the classical statistical mechanics, the law of energy equipartition remarks that the average kinetic energy of **N** particle is equal to the average translational kinetic energy $(3/2)Nk_BT$. Thus, the kinetic energy of a system at equilibrium at a given temperature **T** has the form:

$$E_{\text{kin}} = \frac{1}{2} m \langle v^2 \rangle = \frac{3}{2} N k_B T \tag{1.33}$$

The temperature can then be determined from the total kinetic energy of the system using the following formula:

$$T_K = \frac{1}{3Nk_B} \sum_i m_i v_i^2 \tag{1.34}$$

Note that as the kinetic energy fluctuates in time, the previous relation defines a so-called instantaneous temperature, which is also a time fluctuating quantity [9].

Taylor Expansions and Time Evolution The time evolution of a system takes place along discrete time steps $\mathbf{\Delta t}$. The trajectory of a particle is not the real one but in the single time step. After considering the derivatives for velocity and acceleration as:

$$\frac{\mathrm{d}r}{\mathrm{d}t} = v, \tag{1.35}$$

$$\frac{\mathrm{d}}{\mathrm{d}t}\left(\frac{\mathrm{d}r}{\mathrm{d}t}\right) = \frac{\mathrm{d}^2 r}{\mathrm{d}t^2} = \frac{\mathrm{d}v}{\mathrm{d}t} = a, \tag{1.36}$$

$$\frac{\mathrm{d}}{\mathrm{d}t}\left(\frac{\mathrm{d}v}{\mathrm{d}t}\right) = \frac{\mathrm{d}^2 v}{\mathrm{d}t^2} = \frac{\mathrm{d}a}{\mathrm{d}t} = b \tag{1.37}$$

Thus, the Taylor expansion of coordinates, velocities and accelerations can be written as follows:

$$r(t+\Delta t) = r(t) + \left(\frac{\mathrm{d}r}{\mathrm{d}t}\right)\Delta t + \frac{1}{2}\left(\frac{\mathrm{d}^2 r}{\mathrm{d}t^2}\right)\Delta t^2 + \frac{1}{3!}\left(\frac{\mathrm{d}^3 r}{\mathrm{d}t^3}\right)\Delta t^3 + O(\Delta t^4) \tag{1.38}$$

$$v(t+\Delta t) = v(t) + \left(\frac{\mathrm{d}v}{\mathrm{d}t}\right)\Delta t + \frac{1}{2}\left(\frac{\mathrm{d}^2 v}{\mathrm{d}t^2}\right)\Delta t^2 + O(\Delta t^3) \tag{1.39}$$

$$a(t+\Delta t) = a(t) + \left(\frac{\mathrm{d}a}{\mathrm{d}t}\right)\Delta t + O(\Delta t^2) \tag{1.40}$$

The truncation of Taylor expansions of (1.38) and (1.39) can be done at different levels. Let us consider the derivatives $\mathrm{d}r/\mathrm{d}t = v$ and $\mathrm{d}r^2/\mathrm{d}t^2 = a$ for velocity and acceleration, and approximate the expansions to the first order. Then we receive the **Euler's algorithm**, which is the simplest method for solving the equations of motion:

$$r(t + \Delta t) = r(t) + v(t)\ \Delta t \tag{1.41a}$$

$$v(t + \Delta t) = v(t) + a(t)\ \Delta t \tag{1.41b}$$

The Euler algorithm can determine the new atomic positions $r(t_1)$ starting from given initial coordinates $r(t_0)$ and velocities $v(t_0)$ as well as accelerations $a(t_0)$, which can be calculated from the force field potential of equation (1.29) and then it can be replaced in the second (1.41b):

$$v(t + \Delta t) = v(t) - \frac{1}{m}\left(\frac{\mathrm{d}E}{\mathrm{d}r}\right)\ \Delta t \tag{1.42}$$

1.3.3 Verlet Algorithm

The next level of approximation is to truncate the Taylor expansion in (1.39) to the second order. Then we write down the approximate expansion of $\mathbf{r(t)}$ for a step forward ($+\Delta\mathbf{t}$) and for a step backward ($-\Delta\mathbf{t}$):

$$r(t + \Delta t) = r(t) + v(t)\Delta t + \frac{1}{2}\ a(t)\Delta t^2 \tag{1.43}$$

$$r(t - \Delta t) = r(t) - v(t)\Delta t + \frac{1}{2}\ a(t)\Delta t^2 \tag{1.44}$$

These equations are the basics of Verlet algorithm, which is the most used class of algorithms in molecular dynamics. By adding and subtracting both equations we receive respectively:

$$r(t + \Delta t) = 2r(t) - r(t - \Delta t) + a(t)\Delta t^2 \tag{1.45}$$

$$v(t) = \left[r(t + \Delta t) - r(t - \Delta t)\right] / 2\Delta t \tag{1.46}$$

Verlet algorithm uses the positions and accelerations at time $\mathbf{t}$ and the positions from the time $t - \Delta t$ to calculate the new positions at the time $t + \Delta t$. These new positions of equation (1.45) are of course calculated up to the second order in Δt. The velocity is not explicitly calculated, but it can be derived from the knowledge of the trajectories $r(t + \Delta t)$ and $r(t - \Delta t)$ at every time step $\mathbf{t}$, as given in (1.46). However, alternatively the velocity can be adopted from (1.40) after considering the Taylor expansion to the first order. Resolving (1.40) after $\mathrm{d}a/\mathrm{d}t$ and replacing it by $\mathrm{d}^2v/\mathrm{d}t^2$ in (1.39), we receive:

$$v(t + \Delta t) = v(t) + \frac{1}{2}\left[a(t) + a(t + \Delta t)\right]\Delta t \tag{1.47}$$

1.3.4 Leap-Frog Algorithm

Leapfrog integration is a first order Taylor expansion and similar to Euler integration but more accurate. The Leapfrog integration is equivalent to updating positions $r(t)$ and velocities $v(t)$ at interleaved time points, staggered in such a way that they 'leapfrog' over each other so that the position is updated at integer time steps and the velocity is updated at integer-plus-a-half time steps. However, we take the Euler equation (1.41a) and update the velocity at $\frac{1}{2}\Delta t$ as follows:

$$r(t + \Delta t) = r(t) + v\left(t + \frac{1}{2}\Delta t\right)\Delta t \tag{1.48}$$

The velocity $v(t + \Delta t)$ of (1.41b) can be determined by considering the Taylor expansion equation of equation (1.39) at the first order for $\left(t + \frac{1}{2}\Delta t\right)$ and $\left(t - \frac{1}{2}\Delta t\right)$:

$$v(t + \frac{1}{2}\Delta t) = v(t) + \left(\frac{dv}{dt}\right)\frac{\Delta t}{2} \tag{1.49}$$

$$v(t - \frac{1}{2}\Delta t) = v(t) - \left(\frac{dv}{dt}\right)\frac{\Delta t}{2} \tag{1.50}$$

Subtracting and rearranging both (1.49) and (1.50) then we receive:

$$v(t + \frac{1}{2}\Delta t) = v(t - \frac{1}{2}\Delta t) + a(t)\ \Delta t \tag{1.51}$$

The velocity of equation (1.51) can be calculated at $t + \frac{1}{2}\Delta t$ by means of two quantities, the mid-step velocity $v\left(t - \frac{1}{2}\Delta t\right)$ and the acceleration $a(t)$.

These calculated velocities together with the stored current positions $r(t)$ are used to determine the new positions $r(t + \Delta t)$ of (1.48). In this way, the velocities leap over the positions, then the positions leap over the velocities. Since the velocities are updated at half time steps and leap ahead the positions, the current velocities can be obtained by adding (1.49) and (1.50) to get:

$$v(t) = \frac{1}{2}\left[v\left(t + \frac{1}{2}\Delta t\right) + v\left(t - \frac{1}{2}\Delta t\right)\right] \tag{1.52}$$

1.4 Ab Initio Quantum Chemical Methods

Ab Initio (Latin: from the beginning) methods also called first principle methods are rigorous calculations without using any empirical parameters. The Ab Initio methods distinguish itself from other molecular mechanical methods that they are solely based on the established laws of quantum mechanics. These methods can be applied for small systems (e.g. tens of atoms) or for periodic systems. Because of its high accuracy, the computations are expensive and extensive. Semi-empirical methods are similar to the latter one, but less rigorous, using mostly experimental parameters and more extensive approximations. They are computationally less demanding than the Ab Initio methods and are able calculate systems of some hundreds of atoms. Over the last two decades powerful algorithmic tools and program codes have been developed which are capable to predict structures, energies, reactivities and many properties of high accuracy [10].

1.4.1 Some Basics of Quantum Mechanics

In classical mechanics, the goal was always to find the trajectory of a particle $x(t)$ in relationship with a force acting on this particle and provided by the Newton's second law $F = m\,\ddot{x}(t)$. In quantum mechanics, the goal is to find the wavefunction $\Psi(x, t)$ or precisely its probability density for a particular system. Thus, using the classical description of the total energy as the sum of the kinetic energy and the potential energy, the analogue of Newton's second law is the Schrödinger's equation for a quantum system:

$$\hat{H}\Psi = E\Psi \tag{1.53}$$

where $\hat{H}$ is the Hamiltonian, Ψ is the wavefunction, and E is the total energy of the system. This equation was proposed in 1926 by Erwin Schrödinger, which is the basic non-relativistic wave equation governing the motion of nuclei and electrons in molecules. It is nothing else as a differential (eigenvalue) equation for the energy E and the wavefunction Ψ of a particular state. The kinetic and the potential energies are expressed by the Hamiltonian or the Hamilton operator $\hat{H}$ (see next paragraph), while the wavefunction Ψ depends on the Cartesian and spin coordinates of the component particles. Thus, this equation can be solved using fundamental physical constants and geometry of the molecule as well. In order to determine the energy of heavy elements which incorporate a large number of core electrons, it is necessary to include the relativistic effects and spin-orbit coupling. The relativistic generalization of this equation was proposed in 1928 by Dirac. The standard Ab Initio shorthand notation outlines the theory used Hartree–Fock (HF), Configuration Interaction (CI), and the atomic orbitals (Basis Sets).

Kinetic Energy in Quantum Mechanics: In classical mechanics, the kinetic energy and momentum $\hat{p} = m.v$ are given as:

$$T_{cm} = \frac{1}{2}m \cdot v^2 = \frac{\hat{p}^2}{2m} \tag{1.54}$$

Thus, the correspondence of momentum between the classical and quantum mechanics is given through the operator $\hat{p} = i\hbar\frac{\partial}{\partial x}$. Thus, the kinetic energy operator in quantum mechanics has the following form:

$$\hat{T}_{qm} = \frac{1}{2m}\left[i\hbar\frac{\partial}{\partial x}\right]^2 = -\frac{\hbar^2}{2m}\frac{\partial^2}{\partial x^2} \tag{1.55}$$

For a three-dimensional system becomes the kinetic energy operator the following expression:

$$\hat{T}_{qm} = -\frac{\hbar^2}{2m}\left[\frac{\partial^2}{\partial x^2} + \frac{\partial^2}{\partial y^2} + \frac{\partial^2}{\partial z^2}\right] \tag{1.56}$$

1.4.2 The Hamiltonian

The Hamiltonian is composed of the kinetic and potential energies of the system. Since the molecular system contains electrons (**e**) and nuclei (**n**), which are interacting together, we introduce the terms of the Hamiltonian as the sum of the kinetic energies of nuclei and electrons, the nuclear-nuclear and electron-electron repulsion energies, and the nuclear-electron attraction energies, in a different form of energy operators:

$$\hat{H} = \hat{T}_n + \hat{T}_e + \hat{V}_{nn} + \hat{V}_{ne} + \hat{V}_{ee} \tag{1.57}$$

This is the energy equation describing the components of the kinetic and potential energy. The kinetic and the potential operators are respectively corresponding to:

$$\hat{T}_n = -\frac{1}{2M_A}\sum_A \nabla_A^2 \quad \ldots\ldots\ldots \; kinetic\ energy\ operator\ of\ nuclei \tag{1.58}$$

$$\hat{T}_e = -\frac{1}{2}\sum_i \nabla_i^2 \quad \ldots\ldots\ldots \; kinetic\ energy\ operator\ of\ electrons \tag{1.59}$$

$$\hat{V}_{nn} = \sum_{A<B} \frac{Z_A Z_B}{R_{AB}} \quad \ldots\ldots\ldots \; nuclear - nuclear\ repulsion\ energy \tag{1.60}$$

$$\hat{V}_{ne} = -\sum_{iA} \frac{Z_A}{R_{iA}} \quad \ldots\ldots\ldots\ldots \; nuclear - electron\ attraction\ energy \tag{1.61}$$

$$\hat{V}_{ee} = \sum_{i<j} \frac{1}{r_{ij}} \quad \ldots\ldots\ldots\ldots\ldots \; electron - electron\ repulsion\ energy \tag{1.62}$$

where **M** and **Z** are the mass and atomic number of the atom **A**, ∇_A^2 is the Laplacian operator, given as $\nabla_A^2 = \frac{\partial^2}{\partial^2 x} + \frac{\partial^2}{\partial^2 y} + \frac{\partial^2}{\partial^2 z}$, R_{AB} is the distance between the nucleon **A** and **B**, $\mathbf{r_{iA}}$ is the distance between the electron **i** and the atom **A**, and $\mathbf{r_{ij}}$ is the distance between the electrons **i** and **j**. For example, the Hamilton operator of equation (1.57) for Hydrogen atom reduces simply to the sum of two terms:

$$\hat{H}_{\text{Hydrogen}} = \hat{T}_e + \hat{V}_{ne} = -\frac{1}{2}\nabla_r^2 - \frac{1}{r} \tag{1.63}$$

because the operators $\hat{T}_n = \hat{V}_{nn} = \hat{V}_{ee} = 0$, are equal zero.

1.4.3 Time-Dependent Schrödinger Equation

As mentioned above, the Schrödinger equation is the basis for the quantum mechanics. The solution of this equation is known as the wavefunction. This single-particle three-dimensional time-dependent Schrödinger equation is:

$$i\hbar \frac{\partial \Psi(r,t)}{\partial t} = \hat{H}\,\Psi(r,t) = \left[-\frac{\hbar^2}{2m}\nabla^2 + V(r)\right]\Psi(r,t) \tag{1.64}$$

where $\Psi(r, t)$ is the spatial and temporal dependent wavefunction, $\hat{H}$ is the Hamiltonian, $V(r)$ is the potential. If we write the wavefunction as a product of spatial and temporal terms:

$$\psi(\mathbf{r}, t) = \psi(\mathbf{r}).f(t) \tag{1.65}$$

then inserting (1.65) in (1.64), one yields the following form:

$$i\hbar\,\Psi(r)\frac{\partial f(t)}{\partial t} = f(t)\left[-\frac{\hbar^2}{2m}\nabla^2 + V(r)\right]\Psi(r) \tag{1.66}$$

The left side of (1.66) can be reordered explicitly in terms of time (t), while the right side only in terms of (r).

$$i\hbar\frac{1}{f(t)}\frac{\mathrm{d}f(t)}{\mathrm{d}t} = \frac{1}{\psi(r)}\left[-\frac{\hbar^2}{2m}\nabla^2 + V(r)\right]\psi(r) \tag{1.67}$$

If both sides (dependent upon different variables) are to be equal, then they both must be equal to the same constant. The latter can be called E. Thus, the left side of (1.67) becomes equivalent to the time-dependent Schrödinger equation,

$$i\hbar\frac{\mathrm{d}f(t)}{\mathrm{d}t} = E \cdot f(t) \tag{1.68}$$

while the right side of (1.67) becomes equivalent to the time-independent Schrödinger equation:

$$\left[-\frac{\hbar^2}{2m}\nabla^2 + V(r)\right]\psi(r) = E\ \psi(r) \tag{1.69}$$

The Hamiltonian is a Hermitian operator and the eigenvalues of a Hermitian operator must be real, so E is real. We turn now to the time-dependent equation (1.68), which can be rewritten as:

$$\frac{\mathrm{d}f(t)}{f(t)} = \frac{-i}{\hbar}E \cdot \mathrm{d}t \tag{1.70}$$

This equation is the time-dependent Schrödinger equation. After integration we receive:

$$\int \frac{\mathrm{d}f(t)}{f(t)} = \frac{-i}{\hbar}E \cdot \int \mathrm{d}t \ \Rightarrow \ f(t) = e^{\left(-\frac{iE\cdot t}{\hbar}\right)} \tag{1.71}$$

since $E = h\nu = \hbar\omega$ we get:

$$f(t) = e^{-i\cdot\omega\cdot t} \tag{1.72}$$

This is the solution of the time-dependent Schrödinger equation. If this solution is replaced in (1.65), then one obtains the total wavefunction in the form:

$$\Psi(r,t) = \psi(r) \cdot e^{-i\cdot\omega\cdot t} \tag{1.73}$$

This means that the total wavefunction $\Psi(r,t)$ differs from $\Psi(r)$ only by a phase factor of constant magnitude. One of the consequences is that the quantity $|\Psi(r,t)|^2$ is time independent:

$$|\Psi(r,t)|^2 = \Psi^*(r,t)\ \Psi(r,t) = e^{i\cdot\omega\cdot t}\Psi^*(r)e^{-i\cdot\omega\cdot t}\Psi(r) = \Psi^*(r)\Psi(r) \tag{1.74}$$

The equation (1.73) represents a particular solution of (1.64). The general solution of the Schrödinger equation (1.64) is a linear combination of these particular solutions:

$$\Psi(r,t) = \sum_i c_i\ e^{-i\cdot\omega_i\cdot t}\ \psi(r) \tag{1.75}$$

1.4.4 Hartree–Fock Theory

In quantum chemistry [11], the Hartree–Fock theory (HF) is one of the simplest approximation for solving the many-body Hamiltonian. It seeks the solution of the electronic Schrödinger equation, which results from the time-independent Schrödinger equation (1.69), after invoking the Born–Oppenheimer (BO) approximation, and considering the electronic wavefunction only. The BO approximation implies that the nuclear motion is very slow so that it can be considered as rigid, in opposite to the motion of the electrons, which move much faster than the (vibrational and rotational) motion of nuclei. This means that the term $\hat{T}_n$ in (1.57), which represents the kinetic energy of the nuclei, will be neglected, while the term $\hat{V}_{nn}$, which expresses the repulsion of the nuclei, will be set as constant. Thus, the Hamiltonian is now:

$$\hat{H}_{BO} = \hat{T}_e + \hat{V}_{ne} + \hat{V}_{ee} \tag{1.76}$$

The operators $\hat{T}_e$ and $\hat{V}_{ne}$, which express the kinetic energy and the attraction of an electron to a nuclei, represent dynamic variables that depend only on the position of individual electrons, and therefore they are called one-electron operator:

$$\hat{H}_1 = \hat{T}_e + \hat{V}_{ne} = -\frac{1}{2}\sum_i \nabla_i^2 - \sum_{iA} \frac{Z_A}{R_{iA}} \tag{1.77}$$

The operator $\hat{V}_{ee}$ represents the coulomb interaction between two electrons and therefore it called now the two-electron operator:

$$\hat{H}_2 = \hat{V}_{ee} = \sum_{i<j} \frac{1}{r_{ij}} \tag{1.78}$$

The Electronic Wavefunction: The electronic wavefunction can be obtained by separation the wavefunction into its electronic and nuclear components:

$$\Psi_{\text{Total}} = \Psi_{el} \times \Psi_{nu} \tag{1.79}$$

Thus, the Schrödinger equation (1.69) transforms to the Hartree–Fock equation after considering the Hamilton operator $\hat{H}_{MO}$ of (1.76), and (1.78), as well as the electronic wavefunction Ψ_{el} of (1.79), and the corresponding eigenvalue E of energy:

$$\hat{H}\ \Psi_{el}(r) = E\ \Psi_{el}(r) \tag{1.80}$$

This Hartree–Fock equation is also the basis of the molecular orbital theory, which describes each electron motion by a single-particle function or orbital $\phi(r)$ explicitly independent of the motions of other electrons. The assignment of electrons to orbitals is called the electron configuration of the atom. Hartree proposed firstly that the total wavefunction of the system with N electrons is a product of one-electron orbitals, known as Hartree product:

$$\psi_{\text{Hartree}}(r_1, r_2, \ldots, r_N) = \psi(r_1)\psi(r_2)\cdots\psi(r_N), \tag{1.81}$$

Due to the antisymmetry principle (synonymous with the Pauli exclusion principle) that electrons must be described by wavefunctions, which are antisymmetric with respect to interchange of the coordinates (including spin) of a pair of electrons.

$$\Psi(r_1, r_2, \ldots, r_i, \ldots, r_j, \ldots, r_N) = -\Psi(r_1, r_2, \ldots, r_j, \ldots, r_i, \ldots, r_N) \quad (1.82)$$

The Hartree–Fock method, which is an iterative self-consistent-field procedure, known as (SCF), is based on an anti-symmetrical many-electron wavefunction represented by Slater determinant consisting of one-electron spin orbitals ψ_i:

$$\Psi(r_1, r_2, \ldots, r_N) = \mathscr{A}\left[\psi_1(r_1)\psi_2(r_2)\cdots\psi_N(r_N)\right] \quad (1.83)$$

where $\mathscr{A}$ is the antisymmetrizer, r_i is the set of space and spin coordinates of electron i, and N is the number of the electrons. This is the essential characteristic of the HF method, in which the exact many-electron wavefunction is approximated by an antisymmetrical product of one-electron wavefunctions, besides the electron-electron repulsion, which is treated in an average way. The anti-symmetrical Slater determinant equation (1.83) can be explicitly written as a product of one-electron orbitals:

$$\Psi = \frac{1}{\sqrt{N!}}\begin{vmatrix} \psi_1(r_1), & \psi_1(r_2), & \ldots, & \psi_1(r_N) \\ \psi_2(r_1), & \psi_2(r_2), & \ldots, & \psi_2(r_N) \\ & \ldots & & \\ \psi_N(r_1), & \psi_N(r_2), & \ldots, & \psi_N(r_N) \end{vmatrix} \quad (1.84)$$

The HF spin orbitals ψ_i are described by linear combinations of atomic orbitals ϕ_μ (LCAO), which are generally fixed sets of Slater or Gaussian basis functions:

$$\psi_i = \sum_\mu c_{\mu i}\,\phi_\mu \quad (1.85)$$

where ψ_i is a molecular orbital of number i represented as the sum of μ atomic orbitals ϕ_μ, each multiplied by a corresponding coefficient $c_{\mu i}$, which indicates the weights of the contributions of the μ atomic orbital to the i molecular orbital.

The Fock Operator: Using the bra-ket notation for the wavefunction, (1.80) can be written as follows:

$$\hat{H}\,|\Psi_{el}\rangle = E\,|\Psi_{el}\rangle \quad (1.86)$$

Multiplying the left-hand side by $\langle\Psi_{el}|$, with $\langle\Psi_{el}|\Psi_{el}\rangle = 1$, then one yields:

$$\left\langle\Psi_{el}|\hat{H}|\Psi_{el}\right\rangle = \langle\Psi_{el}|E|\Psi_{el}\rangle = E \quad (1.87)$$

where $\hat{H}$ is the Hamilton operator of equation (1.76), and Ψ is the wavefunction represented by the Slater determinant (1.84). Since the Hamilton operator is the sum of one-electron and two-electron operators, we consider first $\hat{T}_e + \hat{V}_{ne}$, the one-electron operators of equation (1.77), and receive the following equation:

$$E_1 = \left\langle \Psi | \hat{H}_1 | \Psi \right\rangle = \sum_i^n \left\langle \psi_i | -\frac{1}{2}\nabla_i^2 - \sum_A^N \frac{Z_A}{R_{iA}} | \psi_i \right\rangle$$

$$= \sum_i^n \langle i | h_i | i \rangle = \sum_i^n \hat{h}_i \tag{1.88}$$

where n and N are the numbers of electrons and atoms, respectively, $R_{iA} = (r_i - R_A)$ is the distance between the atom A and the electron i. Now one considers the two-electron operator $\hat{H}_2$ of (1.78) and get:

$$E_2 = \left\langle \Psi | \hat{H}_2 | \Psi \right\rangle = \sum_{ij} \left\langle \psi_i(x_1)\psi_j(x_2) | \frac{1}{r_{ij}} | \psi_i(x_1)\psi_j(x_2) \right\rangle$$

$$- \sum_{ij} \left\langle \psi_i(x_1)\psi_j(x_2) | \frac{1}{r_{ij}} | \psi_j(x_1)\psi_i(x_2) \right\rangle$$

$$= \frac{1}{2} \sum_{ij} \underbrace{\langle ij | ij \rangle}_{\text{Coulomb}} - \underbrace{\langle ij | ji \rangle}_{\text{Exchange}} \tag{1.89}$$

Due to the antisymmetrization of the Hartree-product, the two-electron operator V_{ee} of (1.78) splits into a Coulomb expression $\langle ij | ij \rangle = J_{ij}$, which corresponds to the potential V_{ee}^{Coul}, and into an exchange expression $\langle ij | ji \rangle = K_{ij}$, which corresponds to the potential V_{ee}^{Ex}. J_{ij} and K_{ij} are assigned as Coulomb and exchange integrals.

The Exact Hartree–Fock Exchange Energy: As mentioned above, the exchange integral $K_{ij} = \langle ij | ji \rangle$, which presents the exact Hartree–Fock exchange energy, can be explicitly written as a sum over (ij) electrons of all integrals:

$$E_x^{HF} = -\frac{1}{2} \sum_{ij} \left\langle \psi_i(x_1)\psi_j(x_2) | \frac{1}{r_{ij}} | \psi_j(x_1)\psi_i(x_2) \right\rangle \tag{1.90}$$

Thus, the electrostatic potentials of the one-electron & two-electron operators are now:

$$V_c = V_{ne} + V_{ee}^{Coul} + V_x^{HF} \tag{1.91}$$

The electrostatic potential V_C represents the sum of the Coulomb attraction between the nuclei and electrons, the Coulomb repulsion between the electrons, and the exchange potential arose by the antisymmetrization of the Hartree-product.

Finally, when we consider the one-electron and two-electron operators of the equations (1.88) and (1.89), then we receive the Fock operator:

$$\hat{F}_i = \sum_i^n \hat{h}_i + \frac{1}{2} \sum_{ij}^n (J_i - K_i) \tag{1.92}$$

Since the total energy of the electronic system is a function of the many-electron wavefunction (Slater determinant) and of the nuclear positions:

$$E^{HF} = E\left[\Psi\{r, R\}\right] \tag{1.93}$$

one applies the variational principle in order to minimize the expectation value E for fixed nuclei:

$$\partial E/\partial \Psi = 0 \tag{1.94}$$

This procedure leads to the nonlinear Hartree–Fock equations with $\hat{F}_i$ as the Fock operator, after considering the kinetic operator in (1.88) and the electrostatic potential operators of equation (1.91), and with ψ_i the one-electron wavefunctions:

$$\left[-\frac{1}{2}\nabla_i^2 + V_c(x_i)\right]\psi_i(x_i) = \hat{F}_i\ \psi_i(x_i) = \varepsilon_i\ \psi_i(x_i) \tag{1.95}$$

This equation can not be solved exactly, because $\hat{F}_i$ depends on the solutions ψ_i, which are not exactly known. An approximate eigenfunction Ψ can be obtained when $c_{\mu i}$, the expansion coefficients in (1.85), are determined self-consistently, which is nothing else as the self-consistent-field procedure of Hartree–Fock (HF-SCF).

1.4.5 Post Hartree–Fock

In the Hartree–Fock theory, each electron is moving in an average potential field of all nuclei and other remaining electrons, and the prompt influence of electrons that come close together at some point is not taken into account. Electrons repel each other, and try to stay far away from each other. Therefore, their motion is *correlated* and thus this correlation energy reduces the energy of the system because it reduces the electron-electron repulsion. The post Hartree–Fock methods are developed to improve the Hartree–Fock self-consistent field (HF-SCF) method to determine the electron correlation energy. There are many different methods available to approximate and optimize the correlated wavefunction, like the Möller–Plesset (MP) perturbation theory, configuration interaction (CI), or coupled cluster (CC) methods. Also the methods of density functional theory (DFT), which we will discuss in Sect. 1.5, take into account electron correlation, even though in a less systematic and less well defined way than Ab Initio methods.

Configuration Interaction (CI): The Hartree–Fock equations converge to the exact solution through systematic improvements of the many-electron wavefunction. That can happen, for example, by increasing the dimension of the basis sets and, and above all, by applying the configuration interaction (CI) method, which considers the missing correlation energy in the HF approximation. The CI method goes a step further beyond the Hartree–Fock approach. The many-electron wavefunction in the CI method can be expanded into a linear combination of Slater determinants:

$$|\Phi\rangle = |\Psi_0\rangle + \sum_{ra} C_a^r |\Psi_a^r\rangle + \sum_{r<b} \sum_{r<s} C_{ab}^{rs} |\Psi_{ab}^{rs}\rangle \tag{1.96}$$

where Ψ_0 is the Hartree–Fock ground state reference determinant of (1.83), $C_{i\ldots}^{f\ldots}$ are the expansion coefficients, the sets $i \in \{a, b, \ldots\}^{occ}$ and $f \in \{r, s, \ldots\}^{virt}$ indicate the interchange of the initial (occupied) and final (virtual) orbitals, Ψ_a^r, Ψ_{ab}^{rs}, ..., are the singly, doubly, ..., N-tuply excited determinants, which can be constructed with help of the second quantization:

$$|\Psi_a^r\rangle = \hat{a}_r^\dagger \hat{a}_a |\Psi_0\rangle \;\; ; \;\; |\Psi\rangle_{ab}^{rs} = \hat{a}_s^\dagger \hat{a}_b \, \hat{a}_r^\dagger \hat{a}_a |\Psi_0\rangle \tag{1.97}$$

where $\hat{a}_i^\dagger$ and $\hat{a}_i$ are the creation and the annihilation operators, respectively. The operators $\hat{a}_a$ and $\hat{a}_b$, remove the occupied spin orbitals from the Slater determinant and the operators $\hat{a}_r^\dagger$ and $\hat{a}_s^\dagger$, replace them by unoccupied spin orbitals. In other words, the annihilation and creation operators promote or excite the electrons from the occupied orbitals into the unoccupied or virtual orbitals. Subsequently, based on the variational principle, the total energy in the CI approach can be achieved by diagonalizing the full CI matrix in the basis of the N-electron wavefunction Φ in (1.96), which is a linear combination of Slater determinants:

$$E^{CI} = \left\langle \Phi | \hat{H} | \Phi \right\rangle \tag{1.98}$$

The correlation energy can then be obtained after subtraction the Hartree–Fock energy E^{HF} from the total electronic energy E^{CI}:

$$E_c^{CI} = \left\langle \Phi | \hat{H} | \Phi \right\rangle - \left\langle \Psi_0 | \hat{H} | \Psi_0 \right\rangle = E^{CI} - E^{HF} \tag{1.99}$$

Finally, the exact energy of the system consists of the ground and excited states energies, which are the eigenvalues of the Hamiltonian matrix with the elements $\hat{H}_{ij} = \left\langle \psi_i | \hat{H} | \psi_i \right\rangle$. However, since every element of the set $\{|\psi_i\rangle\}$ can be defined by specifying a "configuration" of spin orbitals from which it is formed, this procedure is known as configuration interaction. When the configurations are generated from a single reference determinant allowing single and double (SD) excitations, then one obtains the so-called SD-CI method. Otherwise, when the generation of configurations by SD excitations is done with respect to several reference configurations, the CI expansion is called then the multi-reference-excitation MRD-CI method [12–14]. This method offers in general several advantages, for example, including higher than double excitations and is suitable for the calculation of more than one root of the secular equation.

As the number of electrons increases and the set of basis functions expands, the number of multiply-excited configurations increases dramatically and the CI-Hamiltonian matrix exceeds the dimension of 10^6. Thus, the corresponding Hermitian matrix cannot be computationally stored and therefore cannot be diagonalized. An effective method to overcome these difficulties is the direct CI method, which is proposed by Roos, and Siegbahn [15]. It is based on the calculation of the CI expansion coefficients directly from the given list of molecular one- and two-electron integrals avoiding the construction of a huge CI Hamiltonian matrix.

Coupled Cluster Method (CC): The coupled cluster theory, as developed by Cizek and Shavit et al. [16, 17], provides the exact solution of the time-independent Schrödinger equation (1.87). The wavefunction of the coupled-cluster theory is written as an exponential Ansatz:

$$|\Psi\rangle = e^{\hat{T}}\,|\Psi_0\rangle \tag{1.100}$$

where $|\Psi_0\rangle$ is the reference wavefunction, which is typically a Slater determinant constructed from the Hartree–Fock molecular orbitals, and $\hat{T}$ is the cluster excitation operator which, when acting on $|\Psi_0\rangle$, produces a linear combination of excited determinants from the reference wave function. In the formalism of the second quantization, the excitation operators can be written as a linear combination of single, double, triple, ..., up to N fold excitations for an N electron system:

$$\hat{T} = \hat{T}_1 + \hat{T}_2 + \hat{T}_3 + \cdots + \hat{T}_N \tag{1.101}$$

where

$$\hat{T}_1 = \sum_{a}^{occ}\sum_{r}^{virt} t_r^a \hat{a}_r^\dagger \hat{a}_a \tag{1.102}$$

is the single excitation operator in which the operator $\hat{a}_a$ annihilates an electron from the occupied orbital a, while the operator $\hat{a}_r^\dagger$ creates an electron in the virtual orbital r.

$$\hat{T}_2 = \sum_{ab}^{occ}\sum_{rs}^{virt} t_{rs}^{ab} \hat{a}_r^\dagger \hat{a}_s^\dagger\, \hat{a}_a \hat{a}_b \tag{1.103}$$

is the double excitation operator in which the operators $\hat{a}_a$ and $\hat{a}_b$ annihilate two electrons from the orbitals a and b, while the operators $\hat{a}_s^\dagger$ and $\hat{a}_r^\dagger$ create two electrons in the virtual orbitals r and s, respectively. The single and double excitations have been also applied by the configuration interaction method as in (1.97).

$$\hat{T}_3 = \sum_{\text{abc}}^{occ}\sum_{\text{rst}}^{virt} t_{\text{rst}}^{abc} \hat{a}_r^\dagger \hat{a}_s^\dagger \hat{a}_t^\dagger\, \hat{a}_a \hat{a}_b \hat{a}_c \tag{1.104}$$

is the triple excitation operator in which the operators $\hat{a}_a$, $\hat{a}_b$, and $\hat{a}_c$ annihilate three electrons from the orbitals a, b, and c, while the operators $\hat{a}_r^\dagger$, $\hat{a}_s^\dagger$, and $\hat{a}_t^\dagger$ create three electrons in the virtual orbitals r, s, and t, respectively. The unknown coefficients t_r^a, t_{rs}^{ab}, and t_{rst}^{abc} are to be determined in order to find the coupled cluster wavefunction. In analogous way one can write the quadruple $\hat{T}_4$ excitation and higher excitations.

Taylor Expansion of e-Operator: The exponential operator $e^{\hat{T}}$ of (1.100) can be developed in a Taylor expansion:

$$e^{\hat{T}} = 1 + T + \frac{\hat{T}^2}{2} + \frac{\hat{T}^3}{3!} + \frac{\hat{T}^4}{4!} + \cdots = \sum_{N=0}^{\infty} \frac{\hat{T}^N}{N!} \tag{1.105}$$

After substituting $\hat{T}$ of (1.101) in (1.105) we get:

$$\begin{aligned} e^{\hat{T}} =& 1 + (\hat{T}_1 + \hat{T}_2 + \hat{T}_3 + \cdots) + \frac{1}{2}(\hat{T}_1 + \hat{T}_2 + \hat{T}_3 + \cdots)^2 \\ &+ \frac{1}{3!}(\hat{T}_1 + \hat{T}_2 + \hat{T}_3 + \cdots)^3 + \frac{1}{4!}(\hat{T}_1 + \hat{T}_2 + \hat{T}_3 + \cdots)^4 + \cdots \end{aligned} \tag{1.106}$$

Reordering all terms of excitations:

$$\begin{aligned} e^{\hat{T}} =& 1 + \hat{T}_1 + (\hat{T}_2 + \frac{1}{2}T_1^2) + (\hat{T}_3 + \hat{T}_1\,\hat{T}_2 + \frac{1}{3!}\hat{T}_1^3) \\ &+ (\hat{T}_4 + \frac{1}{2}T_2^2 + \frac{1}{2}T_1^2\hat{T}_2 + \hat{T}_1\,\hat{T}_3 + \frac{1}{4!}\hat{T}_1^4) + \cdots \end{aligned} \tag{1.107}$$

By abbreviation of these terms we receive:

$$\begin{aligned} \hat{C}_1 =& \hat{T}_1 \\ \hat{C}_2 =& \hat{T}_2 + \frac{1}{2}T_1^2 \\ \hat{C}_3 =& \hat{T}_3 + \hat{T}_1\,\hat{T}_2 + \frac{1}{3!}\hat{T}_1^3 \\ \hat{C}_4 =& \hat{T}_4 + \frac{1}{2}T_2^2 + \frac{1}{2}T_1^2\hat{T}_2 + \hat{T}_1\,\hat{T}_3 + \frac{1}{4!}\hat{T}_1^4 \\ &\cdots \end{aligned} \tag{1.108}$$

representing the single, double, triple and quadruple excitations. Thus, the exponential operator (1.105) becomes the new form:

$$e^{\hat{T}} = 1 + \hat{C}_1 + \hat{C}_2 + \hat{C}_3 + \hat{C}_4 + \cdots \tag{1.109}$$

Coupled Cluster Equation: We write the Schrödinger equation:

$$\hat{H}|\Psi\rangle = E|\Psi\rangle \tag{1.110}$$

inserting (1.100) in (1.110) we receive the coupled cluster Schrödinger equation:

$$\hat{H} e^{\hat{T}} |\Psi_0\rangle = E\, e^{\hat{T}} |\Psi\rangle_0 \tag{1.111}$$

Multiplying this equation on the left by $e^{-\hat{T}}$ and $\langle\Psi_0|$, considering $e^{-\hat{T}}\, e^{\hat{T}} = 1$, then we receive the coupled cluster energy E^{CC}:

$$\langle\Phi_0|e^{-\hat{T}} \hat{H}\, e^{\hat{T}} |\Phi_0\rangle = E\langle\Phi_0|\Phi_0\rangle = E^{CC} \tag{1.112}$$

The product $e^{-\hat{T}} \hat{H}\, e^{\hat{T}}$ can be written down using Hadamard's formula in the Lie algebra, also called Baker–Campbell–Hausdorff formula [18]:

$$\widehat{H} = e^{-\hat{T}} \hat{H}\, e^{\hat{T}} = \hat{H} + \left[\hat{H}, \hat{T}\right] + (1/2)\left[\left[\hat{H}, \hat{T}\right], \hat{T}\right] + \cdots \tag{1.113}$$

These resulting equations are a set of non-linear equations which are to be solved in an iterative manner. The coupled cluster correlation energy can be determined in a similar manner to E_c^{CI} of (1.99), by subtracting the Hartree–Fock energy from the coupled cluster energy of equation (1.112):

$$E_c^{CC} = E^{CC} - E^{HF} \tag{1.114}$$

1.5 Density Functional Theory

1.5.1 Kohn–Sham Equations

By contrast, the density functional theory (DFT) employs the single particle electron density $\rho(r)$ as a basis quantity for the investigation of the electronic structure of many-electron systems. In this approach, and according to the first Hohenberg–Kohn theorem [19], that the ground state energy of a system with N electrons and M nuclei is a functional of the ground state electron density function $\rho(r)$:

$$E^{DFT} = E\left[\rho(r)\right] \tag{1.115}$$

where r is the spatial variable. This equation can also tell that the ground state energy of a system can be determined as soon as the ground state density is known. The electron density itself is simply given by:

$$\rho(r) = \sum_i n_i |\psi_i(r)|^2 \tag{1.116}$$

where $\psi_i(r)$ is the one-electron wavefunction, n_i are the occupation numbers. The basic idea of the DFT is that all parts of the energy, kinetic as well as electrostatic, are expressed in terms of electron density. Thus, the electron-nuclear attraction potential V_{en} is given by an external potential $V_{\text{ext}}(r)$, and the electron-electron interaction as well as the kinetic energy of electrons are given by the energy functional $F[\rho(r)]$:

$$E[\rho(r)] = \int V_{\text{ext}}(r)\,\rho(r)\,\mathrm{d}r + F[\rho(r)] \tag{1.117}$$

The $F[\rho]$ is the universal functional of density. It is separated into the classical Coulomb energy V_{ee} and the functional of density $G[\rho(r)]$:

$$F[\rho(r)] = \frac{1}{2}\int \frac{\rho(r)\,\rho(r')}{|r-r'|}\mathrm{d}r\,\mathrm{d}r' + G[\rho(r)] \tag{1.118}$$

The functional $G[\rho(r)]$ contains the terms of the "residual" kinetic energy and the exchange and correlation energy:

$$G[\rho(r)] = T_s[\rho(r)] + E_{xc}[\rho(r)] \tag{1.119}$$

By substitution the functionals $F[\rho]$ and $G[\rho(r)]$ of (1.118) and (1.119) in (1.117), we obtain the energy E as a functional of the density $\rho(r)$:

$$E[\rho(r)] = T_s[\rho(r)] + \int V_{\text{ext}}(r)\,\rho(r)\,\mathrm{d}r + \frac{1}{2}\int \frac{\rho(r)\,\rho(r')}{|r-r'|}\mathrm{d}r\,\mathrm{d}r' + E_{xc}[\rho(r)] \tag{1.120}$$

after adding the Coulomb repulsion of the ions (nuclei) U_{nn}, we obtain the total ground state energy of the system:

$$E[\rho] = T_s[\rho] + V_{en}[\rho] + V_{ee}[\rho] + E_{xc}[\rho] + U_{nn} \tag{1.121}$$

For a fixed external potential $V_{\text{ext}}(r)$, the energy functional is a minimum when it is evaluated at the true density. With the use of the variational principle of Hohenberg and Kohn, with respect to the density $\rho(r)$, we receive from (1.120) the Euler equation:

$$\partial E/\partial\rho = \delta T_s/\delta\rho + V_{\text{ext}}(r) + \Phi_{\text{coul}}(r) + \delta E_{xc}/\delta\rho = \mu \tag{1.122}$$

where the Lagrange multiplier μ is the chemical potential associated with the condition that the number of electrons is kept constant:

$$\int \rho(r)\,\mathrm{d}r = N \tag{1.123}$$

and $\Phi_{\text{coul}}(r)$ is the classical Coulomb potential:

$$\Phi_{\text{coul}}(r) = \int \frac{\rho(r')}{|r - r'|} \, \mathrm{d}r' \tag{1.124}$$

which can be determined via Poisson's equation

$$\nabla^2 \Phi_{\text{coul}}(r) = 4\,\pi\,\rho(r) \tag{1.125}$$

However, (1.122) is exactly obtained if we consider a collection of non-interacting electrons moving in an external effective potential:

$$V_{\text{eff}}(r) = V_{\text{ext}}(r) + \Phi_{\text{coul}}(r) + V_{xc}\left[\rho(r)\right] \tag{1.126}$$

with the exchange-correlation:

$$V_{xc}\left[\rho(r)\right] = \delta E_{xc}\left[\rho(r)\right] / \delta\rho(r) \tag{1.127}$$

The solution of (1.126) can be found using the Schrödinger equation of independent non-interacting particles:

$$\left[-\frac{1}{2}\nabla^2 + V_{\text{eff}}(r)\right] \psi_i(r) = E\,\psi_i(r) \tag{1.128}$$

These are the Kohn–Sham (KS) equations, which can be solved exactly, if the term E_{xc} is known. The wavefunctions ψ_i can be easily obtained by the self-consistent equations and hence the density $\rho(r)$. The convergence of KS equations to the exact solution can be achieved by permanent improvements of the potential V_{xc}. The KS equations include, as in the HF theory, no empirical or adjusted parameters so that the DFT can be considered as Ab Initio theory.

1.5.2 Local (Spin) Density Approximation

The search for accurate approximations of exchange-correlation density functional $E_{xc}[\rho]$ seems to encounter enormous difficulties in the DFT. Kohn and Sham [20] proposed the local density approximation (LDA) for exchange and correlation of the total energy, if the density $\rho(r)$ is varying sufficiently slowly:

$$E_{xc}^{LDA}(\rho) = \int \rho \cdot \varepsilon_{xc}(\rho) \, \mathrm{d}r \tag{1.129}$$

the corresponding exchange-correlation potential $V_{xc}(\rho)$ can be yield upon the variation with respect to the density $\rho(r)$ of (1.127):

$$V_{xc}^{LDA}(\rho) = \frac{d\left[\rho \cdot \varepsilon_{xc}(\rho)\right]}{d\rho} \tag{1.130}$$

where $\varepsilon_{xc}(\rho)$ is the exchange-correlation energy per electron of an interacting homogeneous electron gas of density $\rho(r)$. The energy $\varepsilon_{xc}(\rho)$ can be separated into exchange and correlation energies:

$$\varepsilon_{xc}(\rho) = \varepsilon_x(\rho) + \varepsilon_c(\rho) \tag{1.131}$$

Exchange Energy: After separation of the exchange-correlation energy, and considering the Dirac exchange functional for a single particle [21]:

$$E_x^{Dir}(\rho) = \int \rho \cdot \varepsilon_x(\rho)\,\mathrm{d}r = C_x \int \rho^{4/3}\,\mathrm{d}r \tag{1.132}$$

with the Dirac–Slater exchange potential:

$$\varepsilon_x(\rho) = C_x \cdot \rho^{1/3} \tag{1.133}$$

where

$$C_x = -\frac{3}{4}\left(\frac{3}{\pi}\right)^{1/3} \tag{1.134}$$

substituting (1.133) in (1.130), then the corresponding exchange potential becomes the form:

$$V_x^{LDA}(\rho) = -\left(\frac{3}{\pi}\right)^{1/3} \rho^{1/3} \tag{1.135}$$

Using the generalization of Kohn–Sham scheme of equation (1.129) for spin-polarized systems, which are also called open-shell systems, the exchange-correlation energy $\varepsilon_{xc}(\rho)$ can be split up into functionals of spin up density ρ^α and of spin down density ρ^β, with $\rho = \rho^\alpha + \rho^\beta$, leading to the local spin density (LSD) approximation [22]:

$$E_{xc}^{LDA}\left[\rho^\alpha, \rho^\beta\right] = \int \rho(r) \cdot \varepsilon_{xc}(\rho^\alpha, \rho^\beta)\,\mathrm{d}r \tag{1.136}$$

The terms of the exchange-correlation energy per electron of (1.131) become the spin polarized form:

$$\varepsilon_{xc}(\rho^\alpha, \rho^\beta) = \varepsilon_x(\rho^\alpha, \rho^\beta) + \varepsilon_c(\rho^\alpha, \rho^\beta) \tag{1.137}$$

The exchange energy in the LSD approximation can be separated into density of α and β spin:

$$\begin{aligned} E_x^{LDA}\left[\rho^\alpha, \rho^\beta\right] &= \frac{1}{2}E_x\left[\rho^\alpha, \rho^\alpha\right] + \frac{1}{2}E_x\left[\rho^\beta, \rho^\beta\right] \\ &= \frac{1}{2}E_x^0\left[2\rho^\alpha\right] + \frac{1}{2}E_x^0\left[2\rho^\beta\right] \end{aligned} \tag{1.138}$$

where

$$E_x^0[\rho] = E_x\left[\frac{1}{2}\rho, \frac{1}{2}\rho\right] \tag{1.139}$$

and

$$\frac{1}{2}\rho = \rho^\alpha = \rho^\beta \tag{1.140}$$

using the Dirac potential of equation (1.132) we obtain:

$$E_x^{LDA}\left[\rho^\alpha, \rho^\beta\right] = 2^{1/3}\, C_x \int \left[(\rho^\alpha)^{4/3} + (\rho^\beta)^{4/3}\right] \mathrm{d}r \tag{1.141}$$

Thus the exchange energy per particle of spin σ (α or β) can be written as:

$$\varepsilon_x(\rho^\sigma) = 2^{1/3}\, C_x\, (\rho^\sigma)^{1/3} \tag{1.142}$$

the corresponding exchange potential takes the following form after substituting the functional $\varepsilon_x(\rho^\sigma)$ in (1.136) with respect to (1.130):

$$V_x^\sigma(r) = -2\left(\frac{3}{4\pi}\right)^{1/3} (\rho^\sigma)^{1/3} \tag{1.143}$$

Correlation Energy: Concerning the correlation energy, different potentials have been suggested. Two of these approximations have been proposed by Vosko, Wilk and Nusair (VWN) [23] and also by Gunnersson and Lundqvist [24]. The VWN potentials were made by a very careful analysis of random phase approximation (RPA) and developed a formula for the accurate description of the correlation energy $\varepsilon_c(r)$. This approach is based on the calculations of uniform electron gas, which had been made by Ceperley and Alder [25] using the Quantum Monte Carlo method. The final form of the VWN formula for the correlation energy is:

$$E_c^{LDA} = \int \rho(r_s)\, \varepsilon_c(r_s)\, \mathrm{d}r \tag{1.144}$$

where

$$\begin{aligned}\varepsilon_c(r_s) = A\Bigg\{ & ln\left(\frac{x^2}{X(x)}\right) + \left(\frac{2b}{Q}\right)\tan^{-1}\left(\frac{Q}{2x+b}\right) \\ & - \left(\frac{bx_0}{X(x_0)}\right)\left[ln\left(\frac{(x-x_0)^2}{X(x)}\right)\right. \\ & \left. + \left(\frac{2(b+2x_0)}{Q}\right)\tan^{-1}\left(\frac{Q}{2x+b}\right)\right]\Bigg\}\end{aligned} \tag{1.145}$$

where A is a known constant from the Random Phase Approximation, $x = r_s^{1/2}$, the quantity r_s, with the electron density ρ, is the radius of a sphere whose volume is the effective volume of an electron:

$$r_s = \left(\frac{3}{4\,\pi\,\rho}\right)^{1/3} \tag{1.146}$$

$X(x) = x^2 + bx + c$, and $Q = (4c - b^2)^{1/2}$. The parameters, x_0, b and c are optimized to fit the Ceperley–Alder results for all values $x(r_s, \rho)$ for the spin and non-spin polarized systems.

1.5.3 Non-local Density-Gradient Corrections

However, in the LSD approximation, it is well known that the exchange energy is underestimated and the correlation energy is overestimated. Therefore, the non-local density-gradient corrections to the exchange-correlation energy should be added to the LSD energy, in order to take into account the inhomogeneity of the electron density. The non-local corrections to the exchange- correlation energy can be written in the form [26]:

$$E_{xc} = E_{xc}^{LDA} + E_x^G + E_c^G \tag{1.147}$$

where E_x^G and E_c^G are the gradient corrections to the exchange and correlation energies, respectively. The widely common and used non-local corrections, which have been developed in the last few years, proposed by different theoreticians are already implemented in various program codes, which will be mentioned in Sect. 1.9.

Becke Exchange Energy: The non-local exchange energy correction is proposed by Becke 1988 [27]:

$$E_x^{B88} = -\beta \sum_{\sigma} \int \rho_\sigma^{4/3} \frac{x_\sigma^2}{1 + 6\,\beta\,x_\sigma\,\sin h^{-1} x_\sigma}\,\mathrm{d}r \tag{1.148}$$

with

$$x_\sigma = \frac{|\nabla \rho_\sigma|}{\rho\sigma^{4/3}} \tag{1.149}$$

where β is a constant, x_0 is an dimensionless parameter of spin σ, and $\nabla\rho$ is the gradient of the electron density.

Perdew–Wang Exchange Energy: Another well known approach for the correct exchange energy is the so-called generalized gradient approximation (GGA) of Perdew and Wang [28]:

$$E_x^{PW} = C_x \int \rho^{3/4} F(s)\,\mathrm{d}r \tag{1.150}$$

with

$$F(s) = (1 + 1.296\, s^2 + 14\, s^4 + 0.2\, s^6)^{1/15} \tag{1.151}$$

and

$$s = \frac{|\nabla \rho|}{2k_F \rho}\ , \quad k_F = (3\pi^2 \rho)^{1/3} \tag{1.152}$$

Perdew Correlation Energy: We turn now to the non-local gradient corrections of the correlation energy. The E_c^G energy model is based on the momentum space analysis developed by Langreth and Mehl [29] via the generalized gradient expansion. This model has been later refined and proposed by Perdew [30] to get the functional:

$$E_c^P = \int \rho\, \varepsilon_c(\rho)\, \mathrm{d}r + \int \mathrm{d}^{-1}\, e^{-\Phi}\, C(\rho)\, \frac{|\nabla \rho|^2}{\rho^{4/3}}\, \mathrm{d}r \tag{1.153}$$

where

$$\Phi = 1.745\, \vec{f}\, \left[C(\infty)/C(\rho)\right] \frac{|\nabla \rho|^2}{\rho^{7/6}} \tag{1.154}$$

and

$$d = 2^{1/3}\left[\left(\frac{1+\zeta}{2}\right)^{5/3} + \left(\frac{1-\zeta}{2}\right)^{5/3}\right]^{1/2} \tag{1.155}$$

with

$$\zeta = \frac{(\rho^\alpha - \rho^\beta)}{\rho}\ , \quad \rho = \rho^\alpha + \rho^\beta \tag{1.156}$$

and

$$C(\rho) = 0.001667 + \frac{(0.002568 + \alpha\, r_s + \beta\, r_s^2)}{(1 + \gamma\, r_s + \delta\, r_s^2 + 10^4\, \beta\ r_s^3)} \tag{1.157}$$

and where r_s is that of (1.146), $\alpha = 0.023266$, $\beta = 7.389 \times 10^{-6}$, $\gamma = 8.723$, $\delta = 0.472$, and finally, the cutoff parameter $\vec{f} = 0.11$ has been chosen to fit the exact correlation energy of the neon atom.

Lee–Yang–Parr correlation Energy: Based on the idea of Colle and Salvetti [31], a new GGA correlation energy formula E_c was developed by Lee, Yang and Parr (LYP) [32]. This formula contains several empirical parameters which are fitted to produce the empirical correlation of He:

$$\begin{aligned} E_c^{LYP} = &- a \int \frac{1}{1 + d\, \rho^{-1/3}} \Big\{ \rho + b\, \rho^{-2/3} \Big[C_F \rho^{5/3} \\ &- 2\, t_W + \Big(\frac{1}{9}\, t_W + \frac{1}{18} \nabla^2 \rho\Big) \Big] e^{-c\, \rho^{-1/3}} \Big\}\, \mathrm{d}r \end{aligned} \tag{1.158}$$

and

$$\begin{aligned}E_c^{LYP} = & -a\int \frac{\gamma(r)}{1+d\,\rho^{-1/3}}\Big\{\rho + 2b\,\rho^{-5/3}\Big[2^{2/3}C_F\rho_\alpha^{8/3}\\ & + 2^{2/3}C_F\rho_\beta^{8/3} - \rho\, t_W + \frac{1}{9}\Big(\rho_\alpha\, t_W^\alpha + \rho_\beta\, t_W^\beta\Big)\\ & + \frac{1}{18}\Big(\rho_\alpha\nabla^2\rho_\alpha + \rho_\beta\nabla^2\rho_\beta\Big)\Big]e^{-c\,\rho^{-1/3}}\Big\}\,\mathrm{d}r\end{aligned} \tag{1.159}$$

with

$$C_F = \frac{3}{10}\left(3\pi^2\right)^{2/3}, \quad \gamma(r) = 2\Big[1 - \frac{\rho_\alpha^2(r) + \rho_\alpha^2(r)}{\rho^2(r)}\Big] \tag{1.160}$$

and the parameters $a = 0.049$, $b = 0.132$, $c = 0.2533$, and $d = 0.394$, and

$$t_W = \frac{1}{8}\Big[\frac{|\nabla\rho(r)|^2}{\rho(r)} - \nabla^2\rho(r)\Big] \tag{1.161}$$

is the Weizsäcker kinetic-energy density. The equations (1.158) and (1.159) are the second-order gradient expansion formulas for closed and open shells, respectively.

Wilson–Levy Correlation Functional: Another non-local correlation energy functional is recently developed by Wilson and Levy (WL) [33]:

$$E_c^{WL}\left[\rho_\alpha, \rho_\beta\right] = \int \frac{\left(a\rho + b|\nabla\rho|/\rho^{1/3}\right)\left(1-\zeta^2\right)^{1/2}}{c + d\left(|\nabla\rho_\alpha|/\rho_\alpha^{4/3} + |\nabla\rho_\beta|/\rho_\beta^{4/3}\right) + r_s}\mathrm{d}r \tag{1.162}$$

where ζ and r_s are those of (1.156) and (1.146), respectively. The parameters are $a = -0.7486$, $b = 0.06001$, $c = 3.60073$, $d = 0.9$.

In an bench mark, most of density functional methods above and elsewhere, which include local or nonlocal corrections to exchange and correlation, were extensively studied and tested to atomic systems by San Fabián et al. [34]. They organized different approaches of density functionals into four groups: the local density functionals, the self-interaction corrections, the gradient corrections and finally the methods of the correlation factor. They calculated the correlation energy (CE) of atoms, of the first and the second period, their first ionization potential (IP), and their electron affinities using the density obtained from Ab Initio calculations and then they compared results with experimental data.

1.5.4 Hybrid Functionals

The basic idea behind the hybrid functionals is to mix exchange energies in an exact one like the Hartree–Fock exchange energy of equation (1.90) with those obtained from the non-local density approximations, in order to improve their quality. The general form for hybrid functionals is:

$$E_{xc}^{hybrid} = \mu\big(E_x^{HF} - E_x^{GGA}\big) + E_{xc}^{GGA} \tag{1.163}$$

where μ is the coefficient, which determines the amount of exact-exchange mixing and is fitted semi-empirically. There is a lot of hybrid functionals, which are nowadays in use.

(B3LYP) Functionals: The most popular exchange-correlation functional is the hybrid of three parameter potential of Becke 1988 [27] and Lee–Yang–Parr [32]:

$$\begin{aligned} E_{xc}^{B3LYP} &= E_x^{LDA} + a_0(E_x^{HF} - E_x^{LDA}) + a_x(E_x^{GGA} - E_x^{LDA}) \\ &+ E_c^{LDA} + a_c(E_c^{GGA} - E_c^{LDA}) \end{aligned} \tag{1.164}$$

which can be written as:

$$\begin{aligned} E_{xc}^{B3LYP} &= a_0 E_x^{HF} + (1 - a_0 - a_x)E_x^{LDA} + a_x E_x^{GGA} \\ &+ a_c E_c^{GGA} + (1 - a_c)E_c^{LDA} \end{aligned} \tag{1.165}$$

where $a_0 = 0.2$, $a_x = 0.72$, and $a_c = 0.81$. Substituting these parameters in (1.165), considering E_x^{HF} the exact Hartree–Fock exchange energy of equation (1.90), the Dirac formula E_x^{Dir} of (1.132) for the local density exchange energy term E_x^{LDA} of (1.165), the Becke exchange energy E_x^{B88} of (1.148) for the E_x^{GGA} term, the Lee–Yang–Parr energy E_c^{LYP} of (1.158) and (1.159) for the E_c^{GGA} term, and finally, the Vosko–Wilk–Nusair correlation energy E_c^{VWN} of (1.144) for the E_c^{LDA} term. Then we receive:

$$E_{xc}^{B3LYP} = 0.2\ E_x^{HF} + 0.08\ E_x^{Dir} + 0.72\ E_x^{B88} + 0.81\ E_c^{LYP} + 0.19\ E_c^{VWN} \tag{1.166}$$

(PBE) Functionals: The Perdew–Burke–Ernzerhof (PBE) hybrid functionals [35] are given as follows. The PBE exchange energy is expressed as:

$$E_x^{PBE} = \int \rho(r)\, \varepsilon_x^{unif}(\rho)\, F_x^{PBE}(s)\, \mathrm{d}r \tag{1.167}$$

where $\rho(r)$ is the electron density, $\varepsilon_x^{unif}(\rho)$ is the exchange energy density of a uniform electron gas, and $F_x^{PBE}(s)$ is the PBE exchange enhancement factor given by:

$$F_x^{PBE}(s) = 1 + \kappa - \frac{\kappa}{1 + \frac{\mu\, s^2}{\kappa}} \tag{1.168}$$

where

$$s = \frac{|\nabla\rho(r)|}{2\, k_F\, \rho(r)} \tag{1.169}$$

is the exchange dimensionless reduced density gradient, and

$$k_F = \left(3\pi^2 \rho(r)\right)^{1/3} \tag{1.170}$$

as given in (1.151), with the constants $\kappa = 0.804$, and $\mu = 0.21951$. The PBE correlation energy is given as:

$$E_c^{PBE} = \int \rho(r) \left[\varepsilon_c^{unif}(r_s, \zeta) + H(r_s, \zeta, t)\right] \mathrm{d}r \tag{1.171}$$

where $\varepsilon_c^{unif}(r_s, \zeta)$ is the correlation energy density of the uniform gas, $H(r_s, \zeta, t)$ is a function that takes into account the gradient contribution to the correlation energy, where

$$r_s = \left(\frac{3}{4\,\pi\,\rho(r)}\right)^{1/3} \tag{1.172}$$

is that of (1.146), and

$$\zeta = \frac{\rho^\alpha(r) - \rho^\beta(r)}{\rho(r)} \tag{1.173}$$

similar to (1.156), and

$$t = \frac{|\nabla\rho(r)|}{2\,\phi(\zeta)\,k_s\,\rho(r)} \tag{1.174}$$

with

$$\phi(\zeta) = \frac{1}{2}\left[(1+\zeta)^{2/3} + (1-\zeta)^{2/3}\right], \quad and \;\; k_s = \sqrt{4k_F/\pi} \tag{1.175}$$

ρ^α and ρ^β are the spin-up and spin-down number densities, and k_F is already given in (1.170).

(PBE0) Hybrid Functional: The hybrid functional PBE0 [35] is a mixture of the exchange functional E_x^{PBE} of (1.167) and of the correlation functional E_c^{PBE} of (1.171). These hybrid functionals (PBE) were proposed by Perdew, Burke and Ernzerhof [35], as already mentioned above. This mixture has the following form:

$$E_{xc}^{PBE0} = aE_x^{HF} + (1 - -a)E_x^{PBE} + E_c^{PBE} \tag{1.176}$$

where E_x^{HF} is the exact exchange Hartree–Fock energy, which is already presented in (1.90). The mixing coefficient $a = 1/4$ is determined by perturbation theory [36]. By setting the value of a in (1.176) we obtain:

$$E_{xc}^{PBE0} = 0.25E_x^{HF} + 0.75E_x^{PBE} + E_c^{PBE} \tag{1.177}$$

(TPSS) Functionals: The new functional was predicted by Tao, Perdew, Staroverov, and Scuseria (TPSS) [37]. Without any empirical parameters the TPSS authors constructed the meta-GGA density functionals, which reproduce the exchange-correlation energy satisfying all constraints completely or almost-completely. The main advantages of the both exchange and correlation functionals are the slowly varying densities of one- or two-electron densities so that the molecules and solids are described with high accuracy, as already demonstrated by extensive numerical investigations, beyond the local spin density (LSD) as well as the GGA rungs. However, the primal three steps of the "Jacob's ladder"'s approximations, can be summarized by the formula:

$$E_{xc}[n_\uparrow, n_\downarrow] = \int \mathrm{d}^3r\; n\; \varepsilon_{xc}(n_\uparrow, n_\downarrow, \nabla n_\uparrow, \nabla n_\downarrow, \tau_\uparrow, \tau_\downarrow), \tag{1.178}$$

$$\tau_\sigma(\mathbf{r}) = \sum_i^{occup} \frac{1}{2} \mid \nabla \psi_{i\sigma} \mid^2 \tag{1.179}$$

where $n(r) = n_\uparrow(r) + n_\downarrow(r)$ is the total density, and $\tau_\sigma(\mathbf{r})$ is the kinetic energy density of the occupied Kohn–Sham orbitals $\psi_{i\sigma}(r)$. These orbitals represent the nonlocal functionals of the $n_\sigma(r)$ density. By dropping the ∇n_σ and τ_σ dependences in (1.178), one receives the local spin density (LSD) approximation, which is valid for the uniform electron gas and is in general advantage for solids. However, by removing the τ_σ dependence in (1.178), one receives the generalized gradient approximation (GGA), which has been developed by Perdew–Burke–Ernzerhof (PBE) [35]. However, the GGA approximation possesses just two non-empirical derivations reproducing nearly the exact properties of the exchange-correlation hole and energy. The next step in the ladder is the meta-GGA, which is referred to "TSPP", makes use of all ingredients in (1.178).

However, some very important constraints are added. For example, (a) The TPSS meta-GGA recovers the full fourth-order gradient expansion for the exchange energy in the limit of slowly-varying density $n(\mathbf{r})$. This density was constructed in the sense that the related exchange potential $\delta E_x/\delta n_\sigma(\mathbf{r})$ remains finite at the nucleus of an atom. (b) The next added constraint is the correlation part of the TPSS meta-GGA, predicted by Perdew, Kurth, Zupan, and Blaha (PKZB) [38], which properly vanishes for any one-electron density. (c) The exchange-correlation term in the PBE-GGA approximation is mainly at the low-density limit spin polarization $\zeta = (n_\uparrow - n_\downarrow)/n$ independent.

The meta-GGA functional TPSS exchange energy has the form:

$$E_x^{TPSS}[\,n\,] = \int \mathrm{d}^3rn\; \varepsilon_x^{unif}(n)\; F_x^{TPSS}, \tag{1.180}$$

where $\varepsilon_x^{unif}(n) = -3(3\pi^2 n)^{1/3}/4\pi$ is the exchange energy per particle for the uniform electron gas of density n, and F_x is the exchange enhancement factor showing the nonlocality

$$F_x^{TPSS}(s) = 1 + \kappa - \frac{\kappa}{1 + \frac{\chi}{\kappa}} \tag{1.181}$$

where $\kappa = 0.804$ and $\chi(p, z) \geq 0$ is given as:

$$\chi = \left\{ \left[\frac{10}{81} + c \frac{z^2}{[(1+z^2)^2]} \right] p + \frac{146}{2025} \widetilde{q}_b^2 - \frac{73}{405} \widetilde{q}_b \sqrt{\frac{1}{2} \left[\frac{3}{5} z \right]^2 + \frac{1}{2} p^2} + \frac{1}{\kappa} \left[\frac{10}{80} \right]^2 p^2 + 2\sqrt{e} \frac{10}{81} \left[\frac{3}{5} z \right]^2 + e\mu p^3 \right\} \Big/ \left\{ (1 + \sqrt{e} p)^2 \right\} \tag{1.182}$$

where the constants $c = 1.59096$ and $e = 1.537$ are chosen to yield the correct exchange energy (−0.3125 hartree) for the ground state density of the hydrogen atom. For a density that varies slowly over space, the TPSS F_x recovers the fourth-order gradient expansion:

$$F_x = 1 + \frac{10}{81} p + \frac{146}{2025} q^2 - \frac{73}{405} qp + Dp^2 + O(\nabla^6), \tag{1.183}$$

where $q = \nabla^2 n / 4(3\pi^2)^{2/3} n^{5/3}$ is the reduced Laplacian of the density. In PKZB, the gradient coefficient D is an empirical parameter equal to 0.113, but in TPSS they set $D = 0$, the best numerical estimate for this coefficient. In order to construct F_x, TPSS introduced a variable $\widetilde{q}_b$ which combines p and z:

$$\widetilde{q}_b = (9/20)(\alpha - 1)/[1 + b\,\alpha(\alpha - 1)]^{1/2} + 2p/3, \tag{1.184}$$

where

$$\alpha = (\tau - \tau_W)/\tau^{unif} = (5p/3)(z^{-1} - 1) \geq 0. \tag{1.185}$$

where $\tau^{unif} = \frac{3}{10}(3\pi^2)^{2/3} n^{5/3}$ is the uniform-gas kinetic energy density. When the parameter b in (1.184) is set to zero, $\widetilde{q}_b$ reduces to the dimensionless variable $\widetilde{q}$.

The meta-GGA functional TPSS for correlation energy has the form:

$$E_c^{TPSS}[n_\uparrow, n_\downarrow] = \int d^3r\, n\, \varepsilon_c^{revPKZB} \times [1 + d\varepsilon_c^{revPKZB} (\tau^W/\tau)^3], \tag{1.186}$$

where

$$\varepsilon_c^{revPKZB} = \varepsilon_c^{PBE}(n_\uparrow, n_\downarrow, \nabla n_\uparrow, \nabla n_\downarrow)[1 + C(\zeta, \xi)(\tau^W/\tau)^2] - [1 + C(\zeta, \xi)](\tau^W/\tau)^2 \sum_\sigma \frac{n_\sigma}{n} \widetilde{\varepsilon}_c \tag{1.187}$$

For any $C(\zeta, \xi)$, E_c^{TPSS} of (1.186) properly vanishes for a one-electron density [for which $\tau = \tau^W$ and $\zeta \equiv (n_\uparrow - n_\downarrow)/n = \pm 1$].

(TPSSh) Hybrid Functionals: Rewriting the exchange energy of equation (1.180) we receive:

$$E_x^{TPSS}[\rho_\alpha, \rho_\beta] = \sum_{\sigma=\alpha,\beta} \int \rho_\sigma \varepsilon_x^{LSDA}(\rho_\sigma) F_x(\rho_\sigma, \nabla\rho_\sigma, \tau_\sigma) \mathrm{d}r \tag{1.188}$$

where $\varepsilon_x^{LSDA}(\rho_\sigma) = -(3/4)(6/\pi)^{1/3}\rho_\sigma^{1/3}$ is the σ component of the LSDA exchange energy per particle, and $F_x(\rho_\sigma, \nabla\rho_\sigma, \tau_\sigma)$ is a relatively complicated enhancement factor which reduces to 1 for a uniform density and preserves the LSDA exchange-like scaling. Rewriting the correlation energy of equation (1.186) we receive:

$$E_c^{TPSS}[\rho_\alpha, \rho_\beta] = \int \rho \varepsilon_c^{revPKZB} \left[1 + d\left(\frac{\tau_W}{\tau}\right)^3 \varepsilon_c^{revPKZB}\right] \mathrm{d}r \tag{1.189}$$

Here are $\rho = \rho_\alpha + \rho_\beta$, and $\tau = \tau_\alpha + \tau_\beta$, and $\tau_W = | \nabla\rho |^2 / 8\rho$, and finally $\varepsilon_c^{revPKZB}$ $(\rho_\alpha, \rho_\beta, \nabla\rho_\alpha, \nabla\rho_\beta, \tau)$ which is the revised PKZB correlation energy per particle, and d is a non-empirical constant. However, the global hybrid form of the new meta-GGA functional is obtained by mixing E_x^{TPSS} and E_c^{TPSS} of (1.188) and (1.189) with the exact exchange [39] given as follows:

$$E_{xc}^{TPSSh} = a\, E_x^{exact} + (1-a)E_x^{TPSS} + E_c^{TPSS} \tag{1.190}$$

The TPSSh functional has one empirical parameter whose optimal value $a = 0.10$ was determined by minimizing the mean absolute deviation in the enthalpies of formation of 223 G3/99 molecules using the 6-311++G($3df$, $3pd$) basis. TPSSh satisfies precisely the same exact constraints as the original TPSS.

Jacob's Ladder of Density Functional Approximations: Perdew et al. [40] presented the case for the non-empirical construction of density functional approximations for the exchange-correlation energy by the traditional method of "constraint satisfaction" without fitting to data sets, and present evidence that this approach has been successful on the first three rungs of "Jacob's ladder" or "Jacob's stair" of density functional approximations [local spin-density approximation (LSD), generalized gradient approximation (GGA), and meta-GGA]. They expected that this approach will also prove successful on the fourth and fifth rungs hyper-GGA or (hybrid and generalized random-phase approximation GRPA). However, they realized that this non-empirical construction is a slow and uncertain process. It is hard to get everything right, and impossible to do so on the lower rungs of the ladder: for certain systems and properties, users may want to employ semi-empirical or fitted functionals. In such cases, they recommended functionals that are correct for the uniform electron gas and contain few fitted parameters, such as PBE0 or TPSSh.

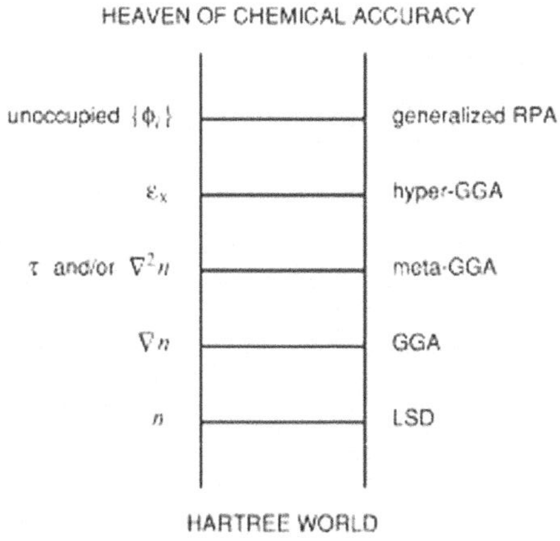

Fig. 1.1 Jacob's stair of density functional approximations to the exchange-correlation energy, where n is the electron density and known as ρ. This figure was reproduced with kind permission of the Journal of Chemical Physics of the American Institute of Physics, from [40]

Visualization of Density Functional Approximations: Thus, in order to visualize the DFT functionals to be a succession of ladder steps, see Fig. 1.1, Perdew et al. [40] formulated the Jacob's stair strategy for constructing improved density functionals for the exchange-correlation energy. These steps can also be expressed in electron density n in relationship to the density functionals as follows:

$$n \sim LSD \curvearrowright \quad \nabla n \sim GGA \curvearrowright \quad \nabla^2 n \sim meta.GGA \curvearrowright \quad \varepsilon_x \sim hyper.GGA \curvearrowright \quad GRPA.. \tag{1.191}$$

Staroverov [26] reported to the above classification, that density-functional approximations constructed using the electron density ρ and no other representative are of rung 1 of the stair and termed as local density approximations (LDA). Using the LDA for spin-polarized systems, then one receives the local spin density (LSD) approximation:

$$E_{xc}^{LSD}[\rho] = \int e_{xc}(\rho)\mathrm{d}r \tag{1.192}$$

Approximations where e_{xc} depends on ρ and g ($g = |\nabla \rho|$ is the modulus of the gradient of density) represent rung 2 and are called generalized-gradient approximations (GGA),

$$E_{xc}^{GGA}[\rho] = \int e_{xc}(\rho, g)\mathrm{d}r \tag{1.193}$$

Rung 3 approximations depend, in addition to ρ and g, on l and/or τ, and are called meta-GGAs (MGGA),

$$E_{xc}^{MGGA}[\rho] = \int e_{xc}(\rho, g, l, \tau)\mathrm{d}r \tag{1.194}$$

where $l = \nabla^2\rho$ is the Laplacian of the density, and $\tau = \frac{1}{2}\sum_k^{occ.} |\nabla\phi_k|^2$ is the Kohn–Sham (non-interacting) kinetic energy density. The functionals of rung 4 involve dependence on a nonlocal ingredient, and are termed hyper-GGAs (HGGA),

$$E_{xc}^{HGGA}[\rho] = \int e_{xc}(\rho, g, l, \tau, e_x^{exact})\mathrm{d}r \tag{1.195}$$

Approximations of rungs 1 through 4 involve only the occupied Kohn–Sham orbitals. As can be seen in (1.191), there is also a fifth rung where one finds approximations that involve occupied and virtual Kohn–Sham orbitals named as generalized random-phase approximation GRPA.

1.6 Semi-empirical Methods

The accuracy of the theoretical quantum chemical methods plays a great role in the calculations. Therefore, one should find a compromise between the accuracy and the size of the investigated system. Thus, there are different approaches in quantum chemistry developed for systems using empirical, semi-empirical potentials, or pseudopotentials. However, a summery of these methods are given.

Introduction: Due to the high cost calculations of integrals in the ab initio quantum chemical calculations, the semi-empirical methods are the simplified versions of the Hartree–Fock theory in which the integrals are approximated by functions with empirical parameters, which in turn are adapted and improved from experimental data. Furthermore, it is supposed that the basis sets do not overlap at all, and the overlap matrix becomes the identity. In addition, the core electrons or say the core orbitals are not included in the calculations since they do not contribute to the chemical reactions or bonds. Therefore only the valence orbitals with minimal basis sets are considered in the calculations. Thus the accuracy of the semi-empirical methods depends mainly on the empirical parameters and their ingredient to the experimental data.

1.6.1 Interatomic Potentials

Introduction: For a system of large number of particles, the quantum chemical methods are not more reliable and thus the molecular dynamics is an alternative approach to overcome the problem with the time and memory. In molecular dynamics one has to assume that each particle is a solid sphere following Newton equations of motion. As one usually neglects gravity, the particles are moving in straight trajectories, being influenced only by their mutual interaction with other particles [41]. The interaction can be modeled using different two-body or many-body potentials. The latter can be written as a series expansion of functional terms depending on the position of one, two, three, etc. atoms at a definite time:

$$V_{\text{Total}}(r_1, r_2, \ldots, r_n) = \sum_i V_1(r_i) + \sum_{ij} V_2(r_i, r_j) + \sum_{ijk} V_3(r_i, r_j, r_k) + \cdots \tag{1.196}$$

Where V_1, V_2, and V_3 are the one-, two-, and three-body potentials, respectively.

1.6.2 Two-Body Potentials

An example for two-body potential V_2, called also pair-potentials, is the well known Lennard-Jones potentials, which is known for modelling inert gases:

$$V(R) = \sum_{ij} v(r_i - r_j), v(r) = 4\varepsilon \left[\left(\frac{\sigma}{r}\right)^{12} - \left(\frac{\sigma}{r}\right)^{6} \right] \tag{1.197}$$

where ε is the depth of the potential well, and σ is the distance at which the intermolecular potential between the two particles is zero. This distance is also called Van der Waals radius. The first part of the above equation, $(\sigma/r)^{12}$ describes the repulsive forces between particles while the latter part of the equation, $(\sigma/r)^6$ denotes attraction.

1.6.3 Three-Body Potentials

The next term in the expansion above is the three-body potential V_3. The terms of the three-body Tersoff potential [42, 43] are usually dependent on the types of atoms i and k. The Tersoff potential for the energy E of a system of atoms can be written as:

$$E = \sum_i E_i = \frac{1}{2} \sum_{i \neq j} V_{ij} \tag{1.198}$$

with

$$V_{ij} = f_c(r_{ij}) \left[a_{ij} f_R(r_{ij}) + b_{ij} f_A(r_{ij}) \right] \tag{1.199}$$

where, f_R is the repulsive two-body term, and f_A is the attractive three-body potential:

$$f_R(r) = \quad Ae^{(-\lambda_1 r)} \tag{1.200}$$

$$f_A(r) = -Be^{(-\lambda_2 r)} \tag{1.201}$$

The f_C is a smooth cutoff function can be written as follows:

$$f_C(r) = \begin{cases} 1 & , r < R - D, \\ \frac{1}{2} - \frac{1}{2}\sin\left(\frac{\pi}{2}\frac{(r-R)}{D}\right) & , R - D < r < R + D \\ 0 & , r > R + D, \end{cases} \tag{1.202}$$

The three-body contributions arise due to the bond-order parameter:

$$b_{ij} = \left(1 + \beta^n \zeta_{ij}^n\right)^{-1/2n}, \tag{1.203}$$

with

$$\zeta_{ij} = \sum_{k \neq i,j} f_C(r_{ik}) g(\theta_{kij}) e^{\lambda_3^3 (r_{ij} - r_{ik})^3}, \tag{1.204}$$

and

$$g(\theta_{kij}) = 1 + \frac{c_i^2}{\mathrm{d}_i^2} - \frac{c_i^2}{\mathrm{d}_i^2 + (h_i - \cos\theta_{kij})^2}, \tag{1.205}$$

θ_{ijk} is the angle between the bonds ij and ik. The function a_{ij} serves the purpose of limiting the range of the repulsive interactions, and was chosen to have the following form:

$$a_{ij} = (1 + \alpha^n \eta_{ij}^n)^{-1/2n} \tag{1.206}$$

with

$$\eta_{ij} = \sum_{k \neq i,j} f_C(r_{ik}) e^{\lambda_3^3 (r_{ij} - r_{ik})^3} \tag{1.207}$$

For example, Dmitriev et al. [44] performed molecular dynamics modeling on amorphous silica using Tersoff potential with the corresponding parameters. For more details see molecular dynamics simulations provided by Nordlund [45]. Finally, the interatomic interaction in both simulations can be calculated in two ways: (i) using classical potential function (this can be pairwise or multi center interaction), (ii) using quantum mechanical methods (this can be ab initio HF, DFT, semi-empirical or tight-binding method, or any other quantum mechanical method).

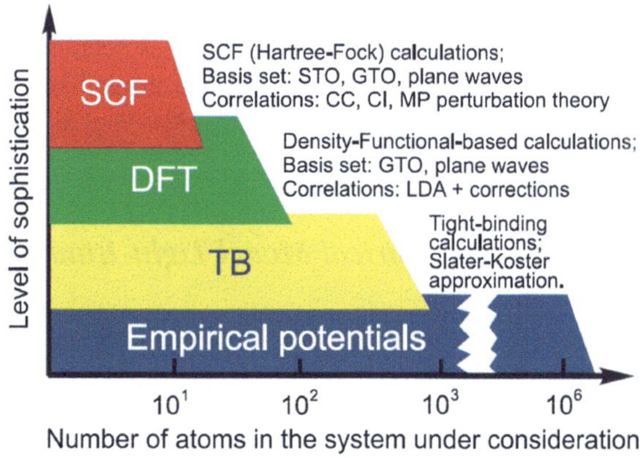

Fig. 1.2 From Sophisticated to non-sophisticated numerical methods as a function of the number of atoms in molecular systems [46]

1.7 Concepts of Semi-empirical Methods

As mentioned above, the semi-empirical methods are attempting to arrive two limitations, namely slow speed and low accuracy. That can be achieved by reducing and parameterizing the integrals in the Hartree–Fock calculation based on experimental data, Therefore, the semi-empirical methods are very fast, because the integrals are already fitted by the experimental parameters of the investigated molecules or solid systems. The graphics of Fig. 1.2 below shows the sophisticated and less sophisticated numerical methods in relation to the number of atoms in the investigated systems.

1.8 Basic Models

As can be seen in the graphics of Fig. 1.2, the Hartree–Fock and density functional calculations present the most sophisticated ab initio methods and are already described in the previous sections. The less accurate and less demanding time methods are the semi-empirical and tight-binding methods appearing at the bottom of the graphics. However, since there are a lot number of semi-empirical methods we will report the most used methods.

1.8.1 Semiempirical Model NDDO

Semiempirical methods have a variety of models. One of these models is named the Neglect of Diatomic Differential Overlap (NDDO) method, which was introduced by John Pople, and is nowadays the most successful semi-empirical method. The overlap matrix in the Hartree–Fock secular equation $|H - ES| = 0$ is replaced with the unit matrix reducing the Hartree–Fock secular equation to $|H - E| = 0$. This method can reproduce well enough the ground state energies and geometries, the dipole moments, the ionization energies as well as the vibrational frequencies.

1.8.2 Semi-empirical Model Tight-Binding

Another model of the semi-empirical methods is the tight-binding approach, which is the simplest quantum mechanical approach to the geometrical and electronic structures of molecular and solid systems. It is less accurate than the first-principle density functional calculations, but however the tight-binding method reproduce attractive and reasonable total electronic energy which can be written as follows:

$$E = \sum_{i=1}^{N} \varepsilon_i + \frac{1}{2} \sum_{\alpha} \sum_{\beta(\neq\alpha)} U(|R_\alpha - R_\beta|) \tag{1.208}$$

where ε_i are the eigenvalues of some of non-self-consistent Schrö-dinger-like equation:

$$\hat{H}\psi_i(r) = [-\frac{1}{2}\nabla^2 + V(r)]\psi_i(r) = \varepsilon_i \psi_i(r) \tag{1.209}$$

where $U(|R_\alpha - R_\beta|)$ is a short-range repulsion between the atoms R_α and R_β. The Schrödinger-like equation above can be solved variationally within a basis of localized atomic-like functions, $\{\phi_\alpha\}$, and so leads to a secular equation:

$$|H - \varepsilon \underline{S}| = 0 \tag{1.210}$$

where

$$H_{\alpha\beta} = \langle\phi_\alpha|\hat{H}|\phi_\beta\rangle \tag{1.211}$$

with the overlap matrix $\underline{S}$

$$S_{\alpha\beta} = \langle\phi_\alpha|\phi_\beta\rangle \tag{1.212}$$

Thus, in order to limit the number of fitting parameters, it is assumed that the inter-atomic matrix elements extend only to first or second neighbors. Many semi-empirical tight-binding models also set the overlap matrix to the identity matrix $S_{\alpha\beta} = \delta_{\alpha\beta}$. Tight-binding is very similar to Hückel and extended Hückel methods used in the quantum chemistry. In comparison to ab initio quantum chemical methods the tight-binding is about two or three orders of magnitude faster.

1.8.3 Semi-empirical Model MNDO

The Modified Neglect of Diatomic Overlap (MNDO) is the most frequently used semiempirical method and is based on the Neglect of Diatomic Differential Overlap (NDDO) integral approximation, as mentioned above. It was developed by Michael Dewar and Walter Thiel in 1977 [47]. In this method, also all parameters are assigned for different atomic types are fitted to reproduce the electronic properties. The two-electron integrals that describe the electron interactions are two-, three- and four-center integrals. In this method, all three- and four-center integrals are ignored, and the remaining integrals are parameterized according to the spectroscopic data of isolated atoms. However, the MNDO method has some deficiencies, like its disability to reproduce the hydrogen bond by a strong interatomic repulsion.

1.8.4 Semi-empirical Models AM1 and PM3

The semi-empirical method Austin Model (AM1), developed by Dewar et al. [48], is based on the NDDO method concerning the two-electron integral approximations. However, in this model the atomic nuclear-nuclear repulsion terms is modified through additional expression, in which non-physical attractive forces that mimic van der Waal interactions. The next semi-empirical model is the parametric method 3 (PM3), developed in 1989 by Stewart [49, 50]. The PM3 methods uses the same Hamiltonian, say the same formalism and Gaussian functions of the AM1 model, but however, in different manner of parametrization. However, both AM1 and PM3 methods produce similar geometrical results, but less accurate total energies and heat of formation.

1.9 Chemistry Software

There is a lot of program soft-wares which are available to calculate molecules as well as periodic solids [51]. The most popular software packages in chemistry are listed as follows:

ABINIT A program package for molecules and periodic systems.

CHARMM A force field molecular dynamics simulation program with broad application to many-particle systems.

CPMD The Car–Parrinello program code is a parallelized plane wave and pseudopotential implementation of density functional theory and Ab Initio molecular dynamics.

Dalton is a powerful molecular electronic structure program.

DeMon stands for Density Montréal, is a software package for density functional theory calculations.

Dirac The Dirac program computes molecular properties using relativistic quantum chemical methods.

GAMESS (UK) is a computational chemistry software program that stands for General Atomic and Molecular Electronic Structure System, developed by Martyn F. Guest et al. in Daresbury. The original code split in 1981 into GAMESS (US) and GAMESS (UK) variants.

GAMMES (US) stands for the General Atomic and Molecular Electronic Structure System and is an Ab Initio quantum chemistry program package, developed by the Mark Gordon research group at the Iowa State University.

GAUSSIAN is a program code for computational chemistry initially released 1970 by John Pople.

GROMACS is a molecular dynamics software package for classical molecular dynamics simulations.

NWChem is a **N**orth**W**est Computational **Chem**istry Package, designed for high performance parallel supercomputers & clusters of conventional workstations.

ORCA is an Ab Initio, DFT & semi-empirical program package.

QUANTUM ESPRESSO It is a computer code for electronic structures based on density functional theory, plane wave and pseudopotentials.

SIESTA is a density-functional program code "**S**panish **I**nitiative for **E**lectronic **S**imulation with **T**housands of **A**toms". It performs efficient electronic structure calculations and ab initio molecular dynamics simulations of molecules and solids.

TRANSIESTA is a non-equilibrium Green function program code, based on density functional theory (DFT). It enables modelling of molecular electronic devices and electron transport properties.

VASP Vienna **A**b-Initio **S**imulation **P**ackage is a program code developed for plane wave and pseudopotentials of density functional theory to calculate molecules and periodic systems.

References

1. F. Jensen, *Introduction to Computational Chemistry*, 2nd edn. (Wiley, Ltd, Sussex, 2007)
2. E.G. Lewars, *Computational Chemistry: Introduction to the Theory and Applications of Molecular and Quantum Mechanics*, 2nd edn. (Springer, Dordrecht, 2011)
3. D. Chahal, Automated, parallel optimization algorithms for stochastic functions. Dissertation, Clemson University, 2011
4. J.A. Nelder, R. Mead, A simplex method for function minimization. Comput. J. **7**, 308–313 (1965)

5. F. Brunet, Contributions to parametric image registration and 3D surface reconstruction. Dissertation, d'Auvergne University, 2010
6. T.W. Shattuck, Molecular mechanics tutorial. Colby College, Waterville, 2008
7. R. Fletcher, *Practical Methods of Optimization*, 2nd edn. (Wiley, New York, 1987)
8. http://www.ch.embnet.org/MD_tutorial/pages/MD.Part1.html. Accessed 12 Feb 2015
9. E. Carlon, Molecular dynamics simulations. Katholieke Universiteit Leuven, 2010
10. I. Boustani, Density functional theory in chemistry. Habilitation thesis, Bergische Universität Wuppertal, Germany, 1998
11. A. Szabo, N.S. Ostlund, *Modern Quantum Chemistry, Introduction to Advanced Electronic Structure Theory* (Dover Publications Inc., Mineola, 1996)
12. R.J. Buenker, S.D. Peyerimhoff, Individualized configuration selection in CI calculations with subsequent energy extrapolation. Theor. Chem. Acta **35**, 33–58 (1974)
13. R.J. Buenker, S.D. Peyerimhoff, Energy extrapolation in CI calculations. Theor. Chem. Acta **39**, 217–225 (1975)
14. R.J. Buenker, Combining perturbation theory techniques with variational CI calculations to study molecular excited states. Int. J. Quantum Chem. **29**, 435–460 (1986)
15. B.O. Roos, P.E.M. Siegbahn, *Methods of Electronic Structure Theory*, ed. by H.F. Schaefer III (Plenum Press, New York, 1977)
16. J. Cizek, On the correlation problem in atomic and molecular systems. Calculation of wavefunction components in Ursell-type expansion using quantum-field theoretical methods. J. Chem. Phys. **45**, 4256 (1966)
17. I. Shavitt, R.J. Bartlett, *Many-Body Methods in Chemistry and Physics: MBPT and Coupled-Cluster Theory* (Cambridge University Press, Cambridge, 2009)
18. M.W. Reinsch, A simple expression for the terms in the Baker-Campbell-Hausdorff series. J. Math. Phys. **41**, 2434–2442 (2000)
19. P. Hohenberg, W. Kohn, Inhomogeneous electron gas. Phys. Rev. **136**, B864–B871 (1964)
20. W. Kohn, L.J. Sham, Self-consistent equations including exchange and correlation effects. Phys. Rev. **140**, A1133–A1138 (1965)
21. R.G. Parr, W. Young, *Density Functional Methods of Atoms and Molecules* (Oxford University Press, New York, 1989), p. 108
22. U. von Barth, L. Hedin, A local exchange-correlation potential for the spin polarized case: I. J. Phys. C **5**, 1629–1642 (1972)
23. S.H. Vosko, L. Wilk, M. Nusair, Accurate spin-dependent electron liquid correlation energies for local density calculations: a critical analysis. Can. J. Phys. **58**, 1200–1211 (1980)
24. O. Gunnersson, B.I. Lundqvist, Exchange and correlation in atoms, molecules, and solids by the spin-density-functional formalism. Phys. Rev. B **13**, 4274 (1976)
25. D.M. Ceperley, B.J. Alder, Ground state of the electron gas by a stochastic method. Phys. Rev. Lett. **45**, 566–569 (1980)
26. V.N. Staroverov, Density-functional approximations for exchange and correlation, in *A Matter of Density: Exploring the Electron Density Concept in the Chemical, Biological, and Materials Sciences*, ed. by N. Sukumar (Wiley, Hoboken, 2013), pp. 125–156
27. A.D. Becke, Density-functional exchange-energy approximation with correct asymptotic behaviour. Phys. Rev. A 38, 3098–3100 (1988); A.D. Becke, A multicenter numerical integration scheme for polyatomic molecules. J. Chem. Phys. **88**, 2547–2553 (1988)
28. J.P. Perdew, Y. Wang, Accurate and simple density functional for the electronic exchange energy: generalized gradient approximation. Phys. Rev. B **12**, 8800–8802 (1986)
29. D.C. Langreth, M.J. Mehl, Beyond the local-density approximation in calculations of ground-state electronic properties. Phys. Rev. B **1984**, 28 (1809)
30. J.P. Perdew, Density-functional approximation for the correlation of the inhomogeneous electron gas. Phys. Rev. B **33**, 8822–8824 (1986)
31. R. Colle, D. Salvetti, Approximate calculation of the correlation energy for the closed shells. Theor. Chem. Acta **37**, 329–334 (1975)
32. C. Lee, W. Yang, R.G. Parr, Development of the Colle-Salvetti correlation-energy formula into a functional of the electron density. Phys. Rev. B **37**, 785–789 (1988)

33. L.C. Wilson, M. Levy, Nonlocal Wigner-like correlation-energy density functional through coordinate scaling. Phys. Rev. B **41**, 12930–12932 (1990)
34. E. San Fabián, F. Moscardo, J.M. Perez-Jorda, Applicability to atoms of a large set of correlation energy functionals. Int. J. Quantum Chem. **52**, 1027–1038 (1994)
35. J.P. Perdew, K. Burke, M. Ernzerhof, Generalized gradient approximation made simple. Phys. Rev. Lett. **77**, 3865–3868 (1996)
36. J. Heyd, G.E. Scuseria, M. Ernzerhof, Hybrid functionals based on a screened Coulomb potential. J. Chem. Phys. **118**, 8207–8215 (2003)
37. J.-M. Tao, J.P. Perdew, V.N. Staroverov, G.E. Scuseria, Climbing the density functional ladder: nonempirical meta-generalized gradient approximation designed for molecules and solids. Phys. Rev. Lett. **91**, 146401-1–146401-4 (2003)
38. J.P. Perdew, S. Kurth, A. Zupan, P. Blaha, Accurate density functional with correct formal properties: a step beyond the generalized gradient approximation. Phys. Rev. Lett. **82**, 2544–2547 (1999)
39. V.N. Staroverov, G.E. Scuseria, J. Tao, J.P. Perdew, Comparative assessment of a new nonempirical density functional: molecules and hydrogen-bonded complexes. J. Chem. Phys. **119**, 12129–12137 (2003)
40. J.P. Perdew, A. Ruzsinszky, J.-M. Tao, V.N. Staroverov, G.E. Scuseria, G.I. Csonka, Prescription for the design and selection of density functional approximations: more constraint satisfaction with fewer fits. J. Chem. Phys. **123**, 062201–062209 (2005)
41. P. Klapetek, Classical nanoscale modeling, in *Quantitative Data Processing in Scanning Probe Microscopy* (2013)
42. J. Tersoff, Empirical interatomic potential for silicon with improved elastic properties. Phys. Rev. B **38**, 9902 (1988)
43. J. Tersoff, Modeling solid-state chemistry: interatomic potentials for multicomponent systems. Phys. Rev. B **39**, 5566(R) (1989)
44. A.I. Dmitriev, A.Y. Nikonov, W. Österle, Molecular dynamics modeling of the sliding performance of an amorphous silica nano-layer-the impact of chosen interatomic potentials. Lubricants **6**, 43 (2018)
45. K. Nordlund, www.acclab.helsinki.fi/~knordlun/moldyn/. Accessed May 2019
46. Y. Wang, http://msse.gatech.edu/CANE/lect03_QMsims_yanwang.pdf. Accessed Jun 2019
47. M.J.S. Dewar, W. Thiel, Ground states of molecules. 38. The MNDO method. Approximations and parameters. J. Am. Chem. Soc. **99**, 4899 (1977)
48. M.J.S. Dewar, E.G. Zoebisch, E.F. Healy, J.J.P. Stewart, Development and use of quantum mechanical molecular models. 76. AM1: a new general purpose quantum mechanical molecular model. J. Am. Chem. Soc. **107**, 3902 (1985)
49. J. Stewart, Optimization of parameters for semiempirical methods I. Method. J. Comput. Chem. **10**, 209 (1989)
50. J. Stewart, Optimization of parameters for semiempirical methods II. Applications. J. Comput. Chem. **10**, 221 (1989)
51. http://www.freechemical.info/freeSoftware/kinds.php. Accessed Feb 2015

Part II
Magic Numbers and Clusters

Chapter 2
Magic Numbers

Abstract Questions, like who mentioned and used the term clusters for the first time, and who found and interpreted the nuclear and electronic shells of clusters, are answered. The definition, history, discovery of electronic shells and synthesis of clusters are presented. The evolution of small clusters to nanoclusters and their function in nanotechnology are discussed. The terms magic numbers in nature and their relationship to clusters are explained. The nuclear, electronic, atomic, and magnetic shells are introduced.

Introduction: Three remarkable events in natural science characterized the Eighties of the last century. The first event happened in 1984, in which W. Knight et al. [1] announced the discovery of the electronic shells in alkali-metal clusters and its relationship to magic numbers. The second one was the discovery of carbon cages called Bucky balls or Buckminster Fullerenes in 1985 by Kroto et al. [2], who was awarded the Nobel Prize in Chemistry in 1996. The third event was the detection of ordered but non-periodic crystalline solids called quasi-crystals in 1984 by Shechtman [3], who was awarded the Nobel Prize in Chemistry in 2011.

Towards Nanodomain: These pioneering events marked the beginning of exploring the world of nanodomain, which was a mystery in the years before. These discoveries inspired and provoked a lot of scientists of different disciplines to improve and develop current and new experimental and theoretical methods to investigate the challenging and fascinating new materials at nanodomain. The snowball of contributions rolled over the years and got larger and larger and released a huge number of publications about the exciting and inconceivable formations and synthesis of nanomaterials. These knowledge led to the prediction and detection of nanostructures, unknown in nature at yet.

Theory and Experiment: By reporting the most important discoveries of nanomaterials we will focus on structures composed of metal or non-metal atoms of the second row of the periodic system like lithium, boron, carbon and nitrogen. For example, structures of small lithium clusters in relationship with electronic structures the magic numbers were proposed in 1987 by Boustani et al. [4]. These theoretically predicted structures were confirmed in 2003 by Kornath et al. [5] experimentally, only after 16

I. Boustani, *Molecular Modelling and Synthesis of Nanomaterials*, Springer Series in Materials Science 290, https://doi.org/10.1007/978-3-030-32726-2_2

years. Boustani predicted in 1994 novel quasi-planar boron clusters [6]. Few years later L. S. Wang et al. [7] confirmed experimentally boron quasi-planar clusters and synthesized these aromatic boron structures. They investigated further on boron clusters and extended their planarity up to B_{32} [8]. Boustani et al. predicted in 1997 [9] via theory the existence of in nature unknown boron nanotubes, and in 1999 [10] that these boron nanotubes are metallic. Ciuparu et al. [11] and Gao et al. [12] succeeded to synthesis single-walled and multi-walled boron nanotubes, respectively.

Aufbau Principle for Clusters: Furthermore, Boustani proposed few years later the "Aufbau Principle" [13] suggesting triangular motifs as building blocks for surfaces of stable boron sheets and nanotubes. Artem Oganov and co-workers [14] confirmed recently the two-dimensional boron allotrope with mass-less Dirac Fermions. New phase of boron crystalline called high pressure γ-boron was predicted and synthesized in 2009 by Oganov et al. [15]. Further discovery on nanostructures in form of single and multi-walled carbon nanotubes was experimentally recorded in 1991 by Iijima [16]. In addition, Novoselov and Geim [17] succeeded in 2004 to synthesis unimaginable single-layer of graphite or one atom thick carbon sheet named graphene. Thus they were awarded the Nobel Prize in Physics in 2010. Inorganic nanotubes were also investigated. Tenne et al. [18] proposed in 1992 polyhedral and cylindrical structures of tungsten disulphide, while Seifert et al. [19] studied in 2000 the structure and electronic properties of MoS_2 Nanotubes.

2.1 Atomic Clusters and Magic Numbers

Clusters (Concept): By definition a cluster is a group of similar things of the same kind, which are close, growing or held together. For example a cluster of grapes, olives, blackberries, cherries or just cornflakes, as presented in Fig. 2.1. Similarly, in physics and chemistry, atomic clusters are aggregates of a countable number of particles such as atoms or molecules containing between two, up to hundred, and a few thousand atoms. They represent a kind of transition state between atom and bulk, and have properties intermediate between those of the isolated monomers and the solids. The species of clusters can be either identical or different.

Types of Clusters: In metals there are two types of valence electrons: bound electrons and free electrons. The bound electrons are glued or localized to their individual atoms, while the free electrons are delocalized and move free between atoms, causing conductivity. Metal clusters are those of the s-block (Fig. 2.2) like alkali and alkaline earth metals. Their electrons are delocalized and bonds are non-directional. Transition metal clusters of the d-block (Fig. 2.2) are those with the valence electrons of d-orbitals. They have greater degree of valency and direct bonds. There are another type of clusters named semi-metals or metalloids. They are placed slanting the line between the metals and non-metals in the periodic system. The six commonly recognized elements of the metalloids are boron, silicon and germanium, arsenic and antimony, and tellurium. They are also called also semiconductors. They have strong

Fig. 2.1 Examples of clusters: cornflakes, grapes, corncob, blackberries, bananas, olives, and cherries

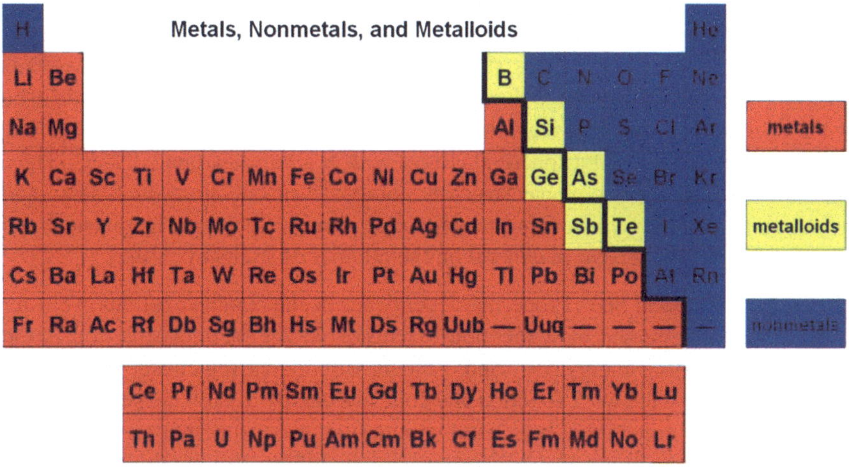

Fig. 2.2 Metals (s-block and d-block), non-metals, and metalloids in the periodic table [20]

directed covalent bonds. Another kind of clusters are those of ionic bonds. They are mostly heteroatomic clusters composed of atoms with large difference in electronegativity, like alkali metal halide clusters Na_xCl_y and Mg_xO_y. Furthermore, there are rare gas atoms called van der Waals clusters. They are formed at low temperature and are bonded very weakly. Finally, clusters of molecules like $(H_2O)_n$, $(C_6H_6)_n$ are weakly bonded either via van der Waals, dipole-dipole, or higher-order multi-polar interactions. Molecular clusters include also inorganic or organometallic clusters, like P_6, $(BH)_{12}^{-2}$ and $Os_6(CO)_{18}$. They are thermodynamically and kinetically stable with respect to coalescence and can exist in solid, liquid and vapor phases.

The Term Cluster: As mentioned above, clusters as a collection of atoms represent the transition phase between single atoms, molecules and solids. The geometrical and electronic properties of the cluster depend strongly on their size. One key question:

at which size can cluster mimic the bulk structure? However, it is very interesting to study metallic, inorganic, semi-conducting, and insulating clusters as a function of size in different fields of physics, chemistry and nanomaterials. However, the term cluster was first used by Benson and Shuttleworth in 1951 [21]. They found that nucleation phenomena depend upon the properties of small clusters of molecules. Once again, the word cluster proposed by F.A. Cotton in the early 1960s [22] referring to some compounds exhibiting metal-metal bonding. In the late 1960s, Burton [23] considered the term clusters for very small crystals, and used the term clustering in dense and vapor phase nucleation. Burton claimed that adequate small clusters can not be considered as solids with a surface, because the most atoms of clusters are nearly laying on the surface building surface atoms.

Properties of Clusters: The fundamental interest on clusters is due to their unique and intrinsic physical and chemical properties and because they present a transition state between molecular and condensed matter. They span a wide range of particle size from very small and nanosize clusters with discrete quantum states up to micro-crystalline with quasi-continuous states. In fact, clusters constitute new materials which have different properties that are distinct from those of discrete solids. The first and singular property of clusters that they have more surface than volume. The surface-to-volume ratio of a spherical cluster with N atoms is around $3N^{-1/3}$. Thus, small cluster have mostly planar structure composed of one atom thick slice, and the atoms of bit larger clusters are nearly on or near the surface. The physical and chemical properties depend in general on the number of atoms in clusters.

Cluster's Function of Size: Properties dependence on the number of atoms in cluster can be simply expressed as a function of cluster size. Functions, like ionization potential and binding energy, vary and oscillate with increasing cluster size so long until they approach the work function and cohesive energy of bulk, respectively. The evolution of molecular orbitals to band structure is also a function of cluster size. In addition, clusters constitute new materials which have different properties that are distinct from those of discrete molecules or condensed matter. The structure of clusters can simulate the growth and phase transition of crystals. Besides the size, the charge of clusters affects very strongly its properties. Removal or addition of one or several electrons can stabilize or destabilize the clusters. Removal of few electrons can lead to Coulomb explosion, as the case of lithium and boron clusters, which will be discussed in the upcoming chapters.

The Terms "Magic Numbers": In mathematics there are some extraordinary numbers, known as mathematical constants, like the Euler number e, the number π, the golden ratio ϕ, and the Fibonacci numbers. However, in modern physics there exist another kind of extraordinary numbers called magic numbers. They are known in the nuclear structures of isotopes, electronic shells in atoms and clusters, atomic shells in condensed or hollow cages.

Ligands and Clusters: The clusters of the main group elements (elements without transition metals), are stabilized mainly by ligands of hydrides. The overlapping of d-orbitals in metal clusters contributes to highly stable clusters inducing the covalent bonds between the valence orbitals. However, based on the polyhedral skeletal electron pair theory proposed by Wade [24, 25], and known as Wade's rules, one can predict the trends in structure stability of metal clusters as well as of boranes. Jemmis electron counting (mno) rules [26, 27] provided additional insight into the relative stability for mono and condensed polyhedral boranes.

2.2 Shell Structures in Atoms and Clusters

2.2.1 *The Shell Structure of Electrons in Atoms*

E. Rutherford proposed in 1911 an atom model which describes the structure of atoms as composition of two parts. The first one is the nuclear core called nucleus with a positive charge, due to the protons inside. The protons and neutrons inside the core build the atomic mass. The other part of the atom are the electrons, which are distributed around the core in different orbitals. They have negative but equal charge to the core. Therefore, the atomic model is similar to the solar system. Neutral atoms have the same number of negatively charged electrons and positively charged protons. Thus, an atom must have equal numbers of protons and electrons, unless it is an ion. The electrons are distributed in electronic main shells. They are divided into sub-atomic shells composed of the atomic orbitals s, p, d, f, g, and h. These sub-shells can hold up to 2, 8, 18, 32, 60, and 72 electrons.

2.2.2 *Shell Structure of Nucleons*

Maria Goeppert-Mayer discovered in 1948 [28] that nuclei have shell structures analogous to electron shell structures in atoms. She won with Eugene Paul Wigner the Nobel Prize in Physics in 1963. Each nucleon in the model of nuclear shell moves in a medial potential well generated by the neighboring nucleons, in similar way to electronic orbitals a potential well generated by the core nucleus in the model of atomic shell. The nuclear orbits form a series of core shells arranged with ascending energy. However, as soon as the outer shells are completely filled, then these nuclei are the most stable and correspond to the so called magic numbers. Figure 2.3 shows a nucleus composed of neutrons and protons and a shell model of neutrons for ^{32}Ne [29].

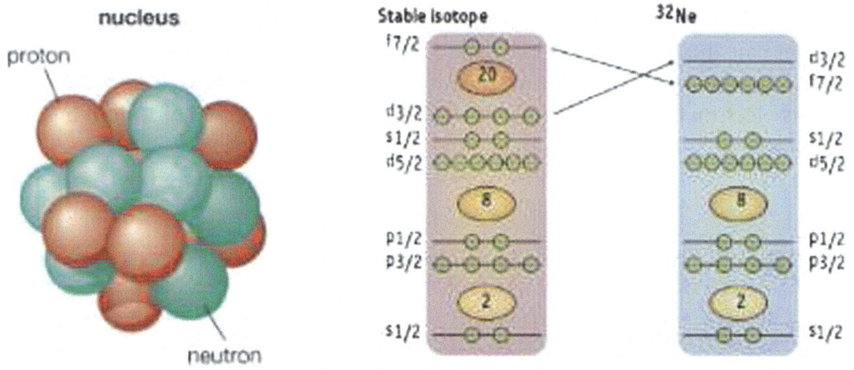

Fig. 2.3 Protons and neutrons of an nucleus, bonded by strong nuclear forces, and related shell model structure [29]

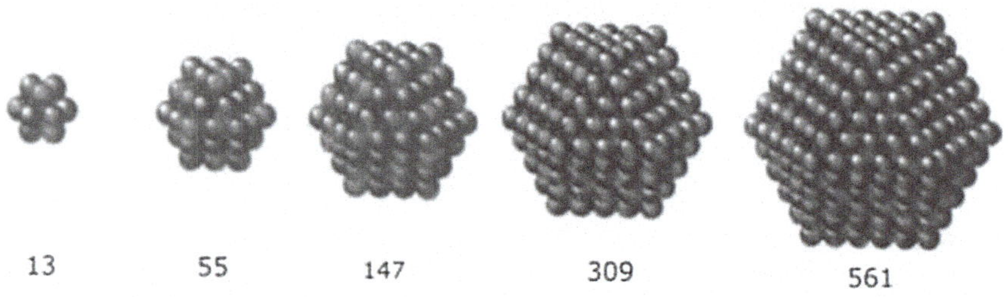

Fig. 2.4 Hexagonal close-packed full-shell magic number clusters of spherical atoms [31]

2.2.3 *Shell Structure of Condensed Clusters*

Shell Structure of Filled Cages: Not only electrons or nucleons but also metal and non-metal atoms can form shell structures. Burton [30] calculated the face-centered-cubic clusters of argon atoms having filled spherical shells of surface atoms. He considered spherical clusters of N argon atoms for $N = 13, 19, 43, \ldots$, up to 675. The shells of a cluster are filled up successively with atoms. The complete filled interior shells are regarded to be rigid, similar to the nuclear cores of atoms. These shells can have icosahedral, hexagonal, or cuboctahedral structures. Figure 2.4 shows hexagonal close-packed full-shell magic number of clusters. The first shell has 12-fold coordination. The number of atoms in each shell nth can be calculated by $(10n^2 + 2)$. However, the total number of atoms in the cluster of filled shells can be determined with $(5/3)n(n + 1)(2n + 1) + 2n + 1)$ [31].

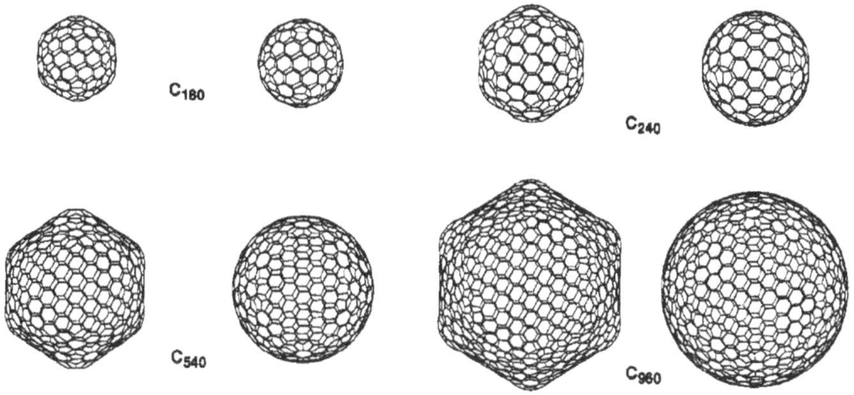

Fig. 2.5 Hollow cages represent shell structures of carbon atoms with the corresponding magic numbers [32]

2.2.4 Shell Structure of Hollow Clusters

The hollow cage of C_{60} fullerene represents the first shell structure of hollow clusters. This cage is of icosahedral structure composed of 60 vertices forming structure of 12, 20, and 90 pentagonal, hexagonal faces, and edges, respectively. Beside the icosahedral, there are another structures, which were found for higher spherical hollow fullerenes of tetrahedral, octahedral, hexahedral and dodecahedral forms. Figure 2.5 shows hollow fullerenes [32].

2.3 Magic Numbers in Nature

2.3.1 Magic Numbers of Electron Shells

The Ionization Energy is known to be the required energy to separate a valence electron from the ground state of an atom, ion, or cluster. Due to the fact that the electrons are distributed in different electronic shells, the energy needed to remove an electron from the outermost shell is called the first ionization energy. The second ionization energy is the energy needed to remove the next electron, and so forth. The first ionization energies of the chemical elements of the periodic system, as can be seen in Fig. 2.6, have an alternating or periodic character with maxima and minima values, as given in [33, 34].

Magic numbers: The maxima in Fig. 2.6 with the electron numbers 2, 10, 18, 36, 54 and 86, which respectively correspond to the noble gas atoms He, Ne, Ar, Kr, Xe, and Rn, are considered to magic numbers. These magic numbers are in turn the number of electrons that fill the quantum electron shells of the noble gas atoms

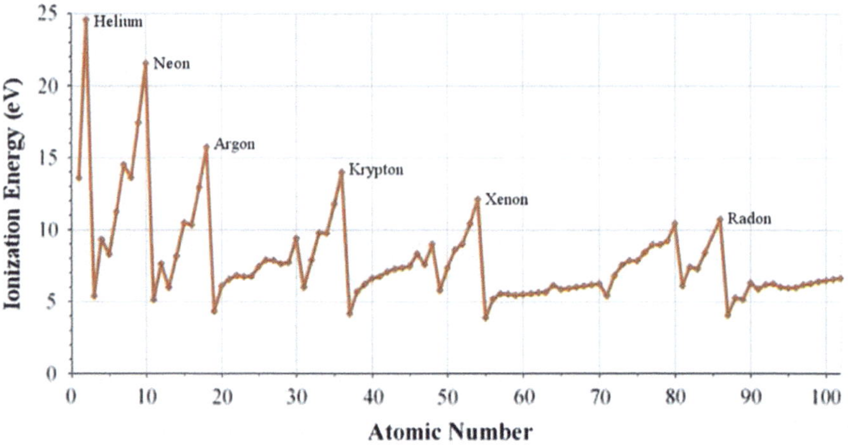

Fig. 2.6 First ionization energies of the elements. The maxima of the diagram 2, 10, 18, 36, 54, and 86 are the magic numbers. They correspond to the noble gas atoms [33]

which are chemically inert and difficult to ionize. This means that filled (closed) quantum shells are highly stable, and consequently more stable that unfilled or open quantum shells. Figure 2.6 shows besides the minima of the ionization energies of the Lanthanide atoms with atomic number $N > 70$, those which refer to the first column of the periodic system H, Li, Na, K, Rb, and Cs, which have only one single electron in the outer electron shell of s-type.

2.3.2 Magic Numbers of Nuclear Shells

Magic Nuclear Shells: In nucleus there exists also nuclear shells similar to the electron shells in atoms, as shown in the previous Sect. 2.3.1. As the electron occupation in electronic shells, protons and neutrons fill shells in nuclei, either partially or completely. Nuclear magic numbers occur for both, neutrons or protons. As soon as the nucleons fill completely the nuclear shells, the nucleus becomes an exceptionally high stability. The experiences have shown that nuclei have high binding energy and those with even number are more stable than those with odd number. These high stable nucleons have half-life of 10^{10} years. Figure 2.7 shows new and classical magic numbers of nucleons 2, 8, 20, 28, 50, 82, and 126 with exceptional stability corresponding to ${}^{4}_{2}He_2$, ${}^{16}_{8}O_8$, ${}^{40}_{20}Ca_{20}$, ${}^{48}_{20}Ca_{28}$, ${}^{54}_{26}Fe_{28}$, ${}^{90}_{40}Zr_{50}$, ${}^{132}_{50}Sn_{82}$, and ${}^{208}_{82}Pb_{126}$. These numbers indicate nucleons of high excitation energy larger than 2.5 MeV [35].

Doubly Magic Nuclear Shells: In general, the number of neutrons in nuclei is surplus compared to the number of protons. However, when their numbers are identical, then they do not undergo radioactive decay and are called doubly magic, as by the element oxygen. However, there are also magic numbers of highly stable isotopes, which have

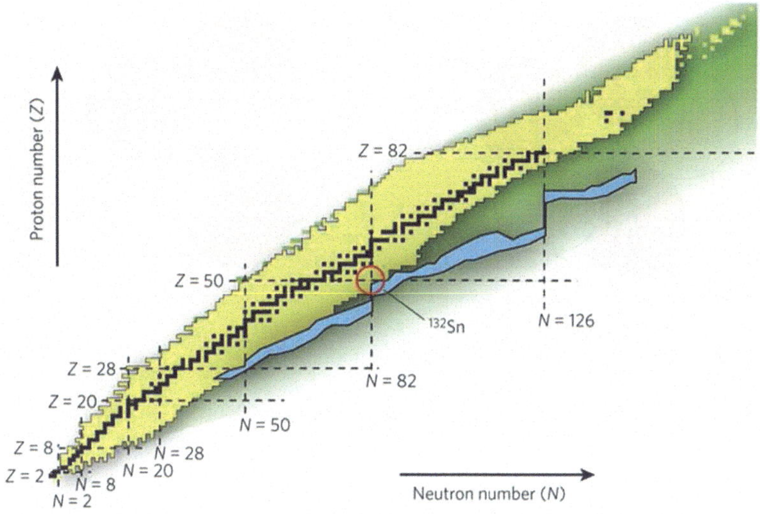

Fig. 2.7 The model of nuclear closed shell structure and the corresponding magic numbers of neutrons and protons, indicated as horizontal and vertical lines, at which the nuclei have exceptional stability, similar to electron shells of atoms of noble gas. The magic numbers 2, 8, 20, 28, 50, 82, and 126 indicate nucleons of high excitation energy larger than 2.5 MeV. This figure was reproduced with kind permission of the Nature Publishing Group [35]

more neutrons in their nuclei and do not decay into other elements, as by Sn and Pb. In general, nucleons with many protons are unstable. They undergo spontaneous positron or beta decay or electron capture to become stable. Also nuclides with more neutrons are unstable and undergo spontaneous beta decay towards stability. Beta decay, in turn, is a type of radioactivity, in which emitted neutrons decay into protons, electrons, and anti-neutrino, because free neutrons are unstable and have a mean lifetime of about 15 min only.

Nuclear Force and Repulsive Protons: Each nucleus of an atom consists of nucleons, which are composed of non-charged neutrons and positively charged protons. Due to the electrostatic or Coulomb laws, similar negatively or positively charged particles repulse each other. Therefore, the protons repulse each other so that the nuclei should be decomposed. However, at a very short distance of approximately 1.8 fm (1 fermi $= 10^{-15}$ m), a very strong nuclear attractive force is induced between the protons including the neutrons and leading to a high stable nuclei, as shown in Fig. 2.8 of [36]. When two protons are very close together the nuclear force is dominant, and when they are far apart the electrostatic repulsive force is dominant. The stability of nucleus, presented as V(r) in Fig. 2.9 [37], is given by the nuclear binding energy E_b, which holds the nucleons together. It is defined as the difference between

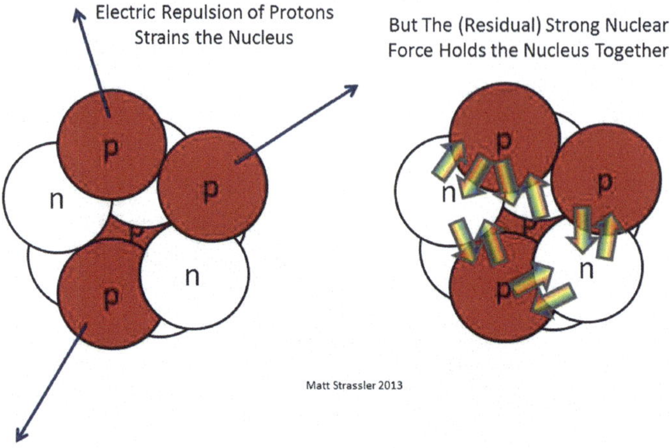

Fig. 2.8 Interplay of Coulomb repulsive and strong nuclear forces between the nucleons [36]

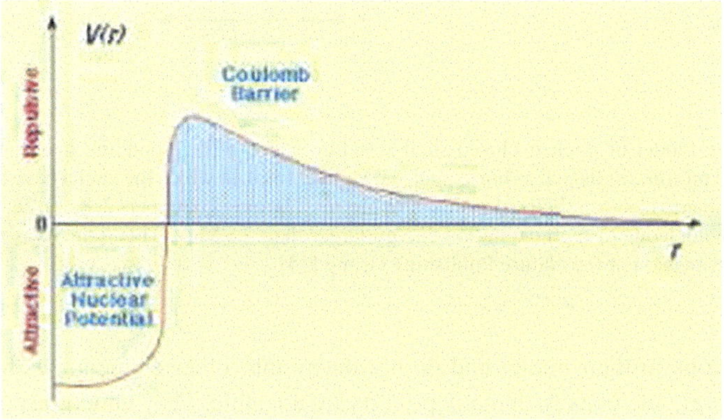

Fig. 2.9 The attractive repulsive potential of the nucleus as a function of intra-nuclear distance V(r). The shorter the distance, the stronger the nuclear force [37]

the mass of nucleus m_N and the total mass of the sum of the individual masses of nucleons $m_{p,n}$, the constituents of a nuclei, protons and neutrons, times the square speed of light: $E_b = [m_N - m_{p,n}]c^2$, or simply $E_b = \Delta mc^2$. Thus, the average of the binding energy per nucleon is about 8 MeV.

Ratio of Distances and Bonds: In order to understand the ratio of distances between the nucleons them-self and between nucleus and electrons, and related binding energy, we take Hydrogen atom H as an example for a comparison. Hydrogen has only one electron moving around the central proton with an average radius of 0.529 Å (1 Ångström = 10^{-10} m), which is known as the Bohr radius a_0. The electron binding (ionization) energy of its single electron in the ground state is about 13.6 eV. One can conclude that the electron-nucleus distance is about 10^5 larger than the proton-proton distance, while the binding energy of protons in nucleus is in the range of mega electron volt or with the factor 10^6 larger than binding electron. The radius of nucleus R is given by $R = r_0 A^{1/3}$, where A is the nucleus number. For example, the nucleus radius of carbon, silicon and iron atoms are 2.66, 3.43 and 4.35 fm, respectively.

2.3.3 Magic Electronic Shells of Clusters

Magic Numbers in Noble Gas Clusters: Echt et al. [38] investigated in 1981 the noble gas experimentally and confirmed for the first time that xenon atomic clusters exhibit magic numbers. The mass spectra of free xenon clusters indicate the magic numbers starting by 13 and up to the magic number 147. It is interesting to observe that some magic numbers like N = 13, 55, and 147 conform the numbers of spheres, which are needed to complete icosahedral shells. However, the occurrence of some magic numbers, like 19, 28, 71, and 87, is due to some extend explanation of former calculations. The mass spectrum as displayed in Fig. 2.10, is divided into two overlapping ranges. The latter Figure shows that the cluster intensity of the mass spectra is decreasing with increasing the cluster size N. It is obviously to see that some cluster sizes increase slightly approaching local maxima, afterwards they decrease and decline. Thus, after the maxima of the magic numbers $N = 13, 19, 25, 71, 87$, and 147, the cluster intensity occurs dramatic drops. However, some numbers like $N = 23, 81, 101$, and 135 could be temporary included in this list. Studying the ratio of intensity dropping at $N = 13, 19$, and 25, it is found to arrive factor 2, in opposite to average decreasing of a few percent.

Magic Numbers in Alkali-Metal Clusters: Small sodium clusters were investigated in 1984 by Knight and co-workers [1]. After production and ionization, the cluster ions were mass-selected. The obtained mass spectra shows particular peaks. Apparently, sodium clusters seem to exhibit shell-like electronic structure, as in atoms. Each peak represents a number of sodium atoms in clusters. The observed peaks are followed by abrupt decrease of the numbers $n = 2, 8, 20, 40, 58$, and 92, which represent electronic shells in Na_n clusters, as can be seen in Fig. 2.11, as reported by de Heer [39]. These numbers correspond to the valence electrons which form the electronic shells of clusters. This means that the valence electrons fill completely the electronic shells. Thus, clusters with closed electronic shells are exceptionally stable and more or less spherical in geometry. The possible electronic shells $1s$, $1p$, $1d$, $2s$, $1f$, $2p$, $1g$, $2d$, $3s$ are proposed by Woods–Saxon potential and Clemenger–Nilsson ellipsoidal shell model.

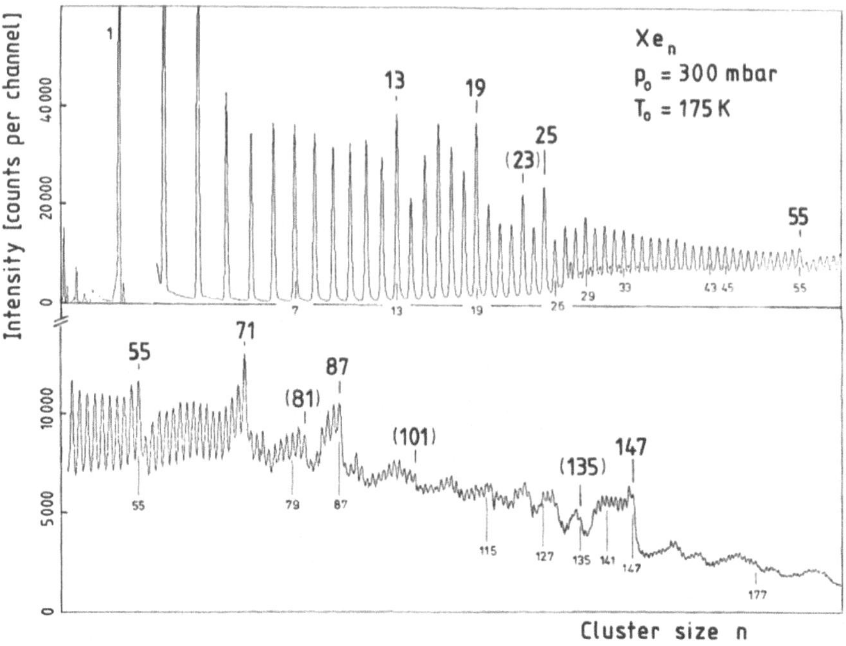

Fig. 2.10 Mass spectrum of xenon clusters. Observed magic numbers are marked in boldface; brackets are used for numbers with less pronounced effects. Numbers below the curve indicate predictions or distinguished sphere packing [38]

2.3.4 Magic Filled Shells of Clusters

Shells of atoms of sodium $(Na)_n$, aluminum $(Al)_n$ & $(Al)_n{}^+$, calcium $(Ca)_n$, strontium (Sr), and Indium $(In)_n^+$ clusters, as well as alkali-halide clusters such as $(NaI)_nNa^+$, $(NaBr)_nNa^+$, $(C_{60})_n^+$, and $C_{60}Ca_x{}^+$ were successfully synthesized and produced by Martin in 1996 [40]. For example, the mass spectra of $C_{60}Ca_x{}^+$, as a fullerene molecule covered by calcium atoms, and the corresponding number of shells are shown Fig. 2.12 [40]. Since a fullerene molecule has 12 pentagonal, and 20 hexagonal faces, for a total of 32. When one metal atom is placed of each of the 32 faces, then it results into the first shell of 32 calcium atoms with high symmetric and presumably high stable icosahedral cluster. Thus, the black circles represent the vertices of an icosahedron, while the white circles are calcium atoms centered on the triangular faces of an icosahedron.

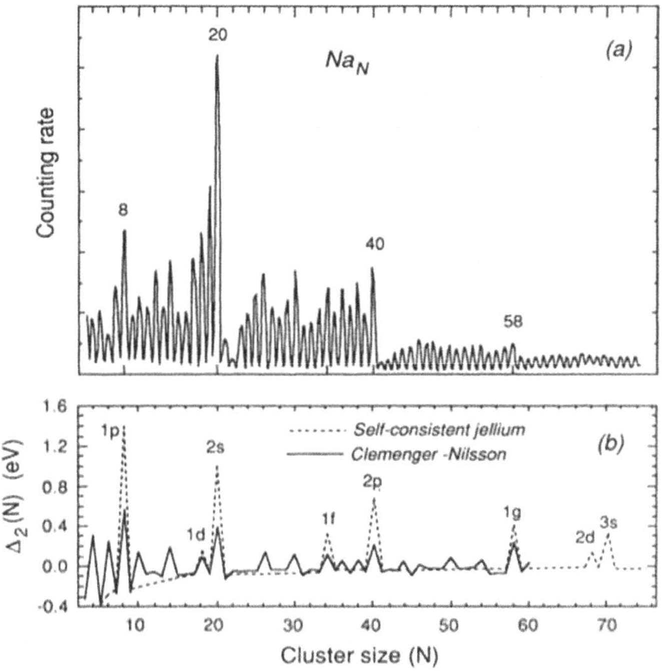

Fig. 2.11 Magic numbers and electronic shells of Na-clusters: **a** experimental abundance spectrum of Na clusters, **b** Woods–Saxon potential (dashed line) and Clemenger–Nilsson ellipsoidal shell model (solid line) [39]

Fig. 2.12 The molecule C_{60} is covered by calcium atoms. Each new complete icosahedral layer of metal atoms added to the C_{60} molecule results in a particularly stable structure [40]

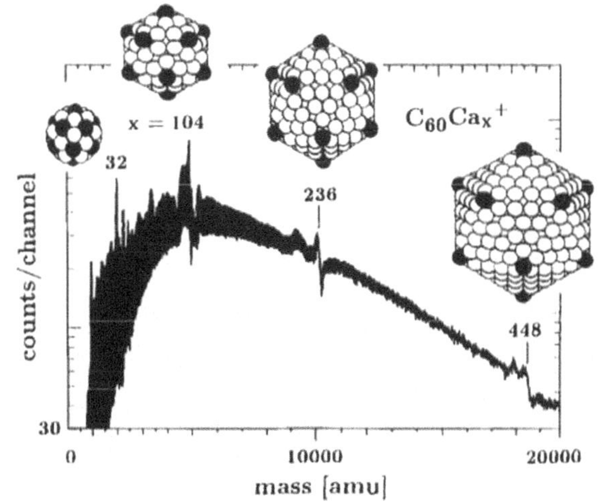

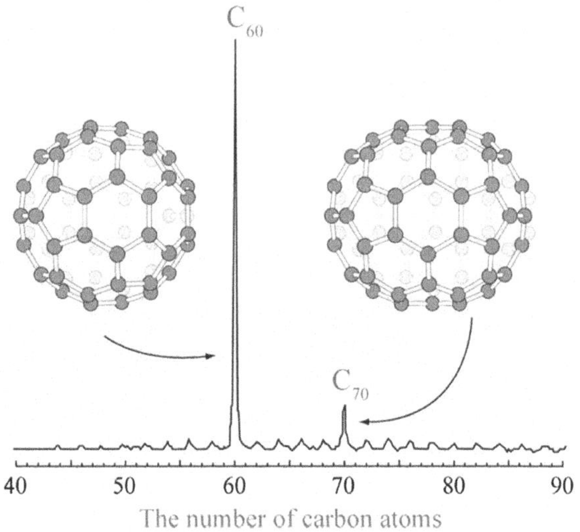

Fig. 2.13 Mass spectra of carbon clusters and the corresponding hollow shells of carbon atoms C_{60} and C_{70} [2]

2.3.5 Magic Hollow Shells of Clusters

The most famous magic number of hollow shell of clusters is the well known Buckminster fullerene C_{60}, which was discovered by Kroto et al. It has an icosahedral symmetry and composed of 12 pentagonal and 20 hexagonal faces. In their experiment, the authors have shown that the mass spectra of carbon clusters have two dominant peaks corresponding to the C_{60} and C_{70} carbon shells, which are shown in Fig. 2.13.

2.3.6 Magnetic Shells of Electrons

Electron Spin as Tiny Magnet: An electron possesses a spin that gives it a magnetic moment; that is, it behaves like a tiny magnet. The spin of electron is already defined as the fourth magnetic quantum number. According to Pauli principle, each orbital can hold a maximum of 2 electrons with different spin orientations. The spin magnetic numbers m_s are $1/2$ and $-1/2$, or spin up and spin down, respectively. Thus, electrons in closed shell systems occur in each orbital pairwise with opposite spins. These opposite spins cause magnetic fields that cancel each other, so that no more net magnet field exists by opposite spins. Alkali earth metals and noble gas are best examples with electron closed shells.

Electron Spin Multiplicity: The spin orientation of valence electrons in atoms, molecules or clusters can be determined through the expression of spin multiplicity $M = 2S + 1$, with $S = \Sigma m_s$, the sum over all electron spins $m_s = \pm\frac{1}{2}$. Here are

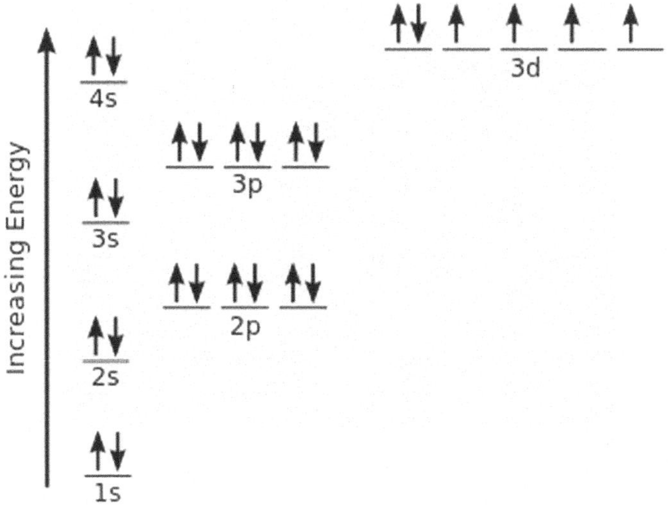

Fig. 2.14 The electron configuration of Fe-atom. The eight valence electrons correspond the 4*s* and 3*d* electronic shells. Due to Hund's rule, the single orbitals must be occupied firstly [41]

some simple examples. Helium atom has two electrons. One electron is with spin up $+\frac{1}{2}$, and one is with spin down $-\frac{1}{2}$. Thus, the multiplicity of a Helium atom, say the spin state is $M = 2(\frac{1}{2} - \frac{1}{2}) + 1 = 1$, assigned as singlet. Therefore, the electrons of Helium are diamagnetic, because the magnetic fields of both spins, up and down, cancel each other. Hydrogen atom has only one electron with $m_s = +\frac{1}{2}$. Thus, the spin of Hydrogen is $M = 2(\frac{1}{2}) + 1 = 2$, and its spin state is 2, or simply doublet with a tiny magnet. The electron configuration of oxygen in the ground state, has two degenerate molecular orbitals each is occupied by one electron. Therefore, the multiplicity of oxygen is $M = 2(\frac{1}{2} + \frac{1}{2}) + 1 = 3$. It is called as triplet state, and hence oxygen is paramagnetic.

Electron Configuration of Fe-Atom: One of the most important atoms with unpaired electrons is iron. At the atomic level, the occupation of electrons in atomic shells of iron atom obeys the Hund's rule. It states that before the electrons occupy the orbital with double electrons, each orbital must firstly be singly-occupied. Thus, as to be seen in Fig. 2.14, iron has at the atomic level four spin up electrons in the 3d shell whose spin magnetism doesn't cancel but is still paramagnetic. The multiplicity of iron atom at the ground state is $M = 2(4 \times \frac{1}{2}) + 1 = 5$, and the spin state is called quintet.

Spin Electrons of Fe_{16} **Planar Cluster**: The Fe_{16} planar cluster is a cut from the BCC(110) surface of iron lattice with the experimental lattice constant 2.86 Å, as presented in Fig. 2.15. It shows that Fe_{16} cluster has different spin orientations denoted by arrows on each iron atom. Since each iron atom of the cluster has four parallel electrons of paramagnetic character, then it is to expect to have 64 parallel electrons

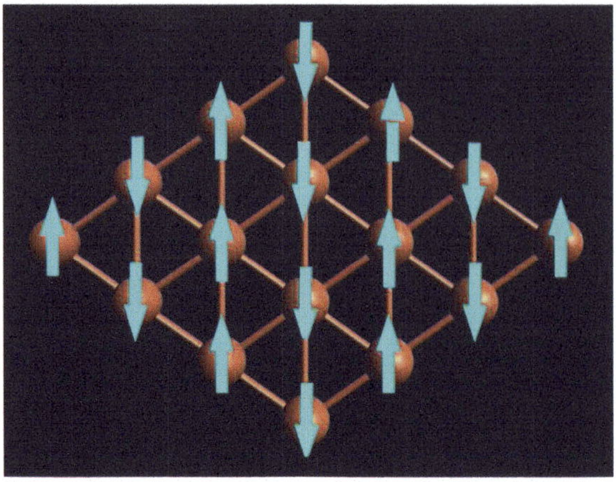

Fig. 2.15 The electron configuration of Fe-atom. The eight valence electrons correspond the 4*s* and 3*d* electronic shells. Due to Hund's rule, the single orbitals must be occupied firstly [41]

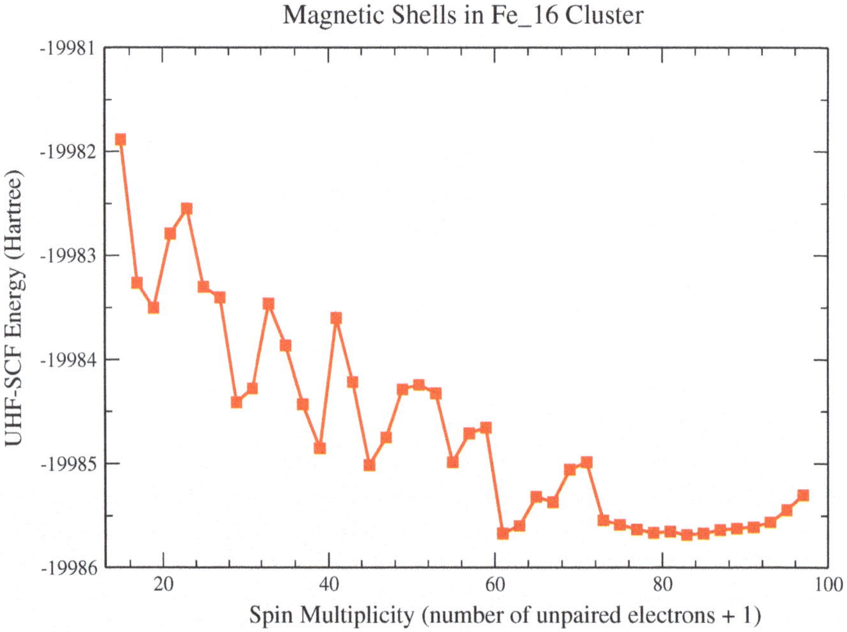

Fig. 2.16 Magnetic Shells indicated by the UHF-SCF energy minima and corresponding to the electron numbers 18, 28, 38, 44, 54, 60, and 82. The energy is a function of the multiplicity of the Fe_{16} cluster [41]

with paramagnetic Fe_{16} cluster. However, the orientation of spins in the Fe_{16} cluster is chaotic, due to the following reasons: (i) the strong interaction of dipoles induced by spins, (ii) the electron spin-spin coupling, (iii) the electron spin-orbit coupling, (iv) the electron-phonon coupling (cluster vibration), (v) the spin fluctuation effect, and finally, (vi) the absence of an external magnetic field. Thus, according to the spins disorientation, the ground spin state of the Fe_{16} planar cluster is unknown.

Spin States Calculations of Fe_{16} Planar Cluster: An attempt to investigate the ground spin state of the Fe_{16} planar cluster was undertaken via ab initio quantum chemical methods [41]. The preliminary calculations, carried out at the UHF-SCF level of theory using the standard STO-3G basis set, were executed for different spin states starting from spin state $M = 15$ towards the high spin state number $M = 97$, where M is the multiplicity, given in Sect. 2.3.6. The obtained energy of the Fe_{16} cluster is depicted in Fig. 2.16, as a function of the multiplicity. The energy function exhibits an oscillating character and possessing energy minima at the numbers 19, 29, 39, 45, 55, 61, 83. They correspond to the number of parallel electrons 18, 28, 38, 44, 54, 60, and 82, as magic magnetic shells.

Nevertheless, calculations at high level of theory under consideration the interaction and coupling effects between the spin electrons, mentioned in the latter paragraph, are desired.

References

1. W.D. Knight, K. Clemenger, W.A. de Heer, W.A. Saunders, M.Y. Chow, M.L. Cohen, Electronic shell structure and abundances of sodium clusters. Phys. Rev. Lett. **52**, 2141 (1984)
2. H.W. Kroto, J.R. Heath, S.C. O'Brien, R.F. Curl, R.E. Smalley, C_{60}: buckminsterfullerene. Nature **318**, 162–163 (1985)
3. D. Shechtman, I. Blech, D. Gratias, J.W. Cahn, Metallic phase with long-range orientational order and no translational symmetry. Phys. Rev. Lett. **53**, 1951–1953 (1984)
4. I. Boustani, W. Pewestorf, P. Fantucci, V. Bonaić-Koutecky̌, J. Koutecky̌, Systematic ab initio configuration-interaction study of alkali-metal clusters: relation between electronic structure and geometry of small Li clusters. Phys. Rev. B **35**, 9437 (1987)
5. A. Kornath, A. Kaufmann, A. Zoermer, R. Ludwig, Raman spectroscopic investigation of small matrix-isolated lithium clusters. J. Chem. Phys. **118**, 6957 (2003)
6. I. Boustani, Systematic LSD investigation on cationic boron clusters $B_n (n = 2 - 14)$. Int. J. Quantum Chem. **52**, 1081 (1994)
7. H.J. Zhai, B. Kiran, J. Li, L.-S. Wang, Hydrocarbon analogues of boron clusters - planarity, aromaticity and antiaromaticity. Nat. Mater. **2**, 827 (2003)
8. Z.A. Piazza, H.-S. Hu, W.-L. Li, Y.-F. Zhao, J. Li, L.-S. Wang, Planar hexagonal B_{36} as a potential basis for extended single-atom layer boron sheets. Nat. Commun. **5**, 3113 (2014)
9. I. Boustani, A. Quandt, Nanotubes of bare boron clusters. Ab initio and density functional theory. Europhys. Lett. **39**, 527 (1997)
10. I. Boustani, A. Quandt, E. Hernández, A. Rubio, New boron based nanostructured materials. J. Chem. Phys. **110**, 3176 (1999)
11. D. Ciuparu, R.F. Klie, Y. Zhu, L. Pfefferle, Synthesis of pure boron single-wall nanotubes. J. Phys. Chem. B **108**, 3967 (2004)

12. F. Liu, C.-M. Shen, Z.-J. Su, X.-L. Ding, S.-Z. Deng, J. Chen, N.-S. Xu, H.-J. Gao, Metal-like single crystalline boron nanotubes: synthesis and it in situ study on electric transport and field emission properties. J. Mater. Chem. **20**, 2197 (2010)
13. I. Boustani, Systematic ab initio investigation of bare boron clusters: determination of the geometry and electronic structures of $B_n (n = 2 - 14)$. Phys. Rev. B **55**, 16426 (1997)
14. X.-F. Zhou, X. Dong, A.R. Oganov, Q. Zhu, Y.-J. Tian, H.-T. Wang, Semi-metallic two-dimensional boron allotrope with massless Dirac fermions. Phys. Rev. Lett. **112**, 085502 (2014)
15. A.R. Oganov, J. Chen, C. Gatti, Y.-M. Ma, T. Yu, Z. Liu, C.W. Glass, Y.-Z. Ma, O.O. Kurakevych, V.L. Solozhenko, Ionic high-pressure form of elemental boron. Nature **457**, 863–867 (2009)
16. S. Iijima, Helical microtubules of graphitic carbon. Nature **354**, 56–58 (1991)
17. K.S. Novoselov, A.K. Geim, S.V. Morozov, D. Jiang, Y. Zhang, S.V. Dubonos, I.V. Grigorieva, A.A. Firsov, Electric field effect in atomically thin carbon films. Science **306**, 666–669 (2004)
18. R. Tenne, L. Margulis, M. Genut, G. Hodes, Polyhedral and cylindrical structures of tungsten disulphide. Nature **360**, 444–446 (1992)
19. S. Seifert, H. Terrones, M. Terrones, G. Jungnickel, T. Frauenheim, Structure and electronic properties of MoS_2 nanotubes. Phys. Rev. Lett. **85**, 146 (2000)
20. http://www.angelo.edu/faculty/kboudrea/periodic/physical_metals_fig1.gif. Accessed 26 Oct 2014
21. G.C. Benson, R. Shuttleworth, The surface energy of small nuclei. J. Chem. Phys. **19**, 130–131 (1951)
22. F.A. Cotton, Metal atom clusters in oxide systems. Inorg. Chem. **3**(9), 1217–1220 (1964)
23. J.J. Burton, Anomalous heat capacity of spherical clusters of atoms. Chem. Phys. Lett. **3**, 594–596 (1969)
24. K. Wade, The structural significance of the number of skeletal bonding electron-pairs in carboranes, the higher boranes and borane anions, and various transition-metal carbonyl cluster compounds. J. Chem. Soc. D 792–793 (1971)
25. K. Wade, Structural and bonding patterns in cluster chemistry. *Advances in Inorganic Chemistry and Radiochemistry*, vol. 18 (1976), pp. 1–66
26. M.M. Balakrishnarajan, E.D. Jemmis, Electronic requirements of polycondensed polyhedral boranes. J. Am. Chem. Soc. **122**, 4516–4517 (2000)
27. E.D. Jemmis, Building relationships between polyhedral boranes and elemental boron. Chemtracts Inorg. Chem. **18**, 620–628 (2005)
28. G. Audi, The history of nuclidic masses and of their evaluation, arXiv:physics/0602050v1. Accessed 1 Oct 2014
29. Rikenresearch: http://www.riken.jp/en/research/rikenresearch/highlights/6041/. Accessed 1 Oct 2014
30. J.J. Burton, Configuration, energy, and heat capacity of small spherical clusters of atoms. J. Chem. Phys. **52**, 345–352 (1970)
31. K.J. Klabunde, R.M. Richards, *Nanoscale Materials in Chemistry*, 2nd edn. (Wiley, Inc., Hoboken, 2009)
32. http://homepage.hispeed.ch/bakowies/graphics/fullerenes_allfullerenes.jpg. Accessed 1 Oct 2014
33. https://en.wikipedia.org/wiki/File:Ionization_energies.png. Accessed 1 Jun 2019
34. N.N. Greenwood, A. Earnshaw, *Chemistry of Elements* (Pergamon Press, Oxford, 1984)
35. P. Cottle, Nuclear physics: doubly magic tin. Nature **465**, 430–431 (2010)
36. http://profmattstrassler.files.wordpress.com/2013/02/nuclear_force.png. Accessed 1 Oct 2014
37. https://encrypted-tbn1.gstatic.com/images?q=tbn:ANd9GcRD1GBCA6O9RIZaapQKVx5TJK_Wbo9xlvn-SSupjvCMalDmqwEz0A. Accessed 1 Oct 2014
38. O. Echt, K. Sattler, E. Recknagel, Magic numbers for sphere packings: experimental verification in free xenon clusters. Phys. Rev. Mod. Lett. **47**, 1121–1124 (1981)
39. W.A. de Heer, The physics of simple metal clusters: experimental aspects and simple models. Rev. Mod. Phys. **65**, 611–676 (1993)
40. T.P. Martin, Shells of atoms. Phys. Rep. **273**, 199–241 (1996)
41. I. Boustani, Ab initio quantum chemical calculations of Fe_{16} planar cluster (2017). Unpublished results

Chapter 3
Alkali Metal Clusters

Abstract Clusters and magic numbers were the major research field mid 80s. Many groups around the world worked theoretically and experimentally to find the relationship between the electronic shells and the magic numbers. The structures of the Lithium and Sodium alkali-Metal, as well as Carbon clusters and their electronic properties were the first investigated agglomerates.

Introduction It is well-known that structures of small molecular aggregates or clusters establish or simulate a transition state between the gas phase and solid state. Therefore, the structural and electronic properties of clusters with a very small number of atoms are of great signification for chemical and physical problems, like nucleation at vapor phase or agglomeration in dense fluid, thermal capacity and temperature change in nanomaterials. In order to handle such kind of problems, it is recommended to simulate small clusters as a cutout from the surface of solid material and reproduce the related properties. Indeed, many investigations have been done for small clusters due to the great computational capacities nowadays which can predict properties with very high accurrence approaching experimental data.

3.1 Alkali Metals

The group 1 of the periodic table consists of 6 elements known as alkali metals, which includes the smallest lithium atom and up to the largest francium atom. However, the most abundant elements of this group are sodium and potassium, which make, respectively, 2.6 and 2.4 percent of the Earth's crust. The other alkali metal elements, like lithium, rubidium and cesium, are considered to be rare. Therefore, lithium is highly demanded element due to its significant important role in different energy storage and particularly its potential applications in batteries. Most of these elements are often stored in mineral oil and are not found in their elemental forms in nature [1].

I. Boustani, *Molecular Modelling and Synthesis of Nanomaterials*, Springer Series in Materials Science 290, https://doi.org/10.1007/978-3-030-32726-2_3

Due to their singly valence electron, they have the largest atomic radii of the elements in their respective periods. They exhibit many physical properties common to metals, and are soft, silver metallic. Their densities are lower than those of other metals. They can easily lose their valence electrons [2]. Though they have low ionization energy and low electronegativity. They are highly reactive and flammable. Due to their high reactivity they exist in nature in combination with other elements. They react readily with non-metals, particularly halogens, like metal-chloride (MCl) minerals: (halite) lithium chloride LiCl, (salt) sodium chloride NaCl, (sylvite) Potassium chloride KCl, and (carnallite) magnesium chloride $MgCl_2$. All of these minerals can easily be extracted and purified due to their high solubility in water (Fig. 2.2).

3.2 Lithium Metal Clusters

3.2.1 Production of Metal Clusters

How are small clusters made? Gas aggregation technique is the most popular technique to generate small atomic clusters. First of all, a sample of the material of the desired clusters in furnace is needed to be heated to reasonable high temperatures under vacuum until the sample is evaporated. The compounds of the evaporated material are composed of separated into individual non bonding atoms. The high temperatures beam of this evaporated gas will be mixed with a cold jet of inert gas, which in turn carries it forward. The low temperature inert gas collides with the atoms of the original materials causing energy loss. These energy lost atoms accumulate together forming aggregates of small clusters with different number of atoms, say with different cluster sizes. In order to determine the mass of these clusters, an electron or laser beam is induced to cross the beam of the obtained clusters causing cluster ionization. These charged clusters can be then analyzed and separated by mass spectrometer [4].

The Mass Spectroscopy is simply a technique that determines the mass of individual atoms, molecules or clusters. It is based on the fact that ions with different masses travel with different velocities. These ions are formed by photo ionization, which is traditionally done by laser. Charged clusters of different sizes have different masses, which can be analyzed by mass spectrometer. The latter separates the masses based on charge to mass ratios and output them to the detector which converts them to a digital output. Figure 3.1 shows a schematic presentation for cluster production, analysis and detection.

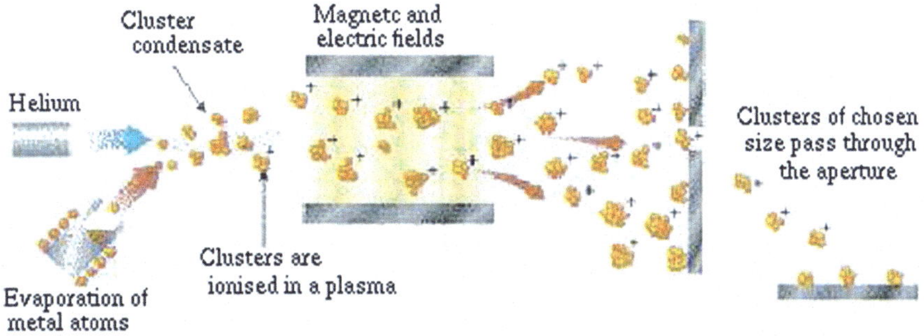

Fig. 3.1 Clusters production [5]

3.2.2 *Lithium Atom*

Lithium metal does not exist in nature in pure phase, but as lithium salts. The root name of lithium is derived from lithos, the Greek word for "stone". Figure 3.2 shows the lithium atom and its crystalline form. Lithium is the lightest used element in the periodic system after hydrogen, which is present in almost all of our life. Its melting point is around 180°C, boiling point 1342°C. It is unique among the minerals, used in synthesis organic compounds and in heat transfer applications as well. Lithium is also included in heat resistant glasses and ceramics. Due to its electrochemical potential, it is very useful for battery anodes. Therefore, the most important use of lithium is in rechargeable batteries, tablets, laptops, and mobile phones. The binary composition of MX, M for alkali-metal and X for halogens atoms is called alkali metal halides. For example, lithium chloride and lithium bromide are highly hygroscopic, and therefore they are used in dehydration as drying agents. In addition, lithium is an ideal element for engineering use like the high-temperature lubrication. The lithium grease provides thicker lubrication, reduces friction, and protect metal-to-metal applications. Despite that lithium is toxic, it has regardless medical applications like in food, drugs, dose-measuring device, and in treatment of mania which is part of bipolar disorder (manic depressive illness). Although it has been used in this area since about 1950, its acceptance has been slow, possibly because it is a natural mineral and not as profitable for the pharmaceutical companies as synthetic drugs [1]. Chemically lithium behaves more closely similar to group (IIa) in the periodic system, and it reacts with most acids giving off hydrogen gas. But however, it does react with oxygen above 100°C but not at room temperature.

Fig. 3.2 The symbol of Lithium atom, the constituents of nucleus and electrons and the corresponding solid [6]

3.2.3 Lithium Clusters

Long before the electronic shells in alkali-metal clusters were discovered, there were only few studies on such metal systems. In the early 1970s, the interest was not focused on clusters but on band-structure calculations by improving the crystal potentials. Lithium is the simplest metal and its crystal structure was the subject of many studies. Hence, Stoll and Preuss [7] studied in 1972 small periodic aggregates and linear chains of Li and Be atoms employing the method of Hartree–Fock molecular-orbital and linear combination of Gaussian orbitals (SCF-MO-LCGO). They have found that small cluster systems can reproduce different properties and characteristic of solids. In addition, they discussed the influence of the shape and size of the cluster systems on these properties. They also found that the binding energy of Li configurations increases with the number of atoms N, monotonous for the chains, but partially alternating for the layers. They also reported that chains are energetically favored in comparison to layers and crystals for the same number of atoms N.

3.2.4 Experimental Evidence

Mass Spectra of Lithium Clusters: Dugourd et al. [8] determined via photoionization technique the mass spectra of Lithium clusters. The selection of these clusters is based on mass analysis using the time of flight (TOF) mass spectrometer, after frequency doubled excimer pumped tunable dye laser ionization. The mass spectra of ionized Li clusters or of Li_n^+ cations is shown in Fig. 3.3 and is recorded for different laser wavelengths. As can be seen, the ionization threshold is decreasing as a function of cluster size. However, the mass spectra exhibit a lack of resolution, which is raised through two isotopes ^{6}Li (7%) and ^{7}Li (93%). Beyond Li_{40}, lithium clusters peaks of

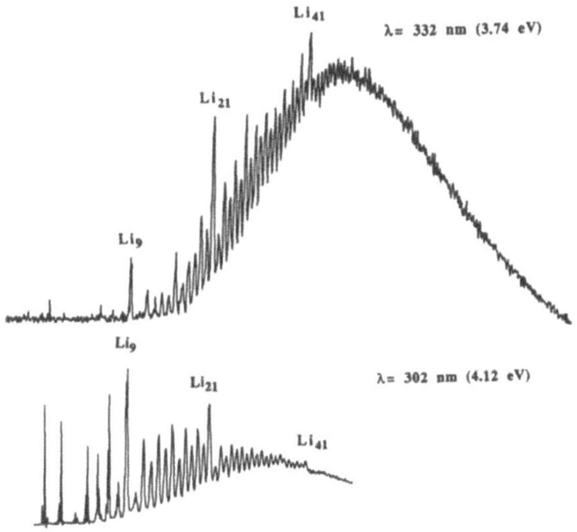

Fig. 3.3 Mass spectra of ionized Li_n clusters as a function of the laser wavelength. The Li_9, Li_{21} and Li_{41} cations have respectively 8, 20 and 40 valence electrons, which are representing completely filled electronic shells [8]

comparable intensity could possess the same mass number, but with different number of atoms. The Li_9, Li_{21} and Li_{41} ionized clusters, which indicate high intensity in the mass spectra, seem to exhibit low ionization threshold. Furthermore, Brechignac et al. [9] determined one year later via photoionization technique also the mass spectra of lithium clusters. The authors used gas-aggregation type source at moderate pressure of around 0.1 torr and by heating via molybdenum oven to produce metal vapor. The latter is diffused in helium stream through 4 mm bore. The metal clusters are transported along by 20–25 torr of helium into a 6 cm long copper tube, which is in turn cooled with liquid nitrogen. Through the cooling along the helium stream the metal clusters waste their formation energy and start to the aggregate. However, the growth of clusters stops at the end of the copper tube, at which the mixture of metal clusters and helium gas enters the three sequent chambers of differential pumping system. The next step is to ionize the neutral stream of metal clusters. They are photoionized by means of a pulsed Nd-YAG laser at a photon energy of around $h\nu = 3.5$ eV. The ionization occurs in between a multi-plate accelerating system of the time-of-flight (TOF) mass spectrometer. The ionized lithium clusters Li_n^+ are accelerated to 2.8 kV streaming by streaming through the drift tube with a length of about 120 cm. A secondary electron multiplier delivers ion signals averaged by a Le Croy 9400 digital oscilloscope. The mass analyzed of mass spectrometer of the Li_n^+ clusters for a given laser wavelength is presented in Fig. 3.4.

The Ionization Potentials (IP) of Lithium Clusters: Broyer and co-workers [10] determined the IP of Li_n clusters experimentally. As shown in Fig. 3.5, the ionization potentials have two main features. The first feature is a clear even-odd alternation, while the second one is a sudden reduction beyond $n = 8$ and $n = 20$ for Li_9 and Li_{21}.

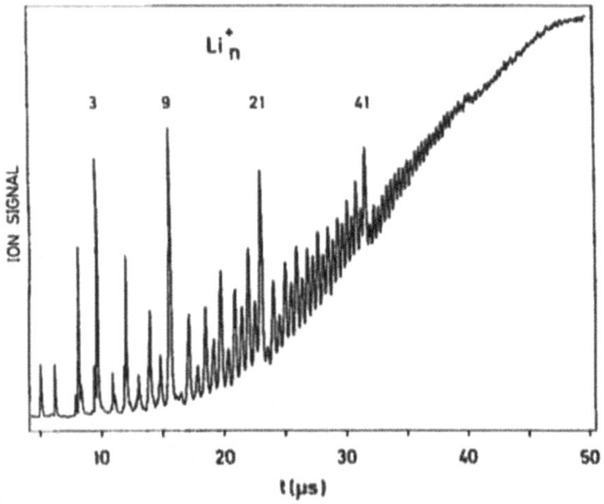

Fig. 3.4 Mass spectra of Li_n^+ clusters as a function of the laser fluence. The shell closing effect is visible as an abrupt change in ion intensities after Li_3^+, Li_9^+, Li_{21}^+, Li_{41}^+ Li_{69}^+ and Li_{93}^+. These cations have respectively 2, 8, 20, 40, 68, and 92 valence electrons, which are representing completely filled electronic shells [9]

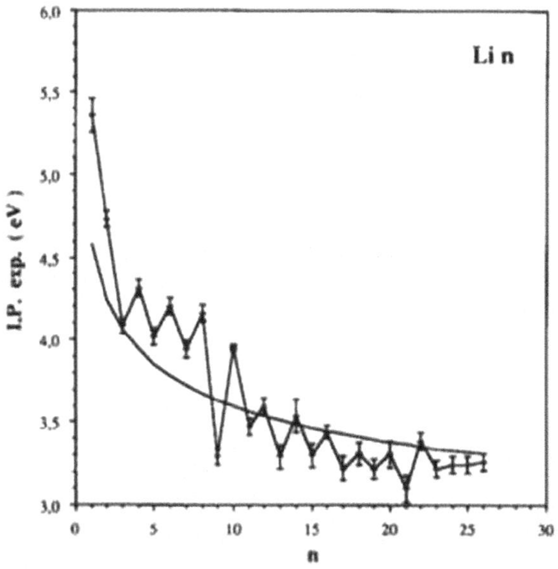

Fig. 3.5 The ionization potential of Li_n clusters (eV), as a function of cluster size n. The low ionization values of Li_9 and Li_{21} clusters are evident. The solid line corresponds to the prediction from the classical droplet model assuming a bulk atomic radius of 1.72 Å, a work function of 2.37 eV and an electron spill-out of 0.746 Å [10]

However, the sudden reduction of the ionization potentials can explained by the shell model of metal clusters: similar to the electronic shells in atom, they can be closed by n = 8 and 20. Thus, the numbers n = 8 and n = 20 correspond to closed electronic shells of lithium clusters. Therefore, the Li_8 and Li_9^+ clusters with 8 electrons, and Li_{20} and Li_{21}^+ clusters with 20 electrons are highly stable. The clusters Li_9, Li_{21} with 9 and 21 electrons seem to be less stable. However, the experimentally determined ionization potential, obtained by means of linear extrapolation, is estimated to be located between vertical and adiabatic ionization potential. Nevertheless, the agreement between the experiment and theory is excellent, within only of 0.2 eV.

However, Dugourd et al. [10] found that the experimental results for small cluster sizes are in good agreement with the ab initio calculations carried out by Boustani et al. [21]. The measured ionization potentials show strong odd-even alternation and pronounced drop in magnitude at $n = 9$ and $n = 21$. In addition, for small cluster sizes n with $n \leq 10$ the odd-even alternation is present also.

Electronic Shells and Magic Numbers in Lithium Clusters: After discovering the electronic shells in Sodium clusters by Knight et al. [11], further alkali-metal clusters were studied by Chou et al. [12]. They calculated the ground state energies of the alkali-metal clusters Li, Na, and K, using the density functional theory as well as the Jellium model. Due to the existence of electronic shell structures in clusters, the authors found that The relative binding energy change δ(N) as a function of cluster size has discontinuities. Therefore, the related magic numbers of closing shells 2, 8, 18, 20, 34, 40, 58, 68, and 92, which correspond to the electronic shells $1s^2$, $1p^6$, $1d^{10}$, $2s^2$, $1f^{14}$, $2p^6$, $1g^{18}$, $2d^{10}$, $3s^2$, and $1h^{22}$, are consistent with the positions of peaks or steps in the mass spectra of sodium. As can be seen in Fig. 3.6, the results of the relative binding energy change δ(N) for lithium and sodium are similar to those for potassium calculated with the Jellium model. The only exception for the shell structure of potassium is at $N = 40$, which gives no rise or change in the total energy.

3.2.5 Li_n Clusters for n $\leq$ 6

Molecular Models of Li_n Clusters (n = 2 and 4): Ray et al. [13] investigated very small Lithium clusters at the ab initio Hartree–Fock and configuration interaction level of theory. They studied different isomers of Li_3 and Li_4 clusters, as presented in Fig. 3.7. They found that an isosceles triangle Li_3 and a rhombus Li_4 are clusters with the most stable structures.

Molecular Models of Li_n Clusters (n = 3 to 6): In 1982, Richts-meier et al. [14] studied the structures of metal clusters M_n ($3 \leq n \leq 6$) of group 1A, applying one of the semi-empirical methods named the diatomic-in-molecules (DIM) approximation using parameterized integrals. They found that the high stable clusters seem to have three-dimensional compact structures. Figure 3.8 shows different isomers of Li_n Clusters for n = 3 to 6. Each structure of these isomers is assigned with the related symmetry point group and the inter-atomic distances, varying between 5.1 and 6.0 a.u. (1 a.u. is around 0.52917 Å).

Jellium Model and Electronic Shells: Rao and Jena [15] carried out the Hrtree-Fock theory particularly the UHF-SCF method at the LCAO-MO procedure as well as the density functional theory (DFT) to calculate the total energies and electronic properties of Li_n clusters for ($n = 1 - 5$). They considered the correlation energy applying the Möller–Plesset approximation to improve the UHF results. They optimized the shapes and stabilities of Lithium clusters using Hellmann–Feynman forces. They confirmed the results of Ray et al. and found that the clusters with isosceles

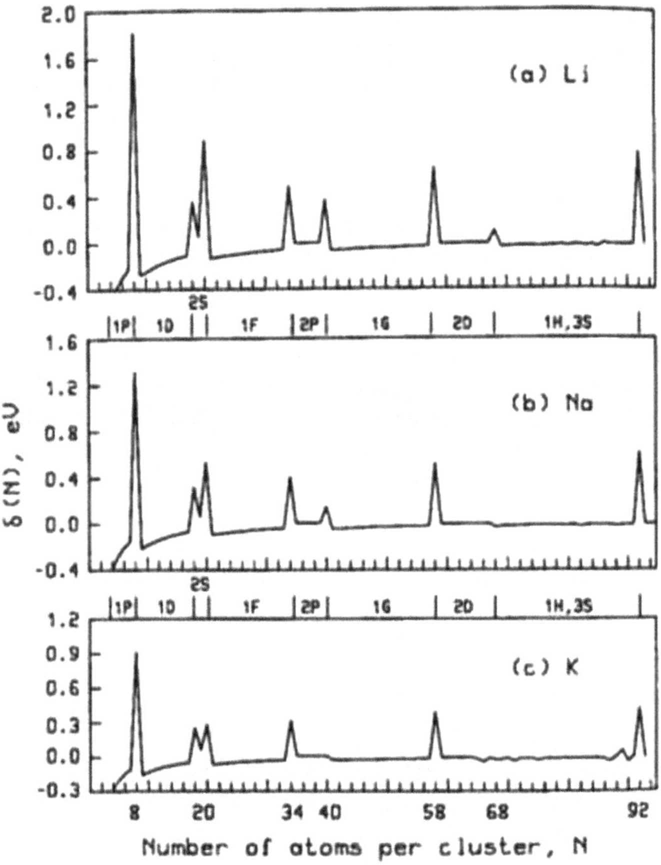

Fig. 3.6 The relative binding energy change $\delta(N)$ versus N for Lithium, Sodium, and Potassium. N is the number atoms in cluster. The peaks 8, 18, 20, 34, 40, 58, 68, and 92 correspond to filled electronic shells $1s^2$, $1p^6$, $1d^{10}$, $2s^2$, $1f^{14}$, $2p^6$, $1g^{18}$, $2d^{10}$, $3s^2$, and $1h^{22}$ [12]

triangle Li_3, rhombus Li_4, and planar three conjuncted triangles Li_5 structures are the most stable isomers of Li clusters, as shown in Fig. 3.9. The authors provided the total energy per atom of Li_n clusters with ($1 \leq n \leq 42$) applying the Jellium model calculations with spin-polarized density functional.

Jellium model is based on the idea that the positive background of the ions is smeared out and the negative charge is regarded as the conduction electron density. Thus, the charges of the system n_{ext} are modeled by

$$n_{ext} = n_0 \, \Theta(R - r) \tag{3.1}$$

where n_0 is the homogeneous spherical density, and R is the radius of a sphere containing N atoms, each atom is occupied with a single electron. By a volume V

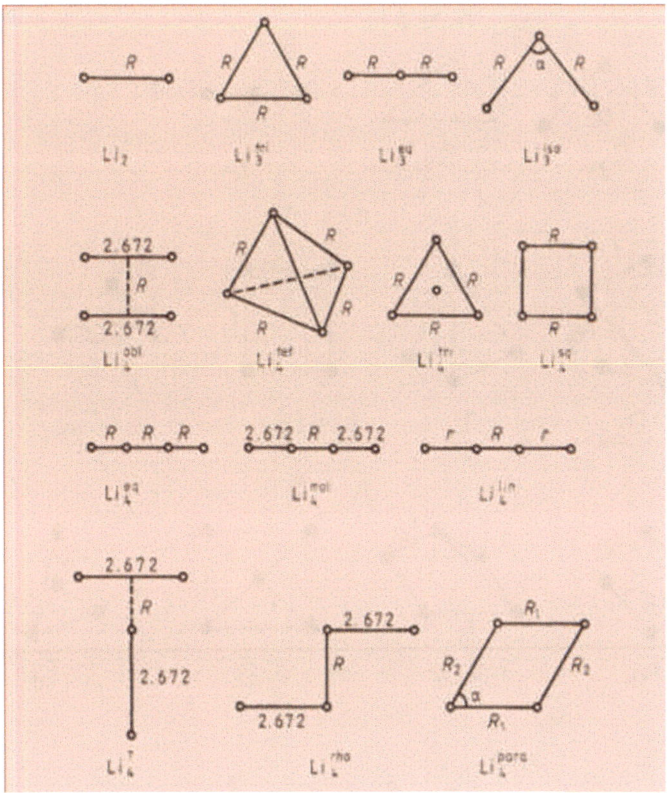

Fig. 3.7 The considered isomers of Lithium clusters, determined at the SCF and CI level of theory [13]

and a total number of atoms or electrons N, the electron density parameter n_0 is given as

$$\frac{V}{N} = \frac{1}{n_0} = \frac{4\pi}{3} (r_s a_0)^3 \tag{3.2}$$

where r_s is the dimensionless Wigner–Seitz or electron density parameter, a_0 is the Bohr radius. The last equation shows that electron density n_0 is a function of the density parameter r_s. The radius of the sphere containing N atoms is given as

$$R = r_s\, a_0\, N^{1/3} \tag{3.3}$$

The authors used for all Jellium calculations of lithium clusters the R_s bulk values. Solving the density functional Kohn–Sham equation self-consistently they obtained the electronic structures of these spherical Jellium clusters. The Jellium model is of course an oversimplification of reality, since the Li clusters up to five Li atoms form

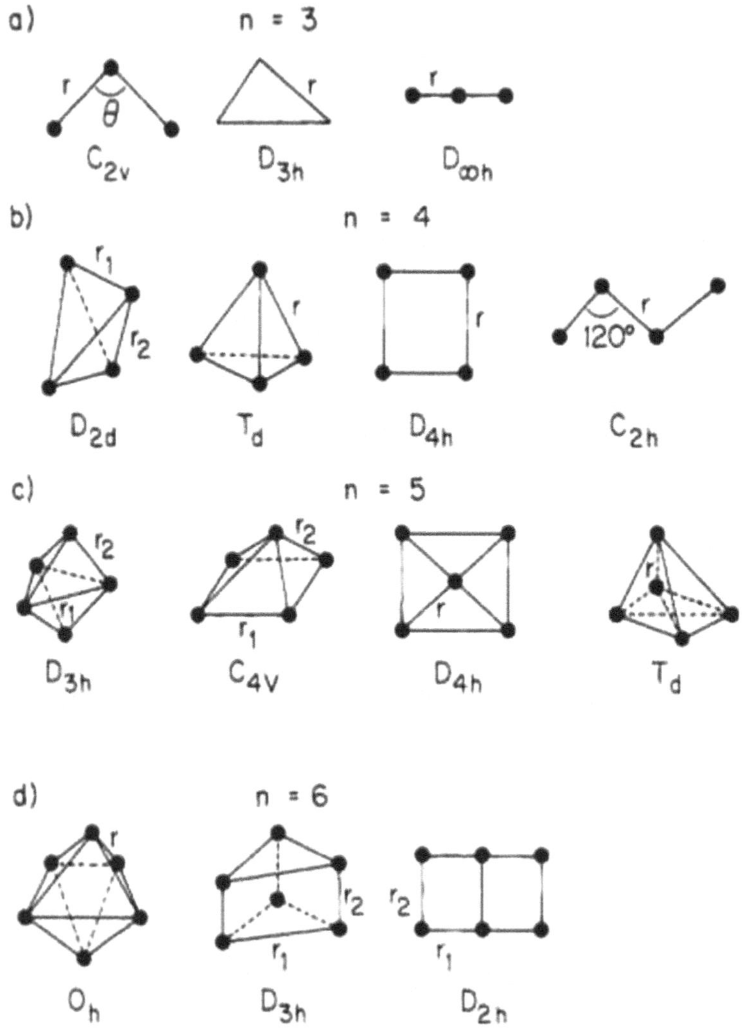

Fig. 3.8 The structures of isomers of Li_n clusters, assigned by the point group symmetry and geometric parameters for: **a** trimers, **b** tetramers, **c** pentamers, and **d** hexamers [14]

two-dimensional structure only. However, Fig. 3.10 below shows definite reductions in the function of the energy per atom for clusters with $N = 2, 8, 20, 34$, and 40. Thus, these numbers correlate with the numbers of valence electrons $N = 2, 8, 20, 34$, and 40 of these Lithium clusters and correspond to the filled electronic shells $1s$, $1s1p$, $1s1p1d2s$, $1s1p1d2s1f$, and $1s1p1d2d1f2p$, respectively. The obtained results for lithium clusters agree ideally with the Jellium calculations for sodium clusters, obtained by Knight et al. [11] having similar experimental characteristics.

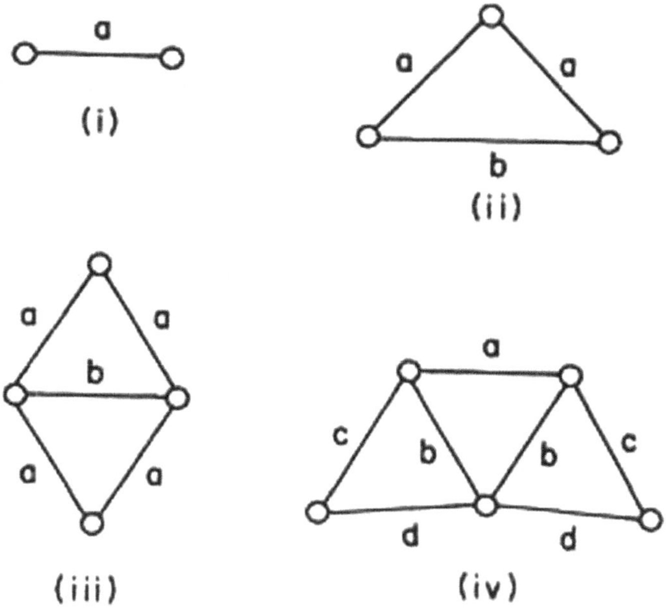

Fig. 3.9 The geometries of Lithium clusters, with the atomic distances (Å) Li_2(a = 2.646), Li_3(a = 2.684, b = 3.266), Li_4(a = 2.938, b = 2.612), Li_5(a = 2.880, b = 2.991, c = 2.939, d = 2.934) [15]

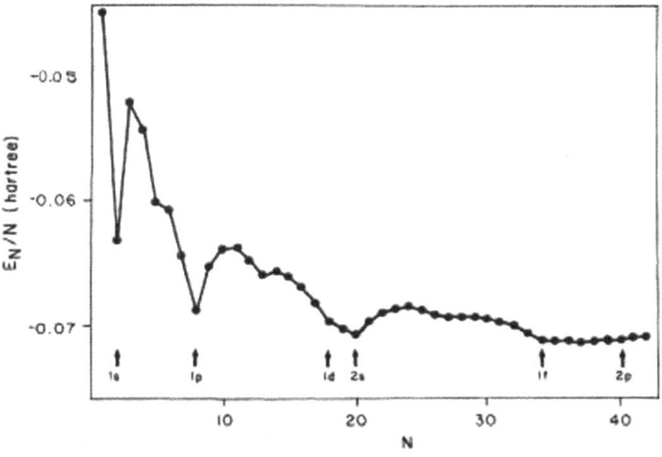

Fig. 3.10 The energy per atom of Lithium clusters as a function of cluster size N. The electronic shell filling has been indicated by 1s, 1p, 1d, 2s, 1f, and 2p [15]

Resonance Structures: Lithium clusters were also investigated by McAdon and Goddard [16] applying the generalized valence-bond method and formulating some general rules to derive the related structural and electronic properties. They studied many isomers of Li_n clusters for (n = 2 to 13). They studied for example the linear chain and ring clusters of Li_8, zigzag and linear chain as well as icosahedral clusters of Li_{13}. In addition, they determined the potential energy curves for the geometric distortion of Li_4 and Li_6 clusters as a function of the bond angle Θ, as shown in Fig. 3.11. It shows also that the delocalization of valence electrons is interpreted through the separate and possible resonance structures in Lithium clusters. However, resonance is a mental method in the valence bond theory that describes the delocalization of electrons within the molecules. The valence bond method of metallic bonding, as discussed by McAdon and Goddard [16], is based on electron localization of particular interurban areas at the middle of bonds between the atoms of 1D, 2D and 3D clusters. The authors found that the ideal bonding should involve valence electrons which are localized in different tetrahedral hollows and the ideal cluster structures should have the maximal numbers of tetrahedra, similar to the 20 tetrahedron, the ingredients of an icosahedron.

Singlet-Triplet Transition of Li_6: The lowest MRD-CI binding energy per atom for the singlet-triplet transition for the Li_6 clusters has been determined by Paccioni et al. [17]. They have calculated the binding energy per atom for Li_6 clusters as a function of the bond angle Θ, during the deformation from the geometry of double squares to the geometry of double rhombuses.

Singlet-Triplet Splitting of Li_6: Furthermore, Koutecký et al. [18] studied the relationship between the geometrical and the electronic properties of the 2D planar and 3D hollow lithium clusters. They found that the 2D and 3D clusters of li_6 isomers exhibit similar stability. The ab initio calculations at the MRD-CI level of lithium hexamers produce lowest energies with degenerate singlet and triplet states. The authors derived some simple criteria concerning the energy gap between the lowest singlet and triplet states. They confirmed that both singlet and triplet ground states of the double squares of Li_6 is lower than the singlet state. In opposite, when the geometry of double squares transforms to the geometry of double rhombuses (section of the fcc (111) Lithium lattice), by varying the angle from 90° to 60°, then the singlet ground state energy of the double rhombuses is lower than the triplet state. In addition, the splitting of singlet-triplet occurs by around $\Theta = 77°$.

The Isomers of Li_6 Clusters: Further investigation of Lithium clusters was carried out by Plavsic et al. They applied ab initio all-electron multi-reference double excitations configuration interaction methods (MRD-CI) studying the structures and related stability of Li_4 and Li_6 clusters. They found that the planar rhombic geometry of Li_4 cluster has the lowest singlet state energy, in agreement with the results of Ray et al. [13] and of Rao and Jena [15]. Both geometries, the capped pentagon Li_6-C_{5v} and the planar cluster Li_6-D_{3h} are the most stable Li clusters, which are in turn similar to the planar section of the fcc Li crystal. However, the spin properties these lithium clusters seem to be very important. The Li_4 and Li_6 clusters with

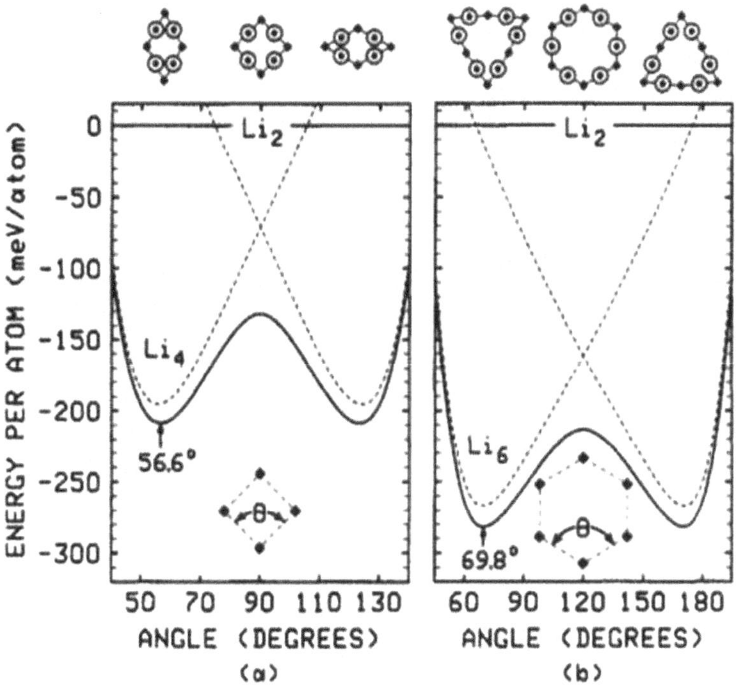

Fig. 3.11 The potential energy curves for geometric distortions of Li_4 and Li_6, as a function of the bond angle Θ. The energy per atom is given relative to Li_2 at equilibrium. The dashed lines represent the separate resonance structures, while solid lines show the energy including all resonance structures. The geometry of Li_4 oscillates between elongated and truncated rhombus over square, while the geometry of Li_6 oscillates between the elongated and truncated sex-angle over regular hexagon. The bold face squares and circles mark the atomic and orbital positions, respectively [16]

closed-shell electronic shells exhibit a singlet ground state lying lower than the lowest triplet state. Furthermore, they also found that closed three-dimensional $Li_4(T_d)$ and $Li_6(O_h)$ clusters with a triplet ground state are less favored. Thus, Fig. 3.12 shows the structures of different isomers of Li_6 clusters, assigned with the corresponding point group symmetry.

The Anions of Li_n^- Clusters for n $\leq$ 7: The anionic lithium clusters were investigated by Alexandrova et al. [20] using a combination of anion photoelectron spectroscopy and ab initio quantum chemical calculations. They also used the gradient embedded genetic algorithm program studying the isomers of each of these anions, which exhibit high stability and likely contribute to the photoelectron spectra. The authors used subsequently different methods like state-of-art ab initio techniques including coupled cluster and multi-reference configuration interaction methods as well as time-dependent density functional theory in order to interpret the experimental spectra. The optimized structures of the anionic Li-clusters are shown in Fig. 3.13.

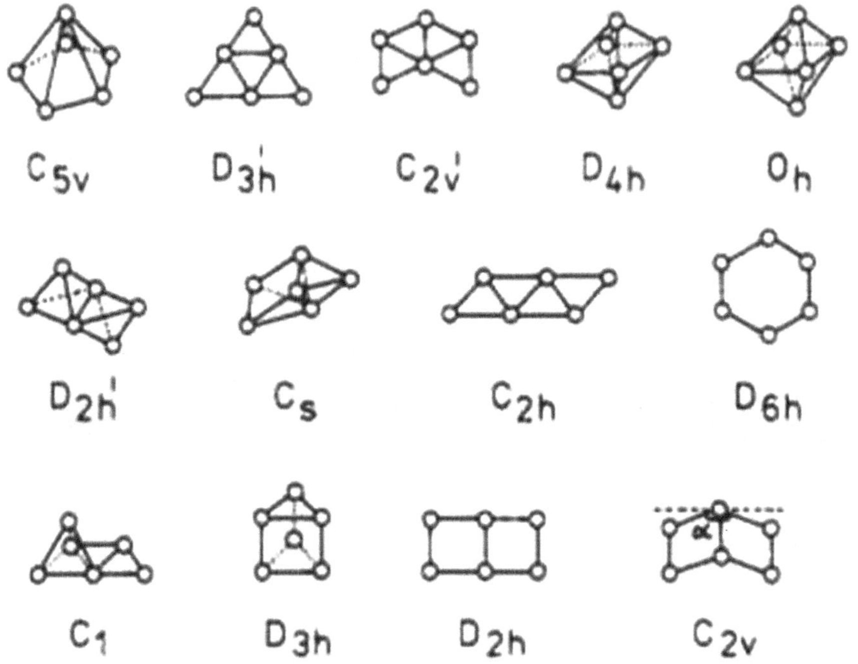

Fig. 3.12 The isomers of Li_6 clusters and their point group symmetry. The first two C_{5v} and D_{3h} clusters are the most stable structures [19]

3.2.6 Li_n Clusters for $n \leq 14$

Molecular Models of Neutral Li_n Clusters: Further studies on neutral and cationic Lithium clusters were carried out systematically by Boustani et al. [21] using ab initio quantum chemical configuration-interaction methods. They found out that there is a strong correlation between the electronic and geometric structures of Li_n Clusters. They found also that the neutral Li_n lithium clusters with ($n \leq 6$) favor planar geometries, which are similar to the altered (111) sections of the fcc-lattice plane. The optimal geometries of larger Li_n clusters for ($6 \leq n \leq 9$) Cavour 3D structures and are composed of condensed deformed tetrahedra. Only the Li_6 clusters with the C_{5v} and D_{3h} structures are energetically degenerate. The structures of neutral clusters, assigned with their point group symmetry, are shown in Fig. 3.14, in correlation with the binding energy per atom (eV), and with their coordination numbers N, as functions of cluster size n, determined at the MRD-CI level of theory. Figure 3.14 shows, the function E_b exhibits a dip at Li_3, and a dip at Li_9, assigned by ▲, besides a maximum at Li_8. Thus, three important features can be recognized in this diagram. The first one is the increasing of the binding energy per atom E_b with increasing the cluster size. The second feature shows that both dips of the binding energy at Li_3 and Li_9 indicate that Li_2 and Li_8 are highly stable, and their stability is connected with

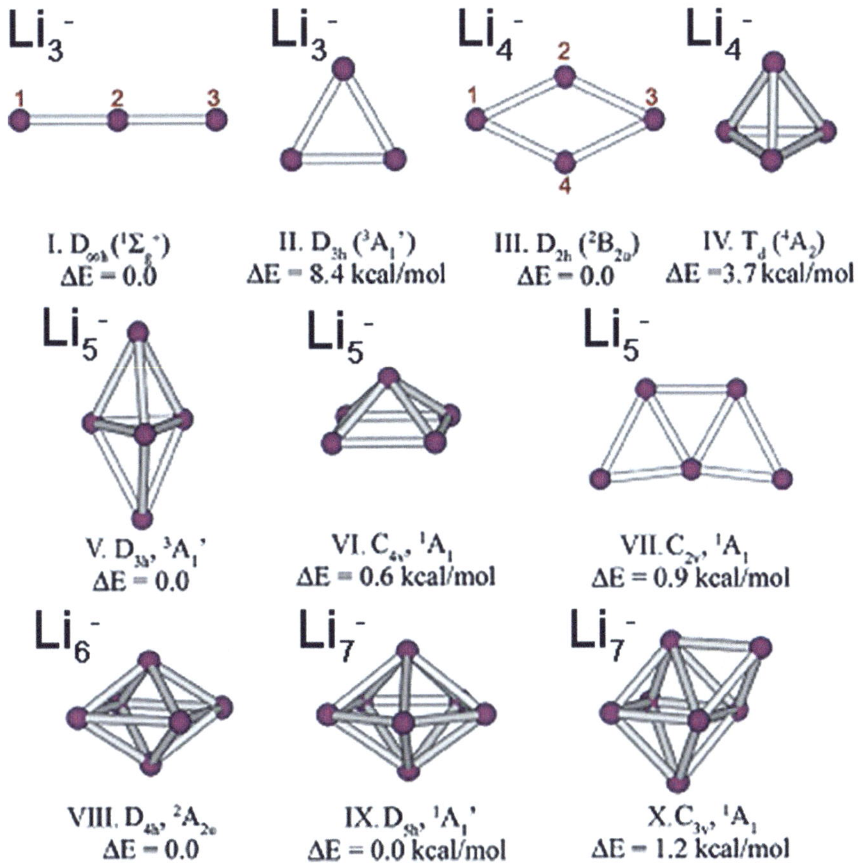

Fig. 3.13 The optimized structures of the anions of lithium clusters Li_n^- for $3 \leq n \leq 7$and their point group symmetry and ground state. ΔE (kcal/mol) gives the lowest energy difference of isomers to the global minimum assigned by ΔE = 0.0. This figure was reproduced with kind permission of the Journal of Chemical Physics of the American Institute of Physics [20]

the magic numbers **2** and **8** corresponding to the **s**, and **p** shells. The third feature shows that the function of the coordination numbers is increasing with increasing the cluster size.

The Neutral, Cationic and Anionic Li_n Clusters: Figure 3.15 shows all cationic and anionic Lithium clusters together with the neutral ones in correlation with the binding energy per atoms as functions of cluster size. The cationic clusters were also determined by Boustani et al. [21]. They found that Lithium cations Li_n^+ prefer planar geometries for (n = 3 and 4). Larger Li_n^+ clusters for ($5 \leq n \leq 10$) favor 3D structures, in similar attitude to the neutral clusters. The anionic clusters Li_n^- were studied by Boustani et al. [24] at the MRD-CI level of theory. They found that the Li_n^- clusters prefer linear geometries for (n = 3 and 4), planar for (n = 5), but high

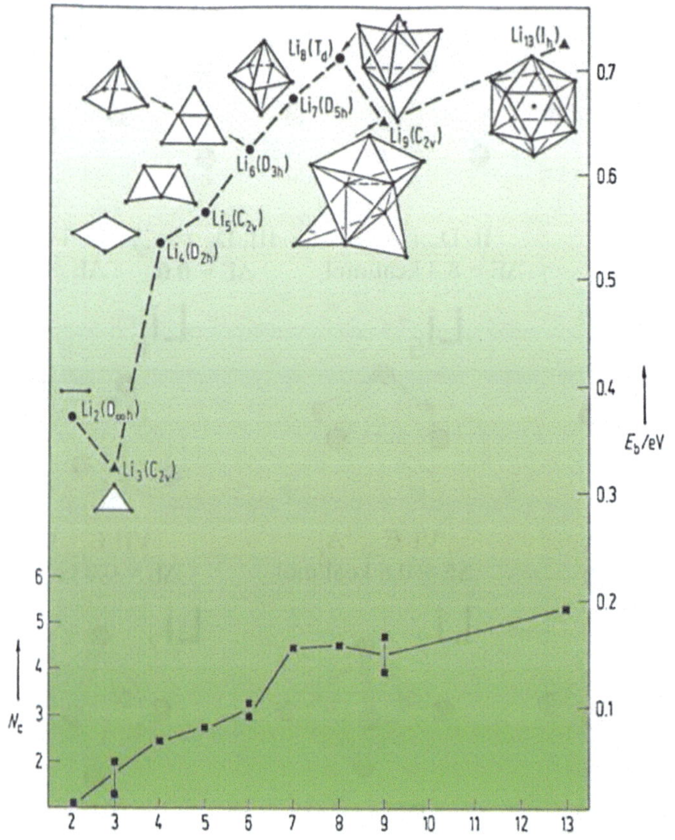

Fig. 3.14 The optimized geometries of neutral lithium clusters Li_n in correlation with their binding energy per atom (eV) E_b (eV), as functions of cluster size n, determined at the MRD-CI level of theory. The points labeled at the bottom with ■ is the average numbers of the nearest neighbors N for the optimized structures [21, 22]

symmetrical geometries for (n = 6, 7 and 8). The latter are octahedral, pentagonal bipyramidal and tetracapped tetrahedral structures with D_{4h}, D_{5h}, and T_d symmetry, respectively. Thus, the optimal geometries of Li_n^- clusters for ($6 \leq n \leq 9$) favor also 3D structures and are composed of condensed deformed tetrahedra, in similar attitude to the neutral lithium clusters.

Development of Tetrahedral 3D Lithium Clusters: All neutral and charged Li_n clusters for ($5 \leq n \leq 10$) favor 3D structures. However, these 3D structures can be developed, by adding atom by atom to trigonal bipyramid to build tetrahedral subunits, as schematically presented in Fig. 3.16. This can also be understood as when one face of a tetrahedron will be capped by an atom to form trigonal bipyramid then further capping and so forth. This can occur by a slight distortion of the pyramids, from which the configuration is built.

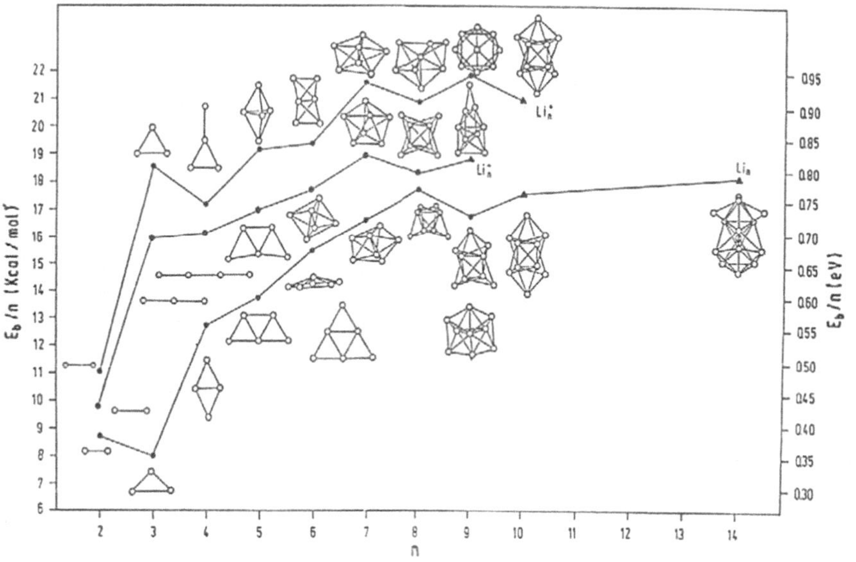

Fig. 3.15 The optimized geometries of neutral, cationic and anionic lithium clusters Li_n, Li_n^+, and Li_n^-, in correlation with their binding energy per atom E_b, E_b^+, and E_b^-, respectively, as functions of cluster size, determined at the MRD-CI level of theory [23]

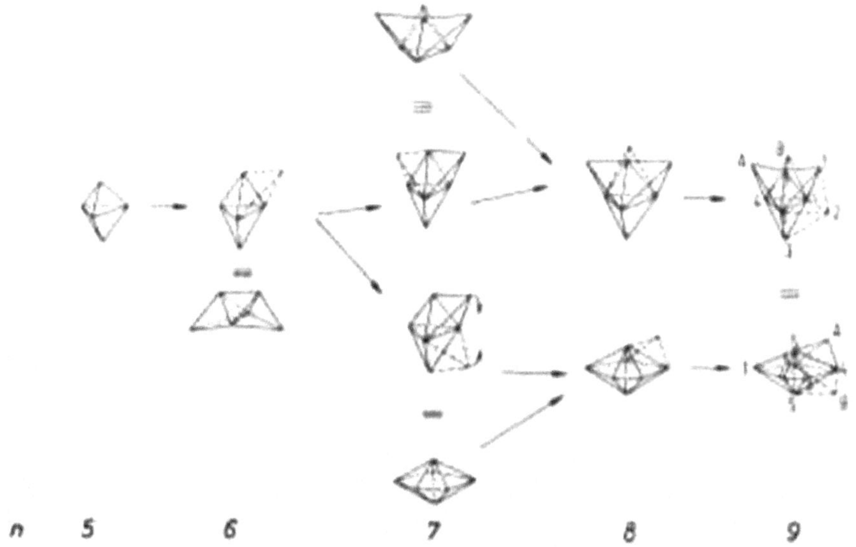

Fig. 3.16 Schematic depiction of the relation between tetrahedral cluster geometries and cluster forms including pentagonal bipyramides as subunits [21]

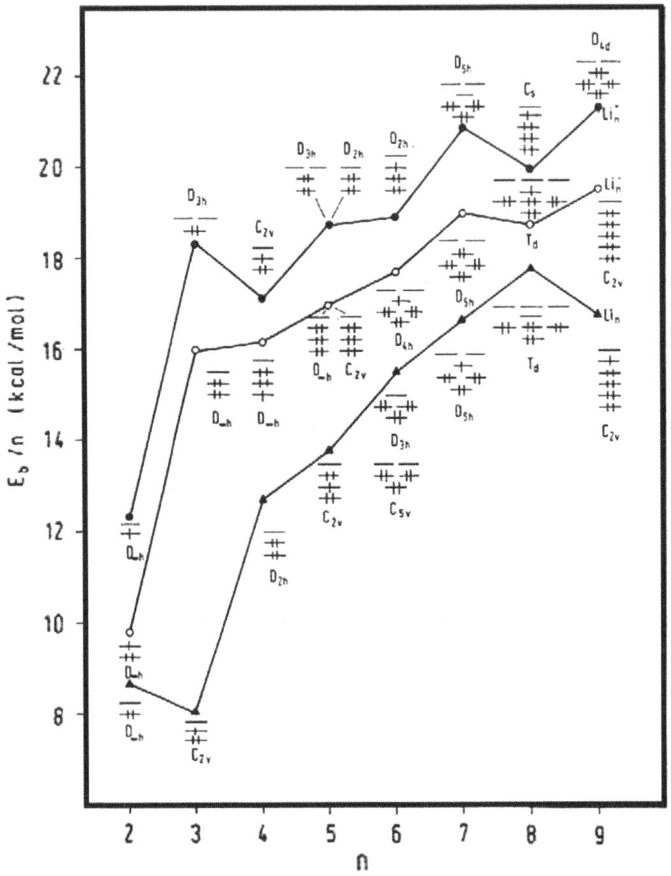

Fig. 3.17 The binding energy per atom of neutral and charged Lithium clusters as functions of cluster size. The electron configurations of molecular orbitals are given schematically and correspond to the geometric structures of Li_n, Li_n^+, and Li_n^- clusters as presented in Fig. 3.15 [24]

The Molecular Orbitals of Lithium Clusters: In order to understand the relationship between the geometrical and electronic structures of Lithium clusters, we present Fig. 3.17 showing the molecular orbitals below or above the E_b, E_b^+, and E_b^- functions, replacing the corresponding geometries of the neutral and charged Lithium clusters, as depicted in Fig. 3.15. The electronic configurations of the neutral Lithium clusters, as shown by the lowest function of Fig. 3.17, are assigned by the point group symmetry and the molecular orbitals as occupied by the valence electrons. As soon as one electron is removed from or added to a neutral cluster, getting respectively cationic or anionic clusters, then there is a change in the point group symmetry and in the molecular occupation.

Correlation between Geometric and Electronic Structures: The quantum chemical investigations of neutral and charged Lithium clusters, carried out by Boustani

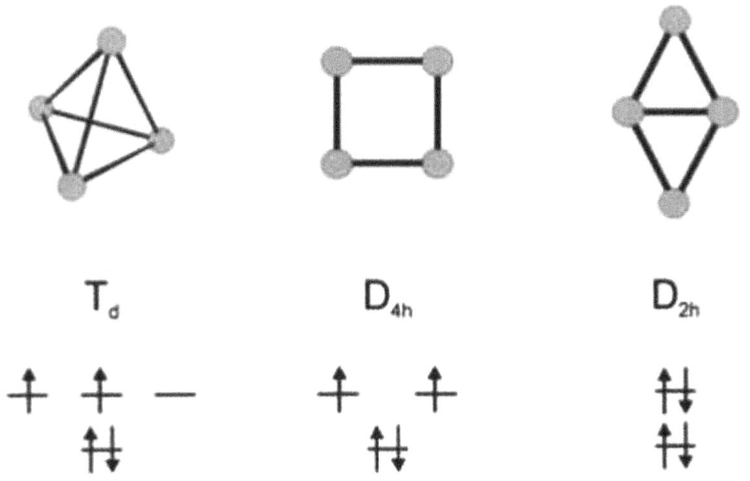

Fig. 3.18 The geometric and electronic configurations of Lithium tetramers. The Jahn–Teller distortion occurs when the degenerate molecular orbitals are not fully, partially, or singly occupied [25]

et al. [21, 23, 24] confirmed that there is a strong correlation between the electronic and geometrical structures of lithium clusters. However, particularly the occupation numbers and the degeneracy of the molecular orbitals of small alkali-metal clusters influence strongly their electronic properties. As by lithium clusters, the degeneracy depends, of course, first upon the cluster symmetry of the structure. Thus, if the degenerate molecular orbitals are not fully or partially occupied, then occurs the Jahn–Teller-Effect (JTE)[1] and distorts the cluster stability and forces the change or deformation of the cluster geometry and consequently leads to lower point group symmetry. This fact can be understood by Li_4 isomers, as illustrated in Fig. 3.18. The degenerate molecular orbitals of the tetrahedron and of the square are occupied by unpaired electrons. Therefore, Jahn–Teller-Effect distorts both structures into a rhombic one with electron pairing forming a singlet ground state [25].

Important Features of Stability: The calculated stability of neutral and charged clusters, determined in form of binding energy per atom, are presented in Figs. 3.15 and 3.17. In both figures, there are two important features which can be recognized. The first feature indicates that the function E_b^+ of the cationic Li_n^+ clusters is higher than the function E_b^- of the anionic Li_n^- clusters, which is in turn higher than the function E_b of the neutral Li_n clusters. Thus, the functions E_b^+, E_b^- and E_b are shifted to each other by an average of 1.0 to 2.0 kcal/mol. The second feature indicates that the maxima of the functions are related to Li_9^+-D_{4d}, Li_7^--D_{5h} and Li_8-T_d clusters and correspond to the magic number 8. The structure of the Li_9^+ cluster is simply a square anti-prism with D_{4d} symmetry, whereas the structure of the Li_7^- cluster is

[1]Jahn–Teller Theorem says: In nonlinear molecule, when degenerate molecular orbitals are asymmetrically or partially occupied, then a distortion will occur and distorts the degeneracy.

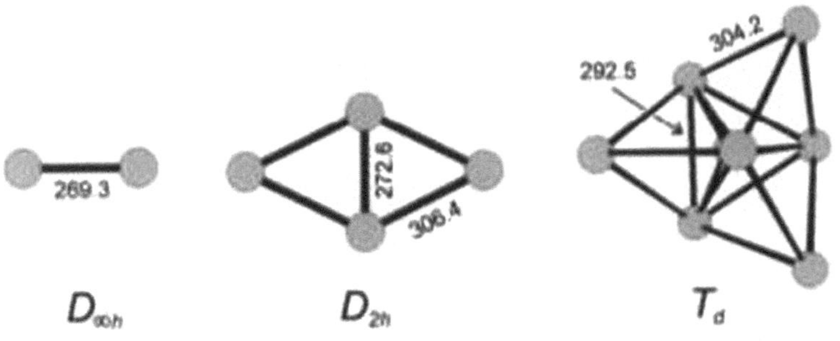

Fig. 3.19 The geometries of Li_2, Li_4, and Li_8 Lithium clusters with the inter-atomic distances (pm), determined at the QCISD/6-311G* level [25]

Fig. 3.20 The electronic configuration of Li_8-T_d shows eight valence electrons filling up the (non)degenerate s-, and p-type molecular orbitals [24]

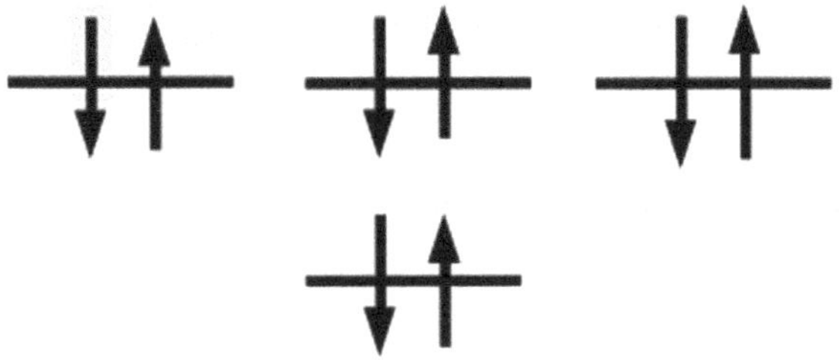

the pentagonal bipyramid of D_{5h} symmetry, and the structure of Li_8 cluster is the tetracapped tetrahedron of T_d symmetry.

The Unique Structure of Magic Cluster Li_8: The structure of the Li_8-T_d cluster was predicted theoretically in 1987 by Boustani et al. [21] for the first time. They found that the Li_8-T_d structure is composed of a central tetrahedron with a bond-length of 3.09 Å, and of tetracapped tetrahedra with a bond-length of 3.11 Å, determined at the MRD-CI/6-31G level of theory. This structure was investigated in 2003 once again by Kornath et al. [25], namely theoretically and experimentally. From the theoretical point of view, they recalculated the atomic distances of the Li_8-T_d cluster at the QCISD/6-311G* level and found that the Li-Li bond-lengths of the inner tetrahedron is 2.925 Å, and of the tetracapped tetrahedra is 3.042 Å, as presented in Fig. 3.19. The results, obtained by Boustani et al. and Kornath et al., differ slightly from each other. The structure of the tetracapped tetrahedron of Li_8-T_d cluster, possesses two particular characteristics: for the tetrahedral subunits there is no more need for deformation due to spatial reasons, while the other feature tells that the eight valence electrons of Li_8-T_d fill up both the non-degenerate *s*-type and the triply degenerate *p*-type molecular orbitals, exhibiting a closed-shell electronic structure , as can be seen in Fig. 3.20.

Raman Spectra of the Magic Cluster Li_8: Kornath et al. [25] investigated Raman spectroscopy of small matrix-isolated Li_2, Li_4, and Li_8 clusters, and confirmed experimentally the tetrahedral structure of Li_8, which was proposed by Boustani et al. [21].

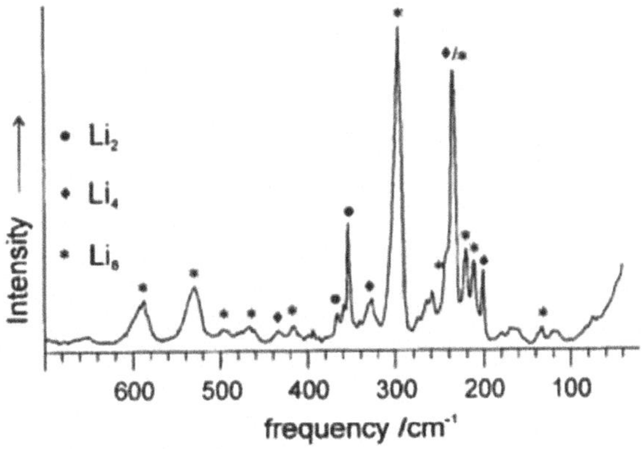

Fig. 3.21 Raman matrix spectra of 7-Lithium clusters in argon at 15 K (experimental conditions: 100 μm layer, laser wavelength 514.5 nm, laser power 100 mV, resolution 1.2 cm^{-1}) [25]

Table 3.1 Raman matrix spectra of 7-Lithium clusters in argon at 15 K (experimental conditions: 100 μm layer, laser wavelength 514.5 nm, laser power 100 mV, resolution 1.2 cm^{-1}) [25]

$^7Li_8$		$^6Li_8$		$\Delta(^6Li/^7Li)$		Assignment	
Expt.	Calc.	Expt.	Calc.	Expt.	Calc.		
83.0	88.2	91.6	95.1	8.6	6.9	nu_4	E
132.8	129.1	...	139.8	...	10.7	nu_8	F_2
201.4	201.5	216.6	218.3	15.2	16.8	ν_7	F_2
221.0	222.2	236.7	239.9	15.7	17.7	ν_2	A_1
241.7	248.0	262.8	268.7	21.1	20.7	ν_3	E
295.3	303.3	318.0	327.1	22.7	23.8	ν_1	A_1
298.5 sh	306.1	320.4 sh	327.4	21.9	21.3	ν_6	F_2
419.2		448.9				$\nu_2 + \nu_7$	F_2
464.7		498.5				$\nu_2 + \nu_3$	E
497.3		532.9				$\nu_1 + \nu_7$	F_2
529.7		567.5				$\nu_1 + \nu_2$	A_1
587.0		631.7				$2\nu_1$	A_1

The observed vibrational frequencies of Lithium clusters, as shown in Fig. 3.21, are compared with the calculated frequencies, obtained at the configuration interaction QCISD/6-311G* level. They determined the Li-Li distances of the octalithium to be 2.925 and 3.042 Å, for the inner tetrahedron and outer tetrahedra, respectively, as given in Fig. 3.19. These results are comparable with the calculated Li-Li distances, 3.09 and 3.11 Å , obtained by Boustani et al. [21] at the MRD-CI/6-311G level of theory.

Raman Spectra Between Theory and Experiment: Kornath et al. [25] compared the observed vibrational frequencies of Li_8 with the calculated ones and presented them in Table 3.1. The observed vibrations in the spectrum, which are laying below 310 cm^{-1} (330 cm^{-1} for $^6Li_8$), are composed of seven Raman active vibrations of

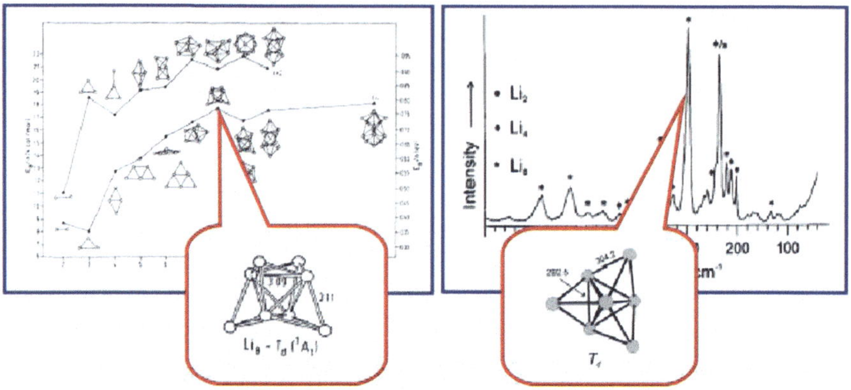

Fig. 3.22 Left pattern: Boustani et al. [21] predicted structure of Li_8 cluster in 1987. Right pattern: Experimentally structure of Li_8 cluster confirmed in 2003 by Kornath et al. [25]

Li_8. Thus, the observed vibration ν_6 is the only one as a shoulder in the intense and the broad ν_1 mode (see Table 3.1), which is in fact the breathing mode of Li_8. One can point out that observed and calculated frequencies are in satisfactory agreement but an excellent agreement for the measured and calculated isotopic shifts. However, the range of spectra above 400 cm^{-1} depicts some overtones plus combination modes. The performance of the latter mentioned vibrational modes is in excellent agreement with the spectroscopic selection rules for a Li_8-T_d cluster with hyper-tetrahedral symmetry.

Theoretical Prediction and Experimental Confirmation: In 1987 predicted Boustani et al. [21] via theory tetrahedral structure of Li_8-T_d (see Sect. 3.2.6) and found that its geometrical structure is composed of four tetrahedra, tetracapped to tetrahedron-shaped core. Around 16 years later Kornath et al. [25] succeeded to observe Li_8 via matrix-isolated lithium clusters in Argon and found the measured Raman spectra of the isolated Li_8 cluster are corresponding to the structure of Li_8-T_d, as shown in Fig. 3.19. The excellent agreement between theoretical prediction and experimental confirmation is presented in Fig. 3.22. Thus, the octalithium Li_8-T_d is the first paradigm of a magic cluster, which was isolated in a noble gas matrix and exhibiting an extremely high stability within the noble gas matrices, as already reported by Kornath et al.

3.2.7 *Li_n Clusters for n $\leq$ 20*

Lithium Clusters For 8 $\leq$ n $\leq$ 20: Fournier et al. [26] investigated Lithium Clusters Li_n for (n = 5 to 20) by density functional theory (DFT) using local spin density (LSD) and gradient-corrected energy functionals (GGA). For the structural optimization they applied the so-called Tabu search algorithm technique. They found that the most stable Li_n isomers can be arranged in two categories. The first group includes pentagonal bipyramid, icosahedron, as well as some related structures which are typical of most pair potentials. The second group was containing structures with centered square anti-prisms which are reminiscent of the bulk bcc structure, which are characterized by two peaks in the pair distribution function. One of them is near 2.60 Å and the other near 3.05 Å. The found structures are described as follows: Li_8-D_d tetra-capped tetrahedron, Li_9-D_{4d} centered square anti-prism, Li_{10}-D_{3h} tricapped centered trigonal prism, Li_{11}-D_{4d} centered square anti-prism capped on both ends, Li_{12}-D_{5h} pentagonal bipyramid with five capping atoms, Li_{13}-I_h icosahedron, Li_{14}-D_{4h} double centered square anti-prism, Li_{15}-C_{4v} double centered square anti-prism plus a top capping atom, Li_{16}-D_{4h} triply capped icosahedron, Li_{17}-D_{3h} double tricapped centered trigonal prism, Li_{18}-C_{5v} double icosahedron with a missing atom, Li_{19}-D_{5h} double icosahedron, and finally Li_{20}-C_{2v} tricapped centered square anti-prism, as depicted in Fig. 3.23.

Fig. 3.23 Optimized structures of Li_n clusters for (n = 8 to 20) [26]

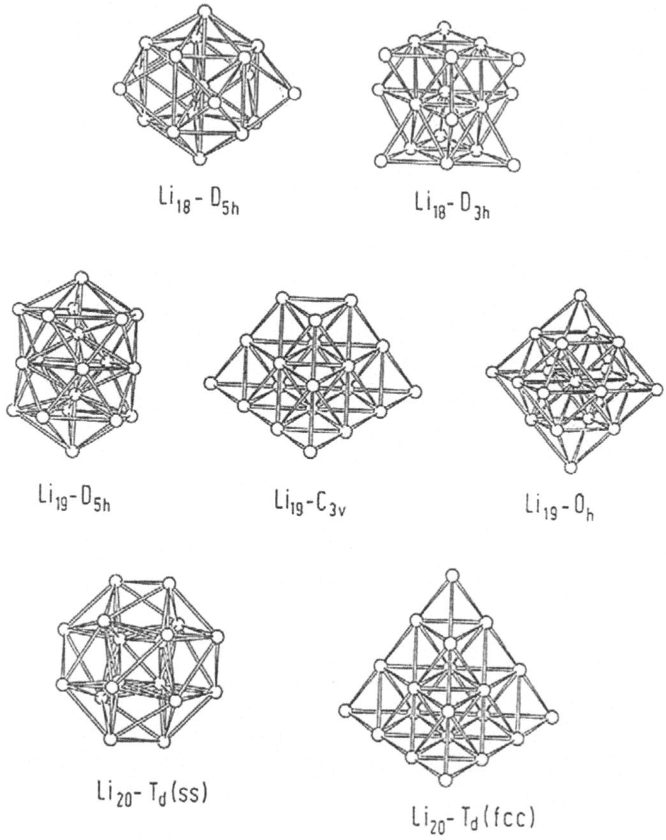

Fig. 3.24 Schematic tetrahedral structures and fcc sections of Lithium crystal of Li_n clusters for (n = 18 − 20) [23]

Lithium Clusters Li_n For 18 $\leq$ n $\leq$ 20: Koutecký et al. [23] investigated at HF-SCF and the MRD-CI level of theory larger Lithium clusters Li_n for (n = 18 − 20). They proposed tetrahedral structures of Lithium clusters and compared their energy with those of the 3D-sections fcc Lithium crystal. They studied two isomers of Li_{18}. The Li_{18}-D_{5h} isomer is composed of double pentagonal bipyramides surrounded by four octahedra, while the second one Li_{18}-D_{3h} is a section of the fcc Lithium crystal, as presented in Fig. 3.24. Three isomers of Li_{19} are proposed. The structure of the first isomer Li_{19}-D_{5h} have the form of double icosahedron. The second and the third isomers Li_{19}-C_{3v} and Li_{19}-O_h are sections or cuts of multi-layered fcc Lithium lattice.

The Magic Cluster Li_{20}: The experimentally found magic number "20" is in the literature interpreted as an indication of the special stability of the neutral Li_{20} cluster. Therefore, Koutecký et al. [23] proposed and studied also two isomers of Lithium clusters Li_{20}. These structures are tetrahedral and sections of fcc Lithium crystal,

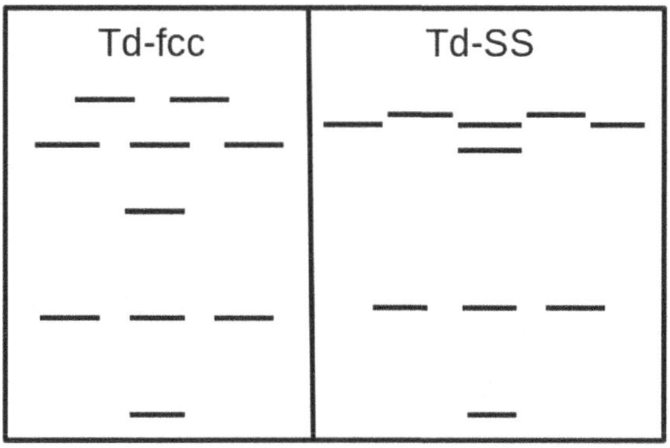

Fig. 3.25 Schematic tetrahedral structures and fcc sections of Lithium crystal of Li_n clusters for (n = 18 – 20). In box are the molecular orbitals of the Li_{20}-T_d-(fcc) cluster of the fcc-crystal-section and the nearly five-fold degenerate molecular orbitals of the spherical Li_{20}-T_d-(SS) structure, simulating the *d* electronic shell [23, 23]

as can be seen in Fig. 3.24. In fact, the structure of Li_{20}-T_d-(SS) cluster has nearly spherical shape and is at the MRD-CI level slightly more stable than the fcc structure Li_{20}-T_d-(fcc). The box of MOs in Fig. 3.25, assigned by T_d-SS, shows the nearly fivefold degeneracy of the highest molecular orbital resembles strongly to the fivefold degenerate of the d-type atomic orbitals.

3.2.8 Li_n Clusters for $n \leq 26$

Lithium Clusters For 18 $\leq$ n $\leq$ 26: Cheng and Yang [27] investigated recently Lithium clusters. They proposed a novel concept of so-called super valence bond. The kernel of this proposal is the superatoms, which can share nuclei and valence pairs for shell closure, which in turn can form delocalized super bonds. They revisited the global minimum energy structures of Lithium clusters Li_n for (n = 2 to 26), using B3LYP/6-311+G* and genetic algorithm method. They determined the energy and structures of these clusters, as shown in the Fig. 3.26. Thus, they confirmed the structures as composed of tetrahedral configurations, similar to those found by Fournier et al. [26].

Molecular Orbital Development of Lithium Clusters: As soon as metal atoms come together and the atomic distance between the atoms is short enough for the atomic orbitals to interact and to overlap, then they form bonding, anti-bonding, and non-bonding molecular orbitals with different energy levels. Since each Lithium atom has one occupied $2s$-type orbital, and three unoccupied $2p$-type orbitals, then for N number of Lithium atoms in a cluster, there are N number of $2s$-type molecular orbitals , and $3N$ number of empty $2p$-type molecular orbitals. Due to the fact that each molecular orbital can hold a maximum of two electrons, or in other words, each $2s$-type orbital is doubly occupied, then only the $N/2$ number of $2s$-type orbitals will be occupied, as shown in Fig. 3.27. By increasing the number of atoms, and accordingly the number of valence electrons, increases the number of occupied $2s$-

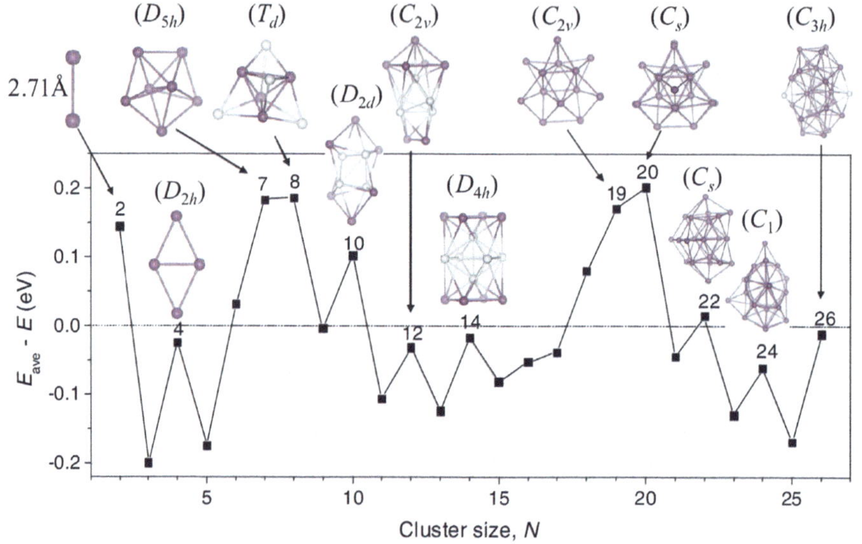

Fig. 3.26 The energies of global minima of Li_n clusters, determined at the B3LYP/6-311+G* level, relative to an average energy E_{eve}. This figure was reproduced with kind permission of the Journal of Chemical Physics of the American Institute of Physics [27]

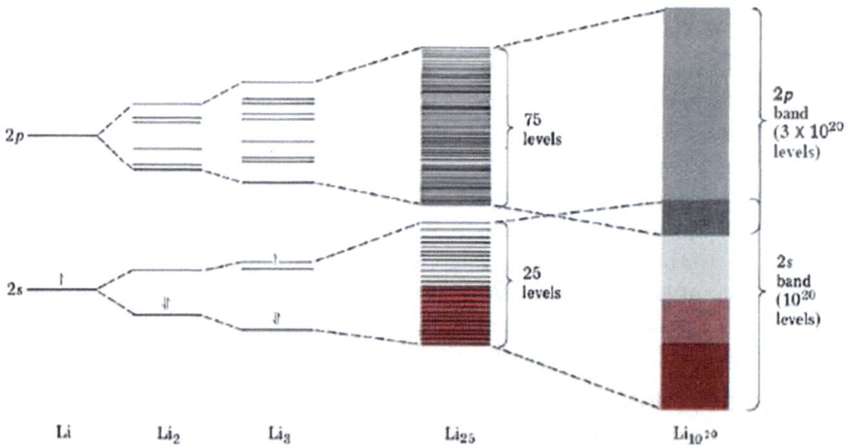

Fig. 3.27 Development of the molecular orbital energies from a single to more valence electrons of Li atoms: 1e$^-$, 2e$^-$, 3e$^-$, ..., 25e$^-$, up to $10^{20}e^-$. Due to the delocalization of valence electrons over increasing numbers of Li atoms. For example, A 1-mg sample of Li would contain nearly 10^{20} atoms. The corresponding orbital energies are so closely spaced that they constitute essentially continuous bands [28]

type and empty $2p$-type orbitals forming respectively the valence $2s$-, and conducting $2p$-bands. Considering around 10^{20} Lithium atoms, less than the (Avogadro)[2] number of atoms in one mole, then there are 10^{20} valence electrons, and as much as number of $2s$-, and $2p$-levels.

3.3 Sodium Metal Clusters

3.3.1 Sodium Atom

The Sodium atom, with the chemical symbol Na, is the second element of the alkali-metals and the eleventh in the periodic system of elements. Sodium shares many properties with Lithium. The only stable isotope contains 12 neutrons, 11 protons and 11 electrons with the electronic configuration $1s^2 2s^2 2p^6 3s^1$, which indicates the full outer shell p by shedding its single valence electron in the third shell $3s$. Sodium only occurs in its ionic form and building ionic bonding. The well known Sodium compound is NaCl salt, as shown in Fig. 3.28. It is well-known in the Earth's crust that sodium is the sixth most abundant element. The crust of Earth contains (2.83%) of Sodium in all its forms. After chloride, the second most abundant element dissolving in seawater, is of course sodium. Therefore, it is possible to estimate that sodium salts found in nature are composed of different compounds, like sodium borate (borax), sodium carbonate (soda), sodium chloride (rock salt), sodium nitrite and sodium sulfate. In addition, sodium salts were found in salty and alkaline lakes, in mineral spring water, and is about (1.05%) in seawater. The production of the salt mineral sodium chloride, in which each gram contains (40%) sodium, and (60%) chlorine, is around 200 million tonnes per year. The salt can be extracted either directly from brine or seawater by solar evaporation, or from rock salt mining using traditional shaft mining by cut and blast mining. Furthermore, many other stars including the sun seem to radiate with a visible light exhibiting high intensity of yellow component indicating that sodium atoms are in a high-energy excited states [31].

3.3.2 Sodium Clusters

The first theoretical studies on sodium clusters, prior mass spectra experiments, were carried out in mid-seventies and in early eighties by Gelb et al. [32], Martins et al. [33], and Ekardt [34], as presented in the subsequent paragraphs.

---

[2]Avogadro's number is that number of constituent particles that are contained in one mole of a substance. This experimentally determined value is about $6.022140857 \times 10^{23}$ particles per mole. This unit can be electrons, atoms or molecules, depending on the nature of the substance.

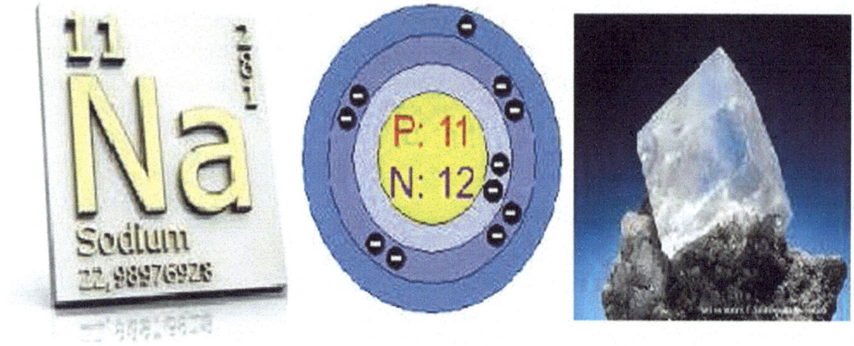

Fig. 3.28 The most stable isotope Sodium, with the chemical symbol Na, contains 12 neutrons, 11 protons and 11 electrons with the corresponding electron configuration $1s^2 2s^2 2p^6 3s^1$ and the well known crystalline salt NaCl [29, 30]

3.3.3 *First Study on Small Na Clusters*

One of the primarily studies on sodium clusters was carried out in 1975 by Gelb et al. [32]. They investigated the electronic structures of small sodium clusters using semi-empirical diatomics-in-molecules approach (DIM). Besides the dimer, they also studied the dimer Na_3 and the isosceles triangle of Na_3, the regular T_d and distorted tetrahedron C_{3v}, the square D_{4h}, the disphenoidal and crossed diatomics D_{2d} of Na_4, the square pyramid C_{4v} and bipyramid D_{3h} of Na_5, the regular octahedron O_h and hexagon D_{6h} of Na_6. The most important result of their calculations is that the closest packed geometries do not necessarily possess the most stable structures or the lowest energy minimum of such kind of clusters. However, the closest packt geometries could be the most stable structures, if the interaction energy could be mounted by a sum of two atoms with spherically symmetric Lennard-Jones potentials. Further stabilization for non-closely packed geometries can achieved by the introduction of the spin symmetries. In addition, due to Jahn–Teller effect, there are many spin splittings appearing on potential energy surface and giving rise to lower symmetry solutions. However, Fig. 3.29 shows the potential curves for Na_5 clusters in the form of regular pentagon D_{5h}, square pyramid C_{4v} and trigonal pyramid D_{3h} structures.

3.3.4 *Electronic Shells of Na Clusters*

The structural and electronic properties of simple metal clusters were also studied by Martins et al. [33] using the formalism of the density functional theory (DFT). In their model, they took into account the electron relaxation effects as well as the lattice structures through initiation of pseudopotentials. Besides the binding energies, the surface energies, and the lattice relaxation, the authors determined the electron

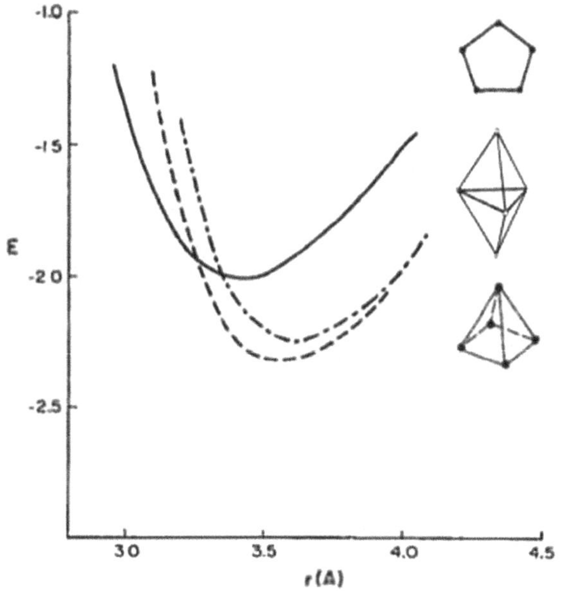

Fig. 3.29 The potential curves for special regular geometries of Sodium cluster Na_5: full line: regular pentagon; dashed line: square pyramid; dot-dashed line: trigonal pyramid. The variable (r) is the bond distance between nearest neighbor atoms in each case. Thus the square pyramid seems the most stable cluster of Na_5 [32]

affinity and the ionization potential. They presented their results for both the bcc lattice as well as for the icosahedral sodium clusters having up to 350 atoms. The most essential achievement in Martins paper [33] in all their models is the orbital degeneracy, scored subsequently by the assumption of spherical effective potential giving rise to propose electron shell structures for Na clusters, as $1s$, $1p$, $1d$, $2s$, $1f$, $2p$, etc., for the first time, similar to the electron shells in atoms, as shown in Fig. 3.30.

3.3.5 Jellium Model for Na Clusters

Ekardt [34] studied in 1984 also the electronic and structural properties of simple metal clusters using the local density approximation (LDA) applied to the spherical Jellium model. He calculated the work function and confirmed the results obtained by Martins et al. [33]. He found also that the shell structure in clusters with the quantum numbers $1p$, $1d$, $2s$, $1f$, $2p$, etc., are caused by the quantum-size effects based on the spherical potential hole. Figure 3.31 presents the development of the charge density of Na_8 cluster, the energy level, and the effective potential of the both completely filled shells of the neutral cluster.

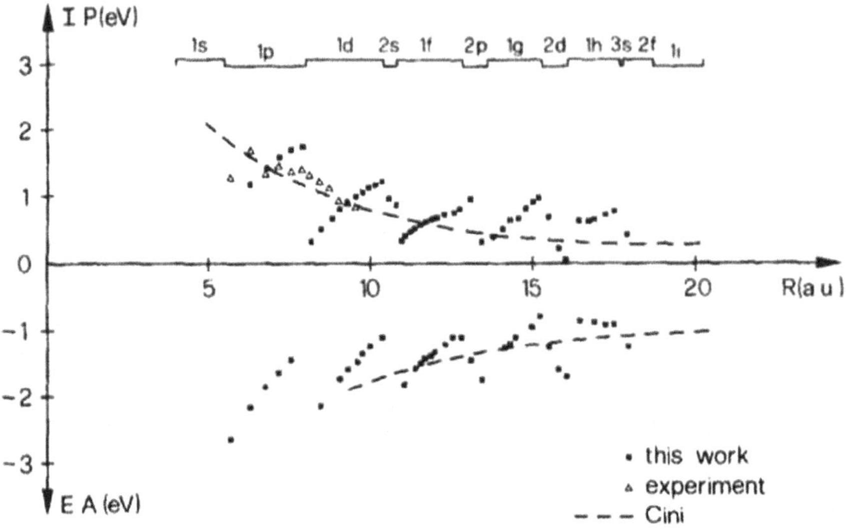

Fig. 3.30 Ionization potential and electron affinity of Na clusters as a function of the particle radius. The squares are the results obtained with the Jellium model, the triangles correspond to the measurements of Herrmann et al. [35]; the dashed curves are the theoretical results of Cini [36]. The highest occupied energy shell for each radius is given at the top of the figure [33]

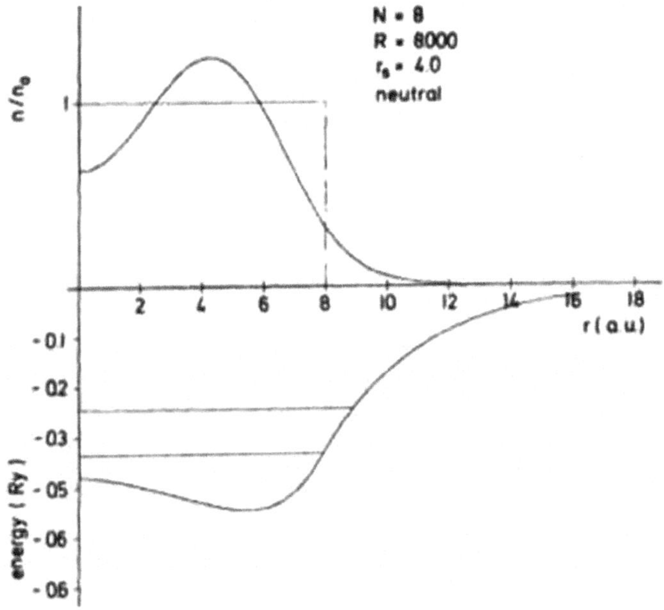

Fig. 3.31 **a** The development of the charge density as a function of size. Given is the total $\langle n \rangle / n_0$ after various shells have completely filled [34]

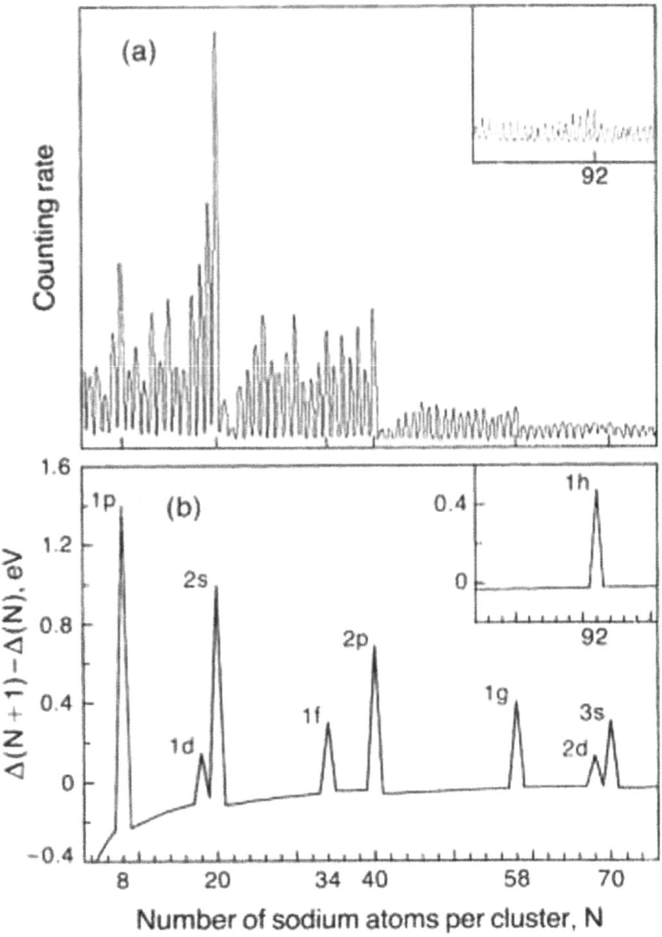

Fig. 3.32 **a** The mass spectra of Sodium clusters Na_N for $N = 4 - 75$. The full scale intensity in the main figure is approximately 20 000 counts/sec. The inset corresponds to $N = 75 - 100$. **b** The calculated change in the electronic energy difference, $\Delta(N + 1) - \Delta(N)$, versus N. The labels of the peaks correspond to the closed-shell orbitals. The peaks 8, 18, 20, 34, 40, 58, 68, and 92 correspond to filled electronic shells $1s^2$, $1p^6$, $1d^{10}$, $2s^2$, $1f^{14}$, $2p^6$, $1g^{18}$, $2d^{10}$, $3s^2$, and $1h^{22}$ [11]

3.3.6 Mass Spectra of Na Clusters

Knight et al. [11] determined in 1984 the mass spectra of sodium clusters of N atoms for ($N = 4 - 100$). The mass spectra were produced in a supersonic expansion carried by an Argon noble gas. These spectra exhibit pronounced peaks at Na_n cluster sizes of $n = 8, 20, 40, 58,$ and 92. These can be interpreted in terms of a one-electron shell model. This model is based on delocalized independent atomic electrons which are bound in spherically symmetric potential well. By accumulation of the eigenvalues of the occupied states one obtains the calculated electronic energy of each cluster $E(N)$ with N atoms. The electronic energy difference between the adjacent sodium clusters is defined as $\Delta E(N) = E(N) - E(N-1)$. The function of this quality against N is plotted in Fig. 3.32. The change in this quality versus N is already shown in Fig. 3.32. The peak positions of the cluster sizes indicate the shell-closing numbers and identifying the positions of the high stable clusters corresponding to the measured abundances with 8, 20, 40 58, 92, as can be seen in Fig. 3.32. The calculations give some peaks at $N = 18, 34, 68$, and 70, which however correspond to weak and total unpronounced peaks in the observed spectra, which are relatively more sensitive to the potential parameters.

3.4 Geometrical Structures of Na_n Clusters

3.4.1 Na_n Clusters for $n \leq 7$

Martins et al. [37] investigated furthermore small sodium clusters Na_n ($n \leq 7$) applying the pseudopotential and the local spin density method and using Hellmann–Feynman forces. The equilibrium geometries were obtained without imposing any symmetry constraint. They started the optimization procedure from randomly generated cluster configurations, then relaxed these geometries under the Hellmann–Feynman forces considering the action on the atoms of each cluster. They found that neutral and cationic sodium clusters up to five atoms have planar structures, the Na_6 cluster is a pentagonal pyramid, and the Na_7 clusters is a pentagonal bipyramid. The structures of Na clusters are similar to the geometries shown in Fig. 3.15.

The Shell Structures in Na_n Clusters: Martins et al. [39] continued their investigations on sodium clusters and shown that the essential issue of the sodium clusters is to recognize that the molecular orbitals, which are mainly composed of linear combination atomic orbitals with 3s character, can be arranged as s-like, p-like, and d-like molecular orbitals, etc., according to their global shape. Hereupon, they found that both Na_8 and Na_8^+ clusters have p-like molecular orbitals. Therefore, the molecular orbitals are to be filled in a definitely clear order when the number of electrons in the clusters increases. The ordering of the energy levels filled by valence electrons in similarity to shell structures found in mass spectrum of sodium clusters are $s, p, d, s, f, p, g, d, h, s, f, \ldots$, etc.

Confirmation the Planar Structures of Na_n Clusters for $n \leq 8$: Further study on neutral and cationic sodium clusters Na_n for $n \leq 8$ were carried out by Bonačić-Koutecký et al. [38] using ab initio quantum chemical configuration-interaction methods. They determined their electronic and geometric structures and found that the neutral sodium clusters up to six atoms have planar structures, the Na_7 is pentagonal pyramid, as can be seen in Fig. 3.33.

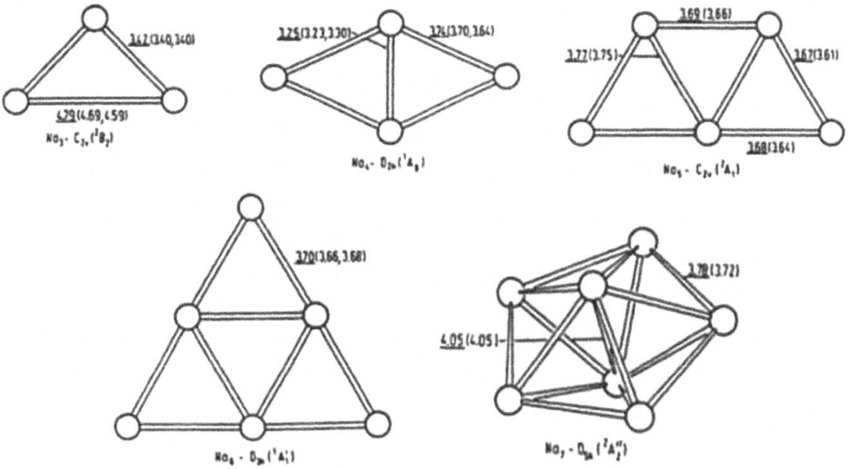

Fig. 3.33 The optimized geometries of neutral sodium clusters Na_n for n = 3 to 7, assigned by the point group symmetry and ground state. The first bond lengths are obtained at the all-electron Hartree–Fock (AE-HF) level. The values in parentheses correspond to effective core potentials (ECP-HF) and configuration interaction (ECP-CI) methods. Distances are given in angstroms [38]

3.4.2 The Structure of the Magic Cluster Na_8

Martins et al. [39] continued their investigations on sodium clusters particularly Na_8 and Na_8^+, studying their structural and electronic properties, based on local-spin-density and self-consistent pseudopotential calculations. They found that the geometric structures of Na_8 and Na_8^+ sodium clusters are of three-dimensional structures. The geometry of the neutral cluster Na_8 has a D_{2d} symmetry, as represented by the left structure in Fig. 3.34. The geometry of the cationic cluster Na_8^+ has a C_{2v} symmetry, which arises through a slight distortion of the Na_8-D_{2d} geometry. However, Bonaĉić-Koutecký et al. [38] found that the structure of Na_8 cluster is composed of a central tetrahedron tetracapped by four tetrahedra with a T_d symmetry, as represented by the right structure in Fig. 3.34. Bonaĉić-Koutecký et al. [40] found that the geometry of the cationic cluster Na_8^+ has a C_{2v} symmetry and is similar to the that one found by Martins et al. [39].

3.4.3 Raman Spectra of the Magic Cluster Na_8

Kornath et al. [41] investigated also Raman spectroscopy of small matrix-isolated Na_2, Na_4, and Na_8 clusters, and confirmed experimentally the tetrahedral structure of Na_8, which was proposed by Bonaĉić-Koutecký et al. [38] The observed vibrational frequencies of Sodium clusters, as shown in Fig. 3.35, are compared with the calcu-

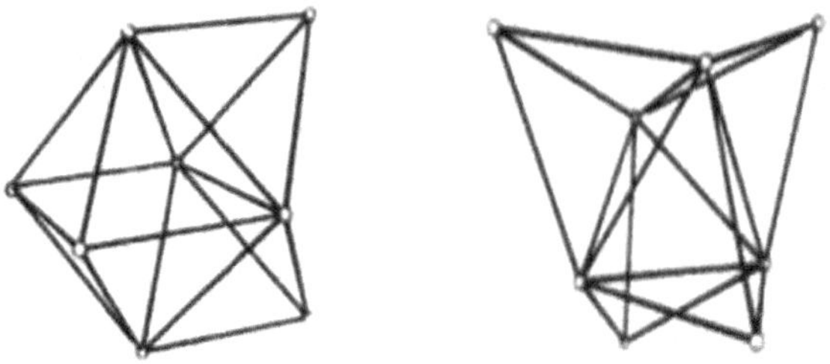

Fig. 3.34 The left structure is the Na_8 cluster with a D_{2d} symmetry. It is composed of an octahedron faced by two tetrahedra, determined by Martins et al. [39]. The right structure is Na_8 cluster with a T_d symmetry. It is composed of a central tetrahedron tetrafaced by four tetrahedra, determined by Bonaĉić-Koutecký et al. [38]

Fig. 3.35 Raman matrix spectra of sodium in argon at 15 K at different stages of cluster formation. (experimental conditions: 100 μm layer; laser wavelength 514.5 nm, laser power 125 mW, resolution 1.2 cm^{-1}) [41]

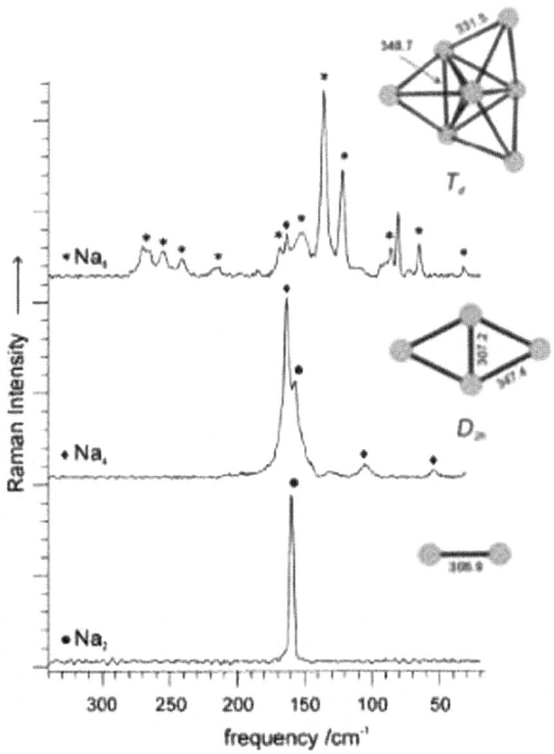

lated frequencies, obtained at the configuration interaction QCISD/6-311G* level. They determined the Na-Na distances to be 3.069 Å for the dimer, 3.072 and 3.474 Å for the rhombus, and 3.487 and 3.315 Å, for the inner tetrahedron and outer tetrahedra of the octalithium, respectively, as given in Fig. 3.35. The distances obtained for Na_8 are comparable with the calculated Na-Na distances, 3.09 and 3.11 Å obtained by Bonaĉiĉ-Koutecký et al. [38], determined at the MRD-CI/6-311G level of theory.

3.4.4 *Basics of Photoabsorption*

Absorption or emission of photons (ultraviolet, infrared, or visible light) by a molecule or cluster depends on electron transitions between different energy levels of the molecular orbitals. The photons are electromagnetic radiations with a definite frequency or wavelength, and the intensity of the absorption varies as a function of the frequency. It is well known that the electronic structures in form of molecular orbitals depend strongly on the geometrical structures of molecules or clusters. Therefore, each molecule or cluster has definite molecular orbitals and thus characteristic photoabsorption spectra. These spectra can be derived either experimentally or theoretically, like time-dependent density functional theory (TDDFT).

3.4.5 *Na_n Clusters for $3 \leq n \leq 8$*

Photoabsorption Spectra of Na_n Clusters for $3 \leq n \leq 8$**:** Greiner and co-workers et al. [42] investigated the photoabsorption spectra of neutral and cationic magnesium and sodium clusters using ab initio framework based on the TDDFT. The results of the TDDFT calculation of photoabsorption of several sodium clusters Na^+_{3-5} and Na_{4-8}, performed with the use of the B3LYP functional, are given in Fig. 3.36 by solid lines. The geometry of sodium clusters, given in the insets, were calculated by Solov'yov et al. [43]. The experimental experimental photoabsorption spectra [44] are plotted by dots. The CI results [38, 40], and the TDDFT results [45], are represented by dashed lines. This comparison shows that TDDFT method is a good alternative to the CI method, providing a good agreement with the experimental data. These results, presented in Fig. 3.36, confirm the predicted structures given in Fig. 3.33.

3.4.6 *Na_n Clusters for $9 \leq n \leq 14$*

Photoabsorption Spectra of Na_n Clusters for $9 \leq n \leq 14$: A series of photoabsorption spectra of Na_n sodium clusters for the range of 9 to 14 was determined by Nieminen and co-workers et al. [46] using the time dependent density func-

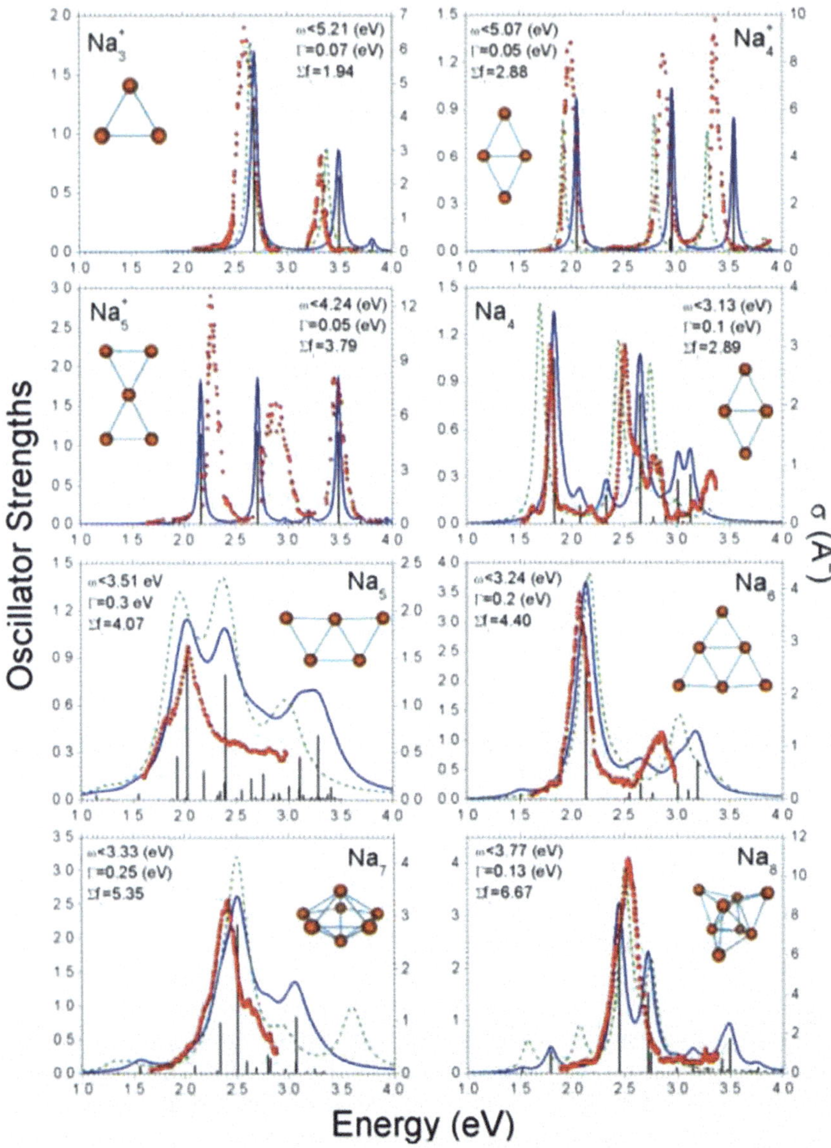

Fig. 3.36 Photoabsorption cross section calculated for Na_{3-5}^{+} and Na_{4-8} using the B3LYP functional (blue/solid lines). Vertical lines show the oscillator strengths for the optically allowed transitions. Cluster geometries calculated by Solov'yov et al. [43] are shown in the insets. The measured (red/dots) and calculated (green/dashed lines) photoabsorption spectra are adapted respectively from [44, 45]

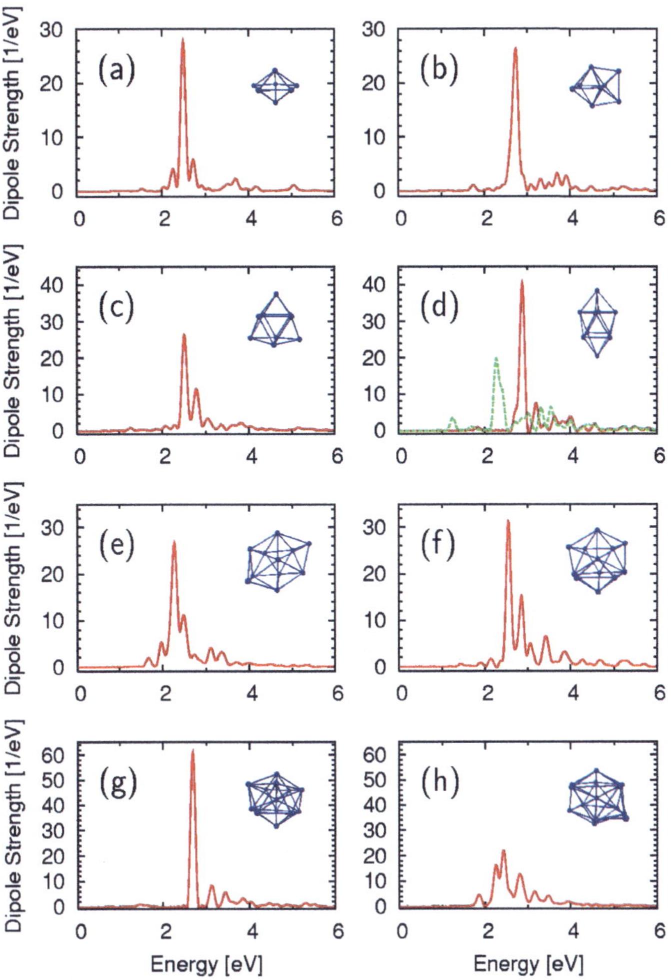

Fig. 3.37 The TDDFT photoabsorption spectra for the clusters Na_7-Na_{14}. Note the different scaling of the axes of ordinates. For Na_{10} (**d**) spectra of a D_{4d} (red/solid) and a trigonal pyramidal (tetrahedral) isomer (green/dashed) are shown. This figure was reproduced with kind permission of the Journal of Chemical Physics of the American Institute of Physics [46]

tional theory scheme, as presented in Fig. 3.37. For some particular sodium clusters, the authors discussed the impact of the cluster geometries on their photoabsorption spectrum. Thus, the photoabsorption spectra of different geometrical isomers of the selected clusters are presented for several cluster sizes. They asserted that the intensities of the photoabsorption spectra are increasing when the cluster size increases. However, the number of peaks keeps constant. In addition, the spectra of very small sodium clusters make some exceptions. Moreover, some spectra exhibit similar photoabsorption bands. This means, that the experimentally observed spectra belong to

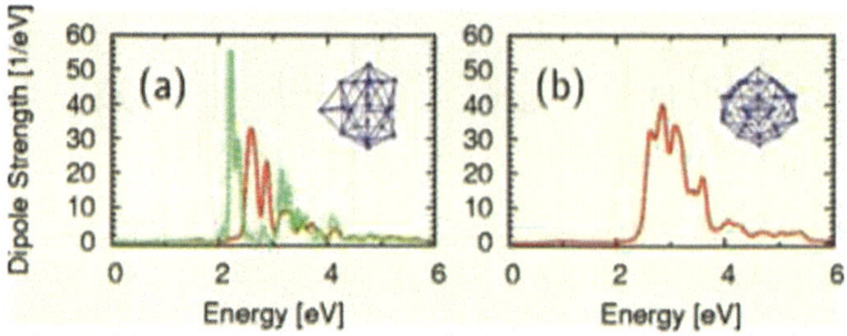

Fig. 3.38 The TDDFT photoabsorption spectra (red/solid curve) for the clusters **a** Na_{20} and **b** Na_{40}. For Na_{20} (**a**), the spectra of two isomers are shown: a capped double-icosahedron (red/solid) and a trigonal pyramid (green/dashed). are shown. This figure was reproduced with kind permission of the Journal of Chemical Physics of the American Institute of Physics [46]

collective plasmon excitations. The authors showed furthermore that their calculated spectra are in excellent agreement with other experimentally or theoretically obtained photoabsorption spectra.

Photoabsorption Spectra of Na_n Clusters for n = 20 and 40: Nieminen and co-workers et al. [46] also calculated the photoabsorption spectra of two larger sizes of sodium clusters Na_{20} and Na_{40} using the TDDFT scheme. Both clusters, as shown in Fig. 3.38. Due to the Jellium model and related electronic closing shells, these clusters are considered as magic clusters. However, Pollack et al. [47] measured two peaks for Na_{20} cluster at 2.42 and 2.78 eV and found that the experimental spectrum corresponds well to the calculated values 2.60 and 2.89 eV. However, it is necessary to ascertain that the TDLDA spectrum (red/solid curve) shows a ratio of 3:2, whereas by Pollack et al. is the ratio 1:3. As can be seen in Fig. 3.38a, the two spectral excitations of Na_{20} correspond to two isomers. The red excitations in the spectrum belong to a capped double-icosahedron which are located at at 2.24 and 2.37 eV, while the green excitations belong to a trigonal pyramid structure of the second isomer. The authors found a feature of crucial importance. It is the fact that the structure of the isomer has a strong influence on the intensity and position of the absorption peaks. Furthermore, they found also that there are some structures with almost degenerate energies and therefore it would be experimentally very difficult, if not impossible, to determine the cluster structure by means of a photoabsorption spectrum. Nevertheless, they presented the experimental photoabsorption spectrum of the Na_{10} and Na_{20} isomers.

3.4.7 Na_n Clusters for $13 \leq n \leq 380$

Where Do Magic Numbers in Na Clusters Come from?: Haberland et al. [48] studied the melting temperatures of Na clusters via calorimetric experiment. They thermalized the positively charged sodium clusters in a heat bath of temperature T,

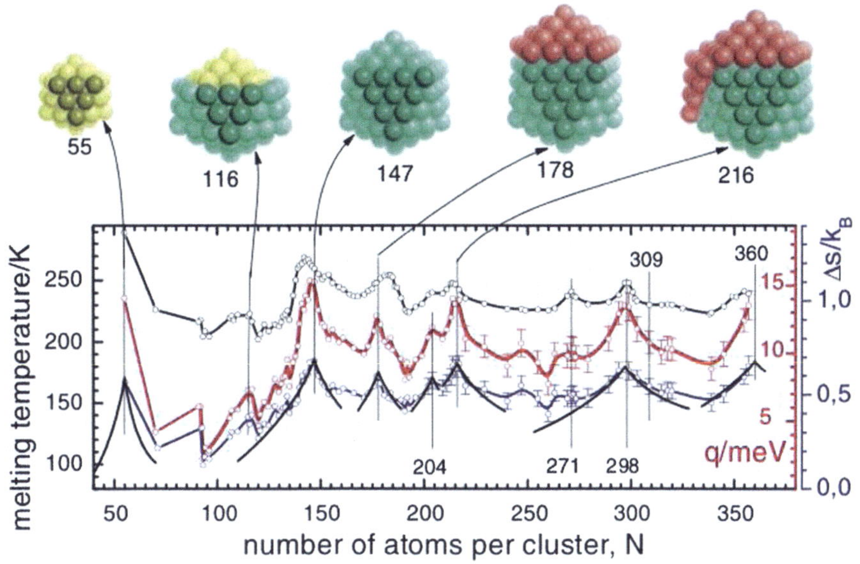

Fig. 3.39 The melting temperatures and the corresponding icosahedral growth pattern of Na clusters (for more details see the paragraph above) [48]

then they mass selected the heated charged clusters via time of flight (TOF), afterwards measured their internal energy E. Thus, they were able to mount the caloric function $E = E(T)$. This function allows to extract the melting temperature (T_{melt}) as well as the latent heat of fusion (q) of the clusters. As expected, they found that the thermal and electronic properties of the mass selected sodium clusters are cluster size dependent. Furthermore, the triple properties, (i) energy, (ii) entropy change upon meting, and (iii) modulation of the photoelectron spectra exhibit maxima, which can be associated with the closed geometrical shells of clusters. In the case when the curves of q and s have different slopes in the neighborhood of the closed shell size then the maxima in the melting temperatures are shifted to different sizes. The upper panel of Fig. 3.39 shows the growth of sodium clusters in icosahedral pattern. The cluster shells are indicated by colored layers. The 2nd layer with yellow, the 3rd layer with green, and the 4th layer with red. The latter figure shows also that the triangular face of the 20 faces of the icosahedron is colored in a deeper shade. Only two structures of the closed shell icosahedra are confirmed and correspond to the magic numbers N = 55 and 147. Further outer layers for other cluster sizes are of course possible and can be constructed. However, the lower panel of the Fig. 3.39 shows three curves with different colors. The upper one, the size depending melting temperature T_{melt}, is colored with black, the middle one, the latent heat of fusion per atom q, is colored with red, and the lower curve, the entropy change upon melting per atom Δs, is colored with blue.[3] The maxima of the Δs and q curves have the

[3] In order to avoid cluttering the figure, error bars are given only for N above 200.

same cluster size N, while the maxima of the upper curve T_{melt} are slightly shifted. Fortunately, some cluster sizes N could be distinguished for definite peaks, and some of them are guessed. As can be seen in the figure, the lower blue curve is partially overlapping with some black solid lines indicating the changing of the calculated entropy upon melting.

Geometric Magic Numbers of Na_n Clusters for $13 \leq n \leq 380$: The investigation of larger sodium clusters with sizes up to $N = 380$ was the next attempt of Noya et al. [49]. They studied the lowest-energy structures of large sodium clusters using the basin-hopping global optimization method, which is based on model potentials required for large sizes of clusters. For the inter-atomic interactions of sodium clusters, they applied two different forms of potentials. They are the Gupta as well as the Murrell-Mottram (MM) potentials (see [49] and references therein). The latter potential of Murrell-Mottram exhibits good transferability because it possesses much more parameters and could be fitted to a wider range of properties. Therefore, this MM potential was anticipated to be more reliable than the other one, but the computations are significantly much expensive. Though, incorporating both potentials can lead to greater reliability in the structural features. However, the considered potentials of Noya et al. [49] support the broad conclusion, that the clusters with large sizes are predominantly icosahedral. This conclusion was derived by Haberland et al. [48] (see previous section), adopted from their recent analysis of the melting

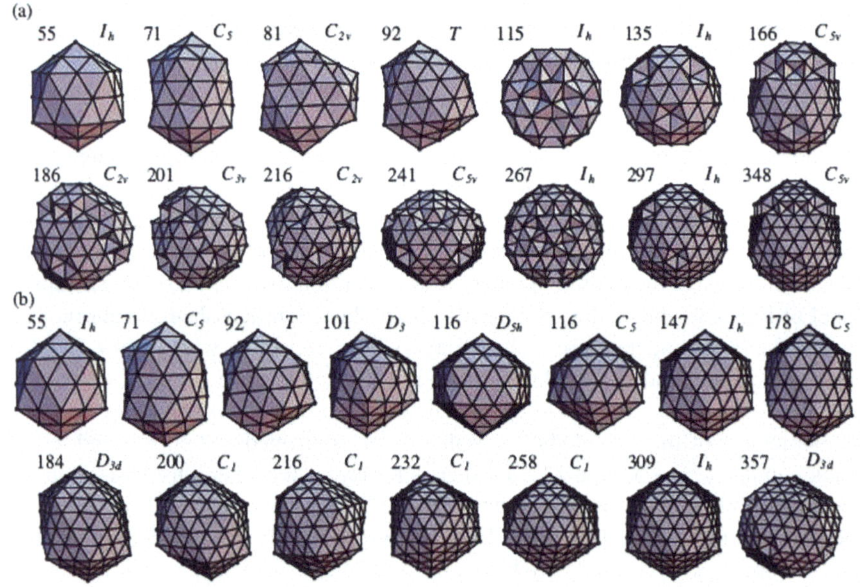

Fig. 3.40 A selection of sodium clusters that show enhanced stabilities for the (**a**) Gupta and (**b**) Murrell-Mottram potentials. Each structure is labelled by the number of Na atoms and its point group symmetry ch3noya

behavior of sodium clusters. By means of the two model potentials, Noya et al. [49] succeeded to locate the lowest-energy structures for all sodium clusters with up to 380 atoms. Finally, the structures of sodium clusters with high stability are presented in Fig. 3.40. Some numbers N, obtained by both potentials, could be interpreted as magic number. However, the results obtained with the Murrell-Mottram potential seem to be reliable, since they indicate an excellent agreement with experimental magic number and with computations at a higher-level of theory.

References

1. M. Halka, B. Nordstrom (eds.), *Alkali and Alkaline Earth Metals* (Facts on File Inc., An imprint of Infobase Publishing, New York, 2010)
2. http://en.wikipedia.org/wiki/Alkali_metal, Accessed 3 Mar 2015
3. http://www.studyvilla.com/Images/periodictable4.jpg, Accessed 25 Feb 2014
4. http://tabish.freeshell.org/physics/Sr/, Accessed 25 Feb 2014
5. http://tabish.freeshell.org/physics/Sr/gas-aggr.gif, Accessed 25 Feb 2014
6. https://shawnhigbee.files.wordpress.com/2013/08/atom.jpg, Accessed 3 Mar 2015
7. H. Stoll, H. Preuss, Hartree-Fock calculation of electron states in small aggregates of lithium and beryllium atoms. Phys. Status Solidi (b) **53**, 519–525 (1972)
8. P. Dugourd, D. Rayane, P. Labastie, B. Pintar, J. Chevaleyre, M. Broyer, L. Wöste, J. Wolf, Size dependence of sodium and lithium clusters ionization potentials. J. Phys. IV **01**, C7-509–C7-512 (1991)
9. C. Brechignac, P. Cahuzac, M. de Frutos, J.P. Roux, K. Bowen, Observation of electronic shells in large lithium clusters, in *Physics and Chemistry of Finite Systems: From Clusters to Crystals*, ed. by P. Jena, vol. I (Kluwer Academic Publishers, Berlin, 1992), pp. 369–374
10. Ph Dugourd, D. Rayane, P. Labastie, B. Vesin, J. Chevaleyre, M. Broyer, Measurements of lithium cluster ionization potentials. Chem. Phys. Lett. **197**, 433–437 (1992)
11. W.D. Knight, K. Clemenger, W.A. de Heer, W.A. Saunders, M.Y. Chow, M.L. Cohen, Electronic shell structure and abundances of sodium clusters. Phys. Rev. Lett. **52**, 2141 (1984)
12. M.Y. Chou, A. Cleland, M.L. Cohen, Total energies, abundances, and electronic shell structure of lithium, sodium, and potassium clusters. Solid State Commun. **52**, 645–648 (1984)
13. A.K. Ray, J.L. Fry, C.W. Myles, A configuration interaction study of small lithium clusters. J. Phys. B: At. Mol. Phys. **18**, 381–386 (1985)
14. S.C. Richtsmeier, D.A. Dixon, J.L. Gole, Structure of group 1A metal clusters, M_n, n = 4–6. J. Phys. Chem. **86**, 3942–3946 (1982)
15. B.K. Rao, P. Jena, Physics of small metal clusters: topology, magnetism, and electronic structure. Phys. Rev. B **32**, 2058–2069 (1985)
16. M.H. McAdon, W.A. Goddard III, Generalized valence bond studies of matallic bonding: naked clusters and applications to bulk metals. J. Phys. Chem. **91**, 2607–2626 (1987)
17. G. Pacchioni, D. Plavsic, J. Koutecký, Ber. Bunsenges. Phys. Chem. **87**, 503 (1983); G. Pacchioni, J. Koutecký, Ber. Bunsenges. Phys. Chem. **88**, 242 1984
18. J. Koutecký, I. Boustani, V. Bonaĉić-Koutecký, Triplet-singlet splitting of the alkali metal clusters: example of the Li_6 clusters. Int. J. Quantum Chem. **38**, 149–161 (1990)
19. D. Plavsic, J. Koutecký, G. Paccioni, V. Bonaĉić-Koutecký, Structure and stability of Li_4 and Li_6. J. Phys. Chem. **87**, 1096–1097 (1983)

20. A.N. Alexandrova, A.I. Boldyrev, X. Li, H.W. Sarkas, J.H. Hendricks, S.T. Arnold, K.H. Bowen, Lithium cluster anions: photoelectron spectroscopy and ab initio calculations. J. Chem. Phys. **134**(044322), 8p (2011)
21. I. Boustani, W. Pewestorf, P. Fantucci, V. Bonac̑ić-Koutecký, J. Koutecký, Systematic ab initio configuration-interaction study of alkali-metal clusters: relation between electronic structure and geometry of small Li clusters. Phys. Rev. B **35**, 9437–9450 (1987)
22. J. Koutecký, P. Fantucci, Theoretical aspects of metal clusters. Chem. Rev. **86**, 539–587 (1986)
23. J. Koutecký, V. Bonac̑ić-Koutecký, I. Boustani, P. Fantucci, W. Pewestorf, Electronic structure and properties of small metal clusters, in *Large Finite Systems*, eds. by J. Jortner, A. Pullman, and B. Pullman (Reidel, Dordrecht, 1987), pp. 303–317
24. I. Boustani, J. Koutecký, Investigation of the electronic and geometric structure of small Li anionic clusters with quantum chemical CI procedure. J. Chem. Phys. **88**, 5657–5662 (1988)
25. A. Kornath, A. Kaufmann, A. Zoemer, R. Ludwig, Raman spectroscopic investigation of small matrix-isolated lithium clusters. J. Chem. Phys. **118**, 6957–6963 (2003)
26. R. Fournier, J.B.Y. Cheng, A. Wong, Theoretical study of lithium clusters. J. Chem. Phys. **119**, 9444–9454 (2003)
27. L.-J. Cheng, J.-L. Yang, Communication: new insight into electronic shells of metal clusters: analogues of simple molecules. J. Chem. Phys. **138**, 141101-1–141101-4 (2013)
28. http://wiki.chemprime.chemeddl.org/images/8/8f/, 10th Line, Accessed 29 Mar 2015
29. http://www.chemicalelements.com/bohr/b0011.gif, Accessed 29 Mar 2015
30. http://ekengrenmeowser.blogspot.de/2013/07/sodium-apex-mineral.html, Accessed 29 Mar 2015
31. http://www.lenntech.com/periodic/elements/na.htm, Accessed 29 Mar 2015
32. A. Gelb, K.D. Jordan, R. Silbey, The electronic structure of small clusters of sodium. Chem. Phys. **9**, 175–182 (1975)
33. J.L. Martins, R. Car, J. Buttet, Variational spherical model of small metallic particles. Surf. Sci. **106**, 265–271 (1981)
34. W. Ekardt, Work function of small metal particles: self-consistent spherical Jellium-background model. Phys. Rev. B **29**, 1558–1564 (1984)
35. A. Herrmann, E. Schumacher, L. Wöste, Preparation and photoionization potentials of molecules of sodium, potassium, and mixed atoms. J. Chem. Phys. **68**, 2327 (1978)
36. M. Cini, Ionization potentials and electron affinities of metal clusters. J. Catal. **37**, 187 (1975)
37. J.L. Martins, J. Buttet, R. Car, Calculation of cluster geometries with the help of Hellmann-Feyman forces. Ber. Bunsenges. Phys. Chem. **88**, 239–240 (1984)
38. V. Bonac̑ić-Koutecký, P. Fantucci, J. Koutecký, ystematic ab initio configuration-interaction study of alkali-metal clusters. II. Relation between electronic structure and geometry of small sodium clusters. Phys. Rev. B **37**, 4369–4374 (1988)
39. J.L. Martins, J. Buttet, R. Car, Electronic and structural properties of sodium clusters. Phys. Rev. B **31**, 1804–1816 (1985)
40. V. Bonac̑ić-Koutecký, I. Boustani, M.F. Guest, J. Koutecký, Ab Initio configuration interaction study of the electronic and geometric structures of small sodium cationic clusters. J. Chem. Phys. **89**, 4861–4866 (1988)
41. A. Kornath, A. Zoermer, R. Ludwig, Formation of the magic cluster Na_8 in noble gas matrixes. Inorg. Chem. **41**, 6206–6210 (2002)
42. I.A. Solov'yov, A.V. Solov'yov, W. Greiner, Optical response of small magnesium clusters. J. Phys. B **37**, L137 (2004)
43. I.A. Solov'yov, A.V. Solov'yov, W. Greiner, Structure and properties of small sodium clusters. Phys. Rev. A **65**, 053203 (2002)
44. H. Haberland, (ed.), *Clusters of Atoms and Molecules, Theory, Experiment and Clusters of Atoms*. Springer Series in Chemical Physics, vol. 52 (Springer, Berlin, 1994)
45. W. Ekardt (ed.), *Metal Clusters* (Wiley, New York, 1999)
46. J.-O. Joswig, L.O. Tunturivuori, R.M. Nieminen, Photoabsorption in sodium clusters on the basis of time-dependent density-functional theory. J. Chem. Phys. **128**(014707), 1–6 (2008)

47. S. Pollack, C.R.C. Wang, M.M. Kappes, On the optical response of Na_20 and its relation to computational prediction. J. Chem. Phys. **94**, 2496 (1991)
48. H. Haberland, T. Hippler, J. Donges, O. Kotsko, M. Schmidt, B. von Issendorff, Melting of sodium clusters: where do the magic numbers come from? Phys. Rev. Lett. **94**, 035701 (2005)
49. E.G. Noya, J.P.K. Doye, D.J. Wales, A. Aguado, Geometric magic numbers of sodium clusters: interpretation of the melting behavior. Eur. Phys. J. D **43**, 57–60 (2007)

Chapter 4
Boron Clusters

Abstract Besides the conventional allotrope of the p-block non-metal element boron, like α-rhombohedral, β-rhombohedral, and γ-orthorhombic solids, and of the p-block non-metal element carbon, like graphite, diamond of many of organic structures, there are a lot of boron and carbon non-conventional structures in the form of clusters, nanotubes, nanocages and nanosheets. These nanostructures present the nanomaterials of both boron and carbon atoms.

4.1 The Element Boron

Introduction: Boron resembles in chemistry and physics its high capability and potentiality to form unbelievable and unexpected networks of molecular and metal-like bonds. In opposite to carbon, boron solid does not exist in nature but however all allotrope of boron, from bulk over nanostructures down to small clusters, are produced and synthesized at the laboratory experimentally. The most structures of those allotrope are composed of B_{12} icosahedra, arranged in different configurations. It is intelligibly and certainly to expect that boron could form, like carbon, some allotrope similar to nanotubes and fullerenes, since the research on boron-based nanostructures and nanomaterials is repidly developing theoretically and experimentally. In the nineties and following years, novel boron structures in form of quasi-planar clusters, nanotubes, cages and sheets named borophene were theoretically predicted and proposed by I. Boustani, the author of this Book. Few years later all of these structures were first theoretically confirmed then experimentally synthesized such as single-walled boron nanotubes as well as single-layered boron sheets known as borophene. However, the mechanical and electronic properties of these novel nanostructures and their potential applications in the nanotechnology are still at the beginning.

Boron versus Carbon: The development of further theoretical methods and computational capacities as well as of novel experimental techniques and measurements delivered not only the structural configuration in atomistic details but also enabled the fabrication and simulation of artificial structures with custom-made electronic and mechanical characteristics creating novel materials unknown in nature. Thus, it

I. Boustani, *Molecular Modelling and Synthesis of Nanomaterials*, Springer Series in Materials Science 290, https://doi.org/10.1007/978-3-030-32726-2_4

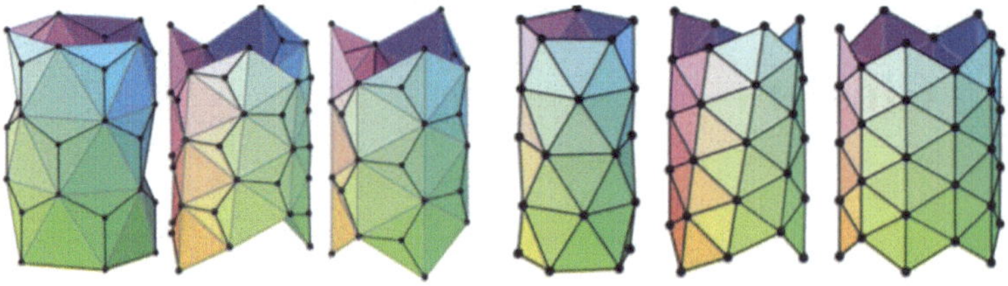

Fig. 4.1 Left set: Polyhedral CNTs (5, 0) zigzag, (4, 3) chiral, (3, 3) armchair, with hexagonal units. Right set: Ployhedral BNTs (5, 0) zigzag, (5, 3) chiral, (5, 5) armchair, with centered hexagonal units [1]

seems to be possible that materials at nanoscale can be routinely designed to achieve the desired properties. As an example for such achievements are the novel structural and electronic properties of carbon. The theoretical calculations and the experimental synthesis of carbon clusters, nanotubes and sheets in form of graphene show that these carbon polymorphism exhibit novel and different physical and chemical properties depending on their size and length scale. These carbon nanostructures are now proposed as candidate materials for a wide variety of applications such as structural (clothes, fire protection and sport equipment), electronic (displays and transistors), chemical (sensors, water filters and hydrogen storage), and mechanical (oscillators, nanotubes membranes and thermal radiation). However, despite their excellent materials and electrical characteristics, CNTs suffer from serious limitations due to a lack of high-purity materials. Therefore, in the recent years attention has also been focused on the counterpart boron nanotubes (BNTs) as alternative to CNTs, which can not only exhibit similar properties as the CNTs, but also has some novel characteristics, that are unique to them. The BNTs are conductive in opposite to the CNTs. To this end, several different nanotubes and novel nanostructures of a wide range of elements and compounds were predicted, synthesized, and characterized. Figure 4.1 presents the BNTs, proposed by Lee et al. [1], as alternative to the CNTs.

Why Boron? In this respect, Boron, followed by the next neighbor carbon in the periodic system (Fig. 4.2), holds a unique promise of substituting the CNTs and also exhibiting its own novel features. The potential candidacy of boron arises from its pecular character the electron deficiency and demonstrates its diverse structural features at cluster level. Boron is the smallest and lightest semi-metal atom. Boron undergoes strong covalent bonds in form of high-strength solids. In fact, β-boron and γ-boron are the hardest elementary crystals after diamonds. Furthermore, some of boron crystals exhibit a series of unique properties. All forms of boron However, the allotropes of boron in nearly all its formations possess very high melting points, namely between 2,200 and 2,300 $^{\circ}$C. They are not reactive and thus have high resistance to chemical attacks. One property of special importance is boron's ability to absorb neutrons. This makes boron useful as the control rods of nuclear reactors for

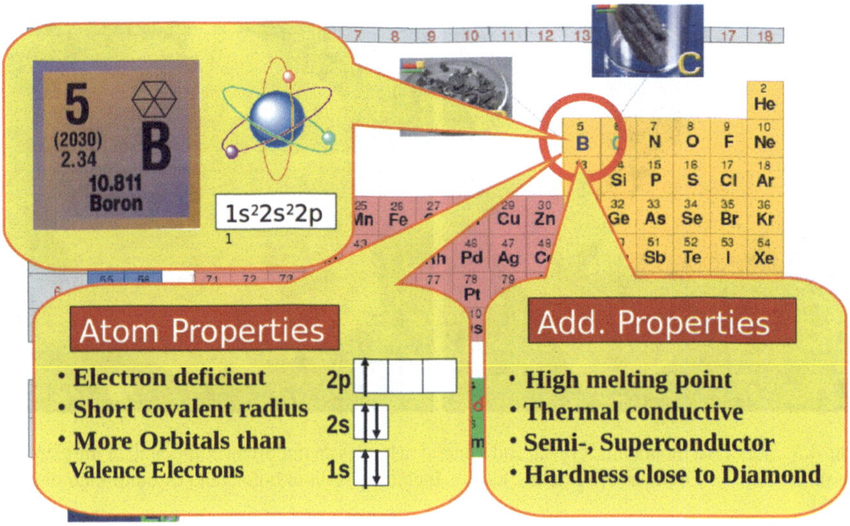

Fig. 4.2 The properties of the element boron in the periodic table [2]

controlling the fission reaction, and important for a medical treatment called boron neutron capture therapy (BNCT), which will be discussed in Sect. 4.1.

4.1.1 History of Boron

Boron Origin: It is a long time ago or thousands of years that boron compound were known starting by Babylonians [3]. Over 4000 years ago they were familiar with importing borax from the Far East of Asia to be used as a chemical cleaning agent named flux for extracting gold. Further applications of boron was medicinal and metallurgic applications as well as mummifying, which was attributed to the ancient Egyptians. On the other side it was reported that, the Arabians, the Tibetans as well as the Chinese were using such kind of materials.

Boron in Borax: The original name of boron is derived from the Arabic word "Al-Buraq", or from the Persian word "Burah" which are names corresponding to the mineral borax. The item boron itself arises from the combination of borax and carbon. In the western of Tibet the mineral borax received the name "Tincal", an item derived from the Sanskrit, the language of ancient India. However, the compound borax glazes ($Na_2[B_4O_5(OH)_4].8H_2O$), as presented in Fig. 4.3, were used in China from AD300. These compounds arrived in AD700 the Arabian chemistry "Al-Kimiya", whereby the Arabic alchemist Jabir Ibn Hayyan was working with. In the 13th century Marco Polo brought some of these glazes to Italy. However, later around the 1600 century, Agricola reported the use of borax mineral as a a flux for metallurgical applications.

Fig. 4.3 Borax is a monoclinic crystal and mineral, which is an important compound of the element boron. It goes by a number of scientific names, including sodium borate and disodium tetraborate [4, 5]

However, around one hundred fifty years later, after detection of the boric acid in hot springs near Florence, Italy, the acid became to be known as sal sedativum and very useful in many medical applications. Sassolite, another borate mineral form of boric acid, was located by Sasso, Italy. From 1827 to 1872 was Sasso, Italy, the major source of European borax. After then this source was replaced by the Americans [6].

Boron Discovery: Boron was discovered by the French chemists Joseph-Louis Gay-Lussac and Louis-Jacques Thénard. At the same year 1808, Sir Humphry Davy, an English chemist, detected boron independently and called this element as boracium. All of them succeeded to isolate boron by a combination of potassium with boric acid (H_3BO_3). The purity of their products was about 50%. Few years later, quite in the year 1824, boron was identified by Jöns Jakob Berzelius proposing it as a new element. Fifty years later, the production of contaminated boron was in full swing besides graphite and diamond. Furthermore, the chemical compound aluminum boride AlB_{12}, (see Fig. 4.4), was recognized to be largely diamond-hard material. Higher purity boron was obtained much later by reduction of boron trioxide with magnesium achieving purities of around 90%. At the beginning of the 20th century, the U.S. American chemist E. Weintraub [8] arguably isolated and synthesized the element boron to a high purity of around 99%. Conventional boron is an amorphous powder with a brown-black color. At low temperatures, crystalline boron is an insulator, but at elevated temperatures becomes boron to be a conductor. The yellow monoclinic crystals of pure boron are extremely hard materials being also semiconductors and resembling with a band gap of 1.50 or 1.56 eV the element silicon.

Steel and Boron: Steels are alloys of iron enriched with other elements, primarily with carbon in an amount of around 2.1% of its weight as the typical alloys of steel. To inhibit corrosion, at least 11% chromium is added to steel forming a layer of

Fig. 4.4 Scanning electron microscopy of AlB_{12} and Al_2O_3 [7]

chromium oxide Cr_2O_3. The steel alloy containing chromium is known as stainless steel. In addition there are different alloys of steel like Fe_2C and Fe_3C, which are called iron carbide. The latter Fe_3C is a compound of 93% iron and 6.6% carbon by weight [9]. Coevally boron was capable to form complex compounds with unusual properties. In 1907 boron was first considered as a potential alloying element in steel. During the next 20 years various investigators studied, and even patented, steels with boron concentrations up to 2% in some cases. It was not recognized until 1921 that such large amounts of boron were responsible for making steel extremely hard and brittle and that even such a minute amount as 0.001 wt% of boron could produce significant effects on steel properties. Only in the thirties and early forties the effect of different elements—including boron—on the hardenability of steels was recognized and boron was found to be by far the most potent [12]. However, the right chemical and physical properties of steel have been determined during the last forty years. With the electrolysis technique of molten potassium fluroborate and potassium chloride (KCl) and by vapor deposition methods it was possible to produce high purity boron. However, one well known compound of iron alloy including boron is the iron boride Fe_2B, which includes square anti-prismatic iron atom coordination containing single boron atoms. Figure 4.5 shows the austeninte steel and the hexagonal structure of the diiron boride.

4.1.2 Energetic Boron

Boron as Rocket Propellent: Boranes, the chemical compounds of boron and hydrogen, does have any equivalents in nature and apart from any analogy to petroleum for boron. At the early of the 20th century took the German chemist in Karlsruhe Alfred Stock the initiative himself and explored the possibility of boron and hydrogen synthesis to configure the so called boron hydrides or simply boranes (Fig. 4.6), in analogy to the well known hydrocarbons. After failed attempts he succeeded to

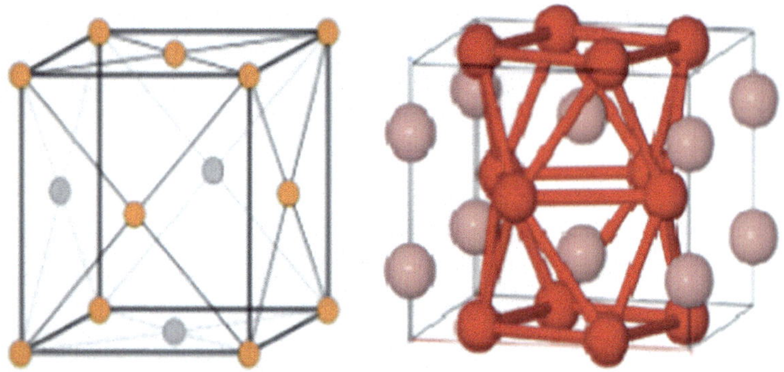

Fig. 4.5 Crystal Structures of Austeninte Steel FeC (fcc) and Di-Iron Boride Fe_2B [10, 11]

Fig. 4.6 Diborane with its hydrogen-bridged structure, while penta- and decaborane are the most stable nido-boranes [14]

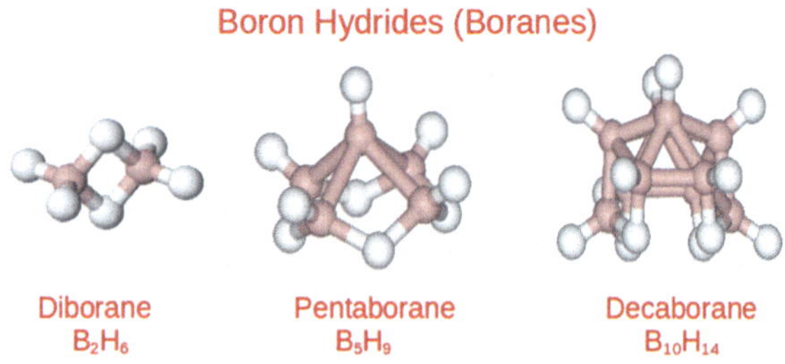

produce small amount of these expensive and highly reactive rare compounds. Until 1950 stayed the chemistry of boranes in darkness and nobody understood neither their structures nor their properties. Afterwards, it was of great pleasure to experience that these synthesized compounds exhibit high thermal properties and can release large quantities of heat and from a sufficient amount of one of these compounds it is possible to make excellent fuels [13].

Rocket Engines and Boron Grains: The "solid-propellant ducted rocket" was in the mid 1970s the focus of interest for developing capable and proficient rocket engines. The search for the ram jet engine, based on the construction and development of a new propellant system, was in progress. Thus, in order to regulate the propulsion of the engine, a control valve was mounted between the gas generator and the combustion chamber. This valve regulates the gas flow to the combustion chamber by an appropriately narrow window, because the generated gas in the system is under high pressure. The distribution of the solid boron 95/97 additive using reliable methods to characterize the grain size is the next step to be determined, due to the fact that coarse boron grains could of course interfere and destroy the engine, if they were disable to be under control. Therefore, the clientele requested solid boron in an amorphous form, containing maximum 2% of grains with sizes laying between 2 and 5 μm only [15].

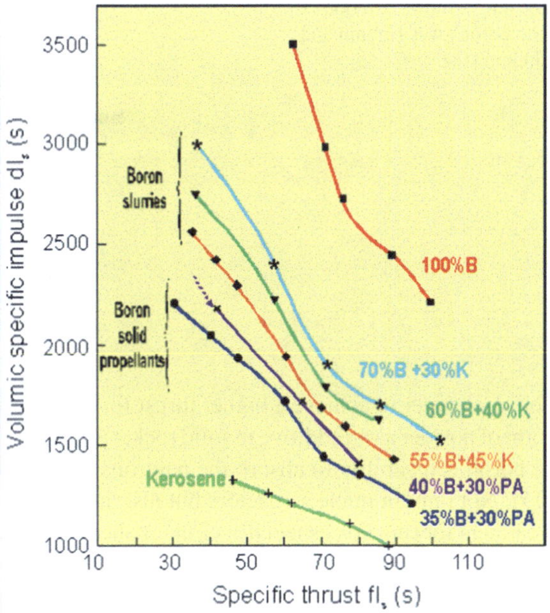

Fig. 4.7 Volumetric specific impulse versus specific thrust for boron kerosene slurry in different ratio [16]

Boron as Mystery Fuel: As mentioned above, a borane is a boron-hydrogen combination, and hydrogen burns with a greater heat of combustion per pound than any other pure material. In a borane, the hydrogen atoms are tied to boron atoms. This combination, also a liquid, contains something like 30.000 BTUs[1] per pound, about the third more heat units (or power) than gasoline or kerosene. This means that in an airplane a borane can deliver a third more thrust than the same amount of JP-4 jet fuel. A borane burns efficiently at high altitudes where the air is so thin that ordinary fuels sputter and flame out. The diagram of the volumetric specific impulse versus specific thrust, presented in Fig. 4.7, shows the I_{sp}[2] for different fuels consisting of boron and slurries of kerosene. The presence of boron in the fuels indicates the increasing impact with increasing boron content during the burning process in ramjet engine, particularly when the content is 100% boron (the upper red curve). This diagram exhibits also the predominant performance of boron over the high power liquid fuel of kerosene, which is assigned by the green line at the bottom of the Fig. 4.7. However, boron and borane additives were used for transfer of atmospheric vehicles, since with these conventional chemical bonds it is possible to increase the energy performance of non-nuclear fuels [16, 17].

Boron as Neutron Absorber: After the World War II the reincarnation of the boron was emerged dealing with completely different products. At the top of the highly challenging industry was not borax but of course elementary boron. As mentioned

[1]BTU (British Thermal Unit) ≈ 1055.06 J ≈ 252 Cal.

[2]I_{sp} is the specific impulse, a measure of the efficiency of rocket and jet engines.

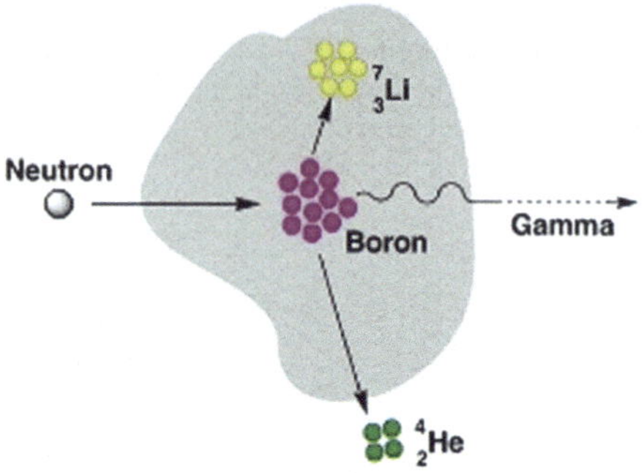

Fig. 4.8 Neutron reaction with boron gives Helium and Lithium [18]

above, in order to achieve a higher thrust force it was very necessary to use boron in form of powder as an additive in solid rocket propellants. Further essential application of boron is its ability to absorb the neutrons in the reactors of nuclear power plants [15]. Not only in nuclear reactors but also in medicine is boron of great interest. A treatment for cancers, especially virulent ones of the brain, in which a person who has been injected with a boron compound that concentrates in cancerous cells is exposed to neutron irradiation causing the boron compound to emit α-particles that destroy the cancer cells. This concept is known as boron neutron capture therapy (BNCT), which is represented in Fig. 4.8 and based on the typical neutron absorption reaction as:

$$^{10}_{5}\mathrm{B} + {}^{1}_{0}\mathrm{n} \rightarrow {}^{11}_{5}\mathrm{B} \rightarrow {}^{7}_{3}\mathrm{Li} + {}^{4}_{2}\mathrm{He} + 2.8\,\mathrm{MeV} \tag{4.1}$$

After the integration of the stable isotope ^{10}B into the compound of the quested tumor. This compound will be then introduced into blood, which in turn and after a given time accumulates in the tumor. After then, the irradiation of epithermal neutrons are captured by the ^{10}B isotope inside the tumor causing transformation into ^{11}B isotope. Capturing neutrons causes the instability of the boron isotope ^{11}B and leads its nuclei to break apart provoking the emission of α-radiation (the nucleus of Helium atom only) and recoiling ^{7}Li nuclei, as shown in (4.1).

4.1.3 Boron Materials

Boron Minerals: In opposite to carbon, the element boron does not exist in nature sheerly, but in compositions of rocks, soil, coal, and water. Boron is mostly integrated in compounds with oxygen called borates. In nature there are more than 230 minerals, which include boron. Some of these boron containing minerals are highly soluble in water. For example, boron can consolidate easily with hydrological systems, which in turn can be coupled to plant and animal organic systems. Consequently, boron would also have effects on human health after merging with plants, animals and drinking

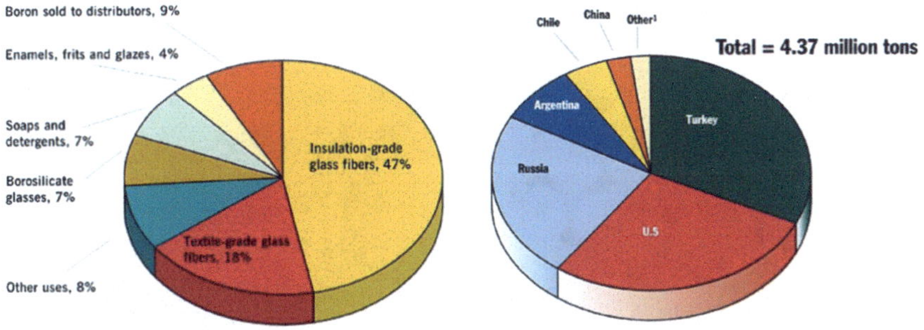

Fig. 4.9 Left: The U.S.A. boron consumption by the end use in 2000. Right: the world production of boron minerals in 1999 [20, 21]

water. Boron can burn with Oxygen only at higher temperatures, forming boron oxide (B_2O_3), but by reaction with Nitrogen it forms boron nitride (BN). By reactions with metals boron forms the so called refractory borides [19]. Due to its high reactivity, boron precludes its incidence in the free state. The occurrence of some boron compositions can be listed as follows. Boracite ($Mg_3B_7O_{13}Cl$), was found in Germany. Ulexite was also found in Bolivia, Chile and Peru. Ulexite ($CaNaB_5O_6(OH)_6.5H_2O$); and Kernite (rasorite, $Na_2B_4O_7.4H_2O$), were found in the United States. Sassolite (H_3BO_3), was found in Sasso, Italy, as already referenced in Sect. 4.1.1.

Boron Sources: The mineral kernite is the world's major source of boron located in the Mojave Desert of California, USA. For many years, the colemanite mineral was supplied by Turkey being the main producer of boric acid to Europe. In Kirka, Turkey, the mineral sodium borates were found in 1960. Other deposits of boron compounds were also found in Anatolia, Turkey. Therefore, Turkey became to be the largest producer of borate products worldwide, exporting colemanite, ulexite, borax pentahydrate, ($Na_2B_4O_7.5H_20$), anhydrous borax ($Na_2B_4O_7$), refined borax decahydrate, and tincal, in form of mineral concentrates. Thus, the world production and consumption of boron is presented in Fig. 4.9. As can be seen, the major production of boron minerals are Turkey, U.S. and Russia. By the boron consumption in the USA can be seen that the glass industry remains the largest domestic market in 2000, booking the main consumption of boron attaining 76%. Large amounts of boron compounds were consumed by further kinds of glass, like the textile-grade glass fibers with 18%, and insulation-grade glass fibers with 47%. Furthermore, small amounts of boron composites were also consumed like borosilicate glasses to 7%, soaps and detergents to 7%, and frits and glazes to 4% only.

Boric Acid: Sassolite or boric acid ($B[OH]_3$) (Fig. 4.10) is a mineral in form of white and yellow crystalline solids, known as hydrogen borate. It is a soluble in hot water to form a weak acid, used as a mild antiseptic, flame retardant, insecticide and in manufacture of borax and other borate salts [22]. The common use of the hydrogen borate, the boric acid, is the swimming-pool type of nuclear reactors, where boron serves as a neutron absorber. Further applications of boric acid is in electroplating

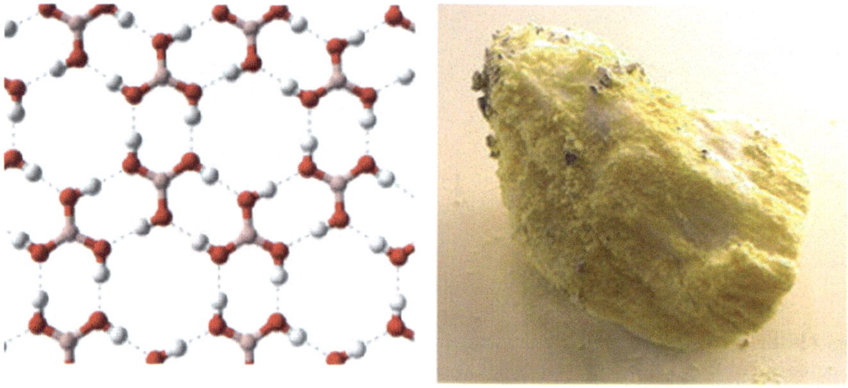

Fig. 4.10 Hydrogen bonding in boric acid $B[OH]_3$ allows the molecules to form parallel layers in a solid state building yellow sassolite crystals [23, 24]

baths, in which deposition of nickel takes place. But however, boric aced is used mostly as an eye washer. As a source of boron in the fused salt electrolysis, the anhydride of boric acid is also used for the preparation of elemental boron.

Superhard Materials: Light elements like boron, carbon, nitrogen and oxygen are the normal compositions of these materials. These elements, with their well known intrinsically strong and covalent directed bonds, can tight networks of three-dimensional character with extreme hardness and adamant resistance to external shear strain. This structure creates a unique combination of properties including low density, high hardness as well as high chemical stability and excellent wear resistance, which allows such materials to find extensive use as nozzles, pump impellers, grinding wheels, drill bit components, abrasives and cutting tools. Potential industrial applications have in recent years brought about intense studies of the physical and chemical properties of these materials [25].

Boron Carbide and Nitride: Boron carbide B_4C, Fig. 4.11a, is a ceramic material with an extremely hard character. It is mostly used in tank armor but also in wear. The chemical reaction of boron trioxide (B_2O_3) with carbon leads to the boron carbide powder B_4C. The Vickers hardness of boron carbide B_4C is about 38 GPa,[3] and considered to be the next hardest known materials after diamond and cubic boron nitride. It is thermally conductive and a semiconductor of *p*-type with a band gap of about 0.84 eV. Cubic boron nitride (c-BN), Fig. 4.11 (right), is a superhard material having no natural analogs. The Vickers hardness of c-BN is 48 GPa, ranking the second behind diamond, whose Vickers hardness is in the range of 70–120 GPa. The c-BN is an insulating material with electrical resistivity $\rho > 10^{15}\ \Omega$ and a band gap of around 10 eV. The crystal lattice of c-BN is similar to that of diamond. The c-BN ceramic is normally produced at temperatures above 1600 °C by a reaction of boric acid with boron oxide and nitrogen as well. Cubic BN is chemically inert and

[3] The Pascal (symbol: Pa) is the unit of pressure, internal pressure, stress, and Young's modulus.

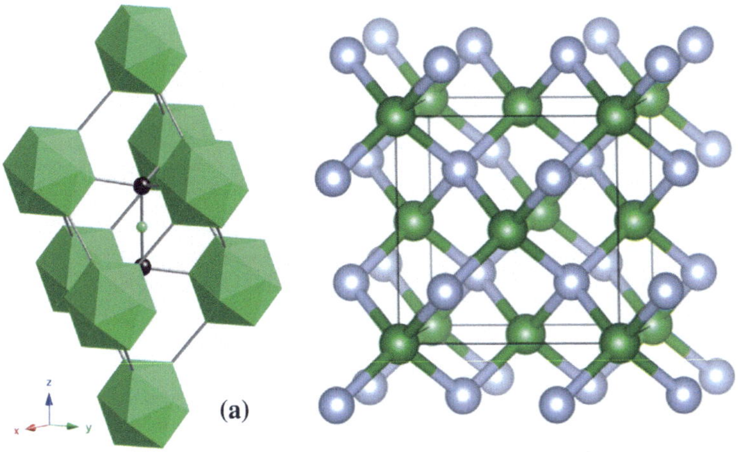

Fig. 4.11 Left structure: Boron Carbide, which is composed of C-B-C chain surrounded by eight B_{12} icosahedra forming rhombohedral lattice. Right structure: Diamond-like Cubic Boron Nitride, in which each nitrogen atom is covalently bonded to four carbon atoms and vice versa [26, 27]

non-toxic. It has high thermal conductivity and high melting temperature of 2973 °C, excellent wear resistance and good chemical inertness. It is also used as an abrasive, wear-resistant coating and for cutting tools [28].

Boron Suboxide: In 1998, Hubert et al. [29] succeeded to synthesis a grain of boron suboxide (B_6O), which is obtained in a scanning electron microscope pseudo color image. After mixing boron and boron oxide, these compounds were heated to 1700 °C and settled under a high atmospheric pressure of about 4 GPa forming boron suboxide (B_6O) particles with an orange-red color, with near-perfect, regular icosahedra. As shown in Fig. 4.12, these structures are similar to some particles with multiply-twinned form monitored in some cubic materials. The structure of these icosahedral particles is nothing else as the Mackay packing of icosahedral B_{12} units, and therefore it has a long-ranged order beyond the translational symmetry. The boron suboxide crystals is hardest known oxide. It is considered to be one of the hardest material in the world. The Vickers hardness H_V for polycrystalline B_6O ceramics may vary from 30 to 45 GPa [31]. Similar to boron carbide B_4C ranks the hardness of B_6O ceramic direct after diamonds and cubic boron nitride. Due to the extremely high hardness of B_6O ceramic, the promising potential opens the door to new class of applications like wear and hardfacing materials, cutting and polishing tool, and abrasive substance in working industry. Furthermore, due to its low chemical reactivity as well as to its mechanical features, this material could potentially substitute the tungsten carbide compound in high-wear resistance utilizations. Beside its hardness property, boron suboxide can be applied in the electronics due to its particular semiconductor properties [30].

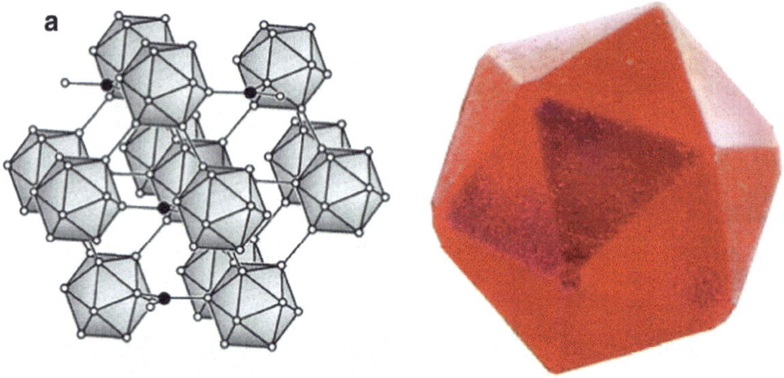

Fig. 4.12 Each oxygen atom O (•) in the crystal structure of B_6O is bonded to B (∘) atoms in three separated icosahedra. The centers of the outer 12 icosahedra are at the vertices of a cuboctahedron. The image of the orange-red crystals has a perfect icosahedral shape, which has 20 triangular faces and 12 corners with I_h symmetry [29, 30]

Fig. 4.13 The symbol and form of boron atom as well as the vital role of boron in cell walls of plants [32]

4.1.4 Chemistry of Boron

Atomic Boron: Boron is the direct neighbour to carbon in the periodic system, with the atomic number 5, Fig. 4.13. It is known to be hard, brittle, lustrous black semi-metal. It is also conversant with its directed covalent multi-center bonds with short covalent radius. Due to its electron deficient character as well as thermic performance boron holds unique chemical and physical properties. The electron configuration of boron is $1s^22s^22p^1$. The second electronic shell has three valence electron, and the *p*-type orbitals are occupied by only one valence electron, which needs 8.29 eV to be ionized. Therefore, in an electrovalent bond, boron does not donate or accept electrons so easily but it goes in covalent bonds.

Boron Isotopes: Up to date around 14 boron isotopes have been discovered. The most stable non-radioactive boron isotopes, as shown in Fig. 4.14, occurs naturally are ^{10}B and ^{11}B. The nucleus of ^{10}B is composed of five protons and five neutrons. For thermal neutron absorption ^{11}B exhibits a high cross section, while the ^{11}B nucleus has five protons and six neutrons, which is compatible as a fuel for a neutronic fusion

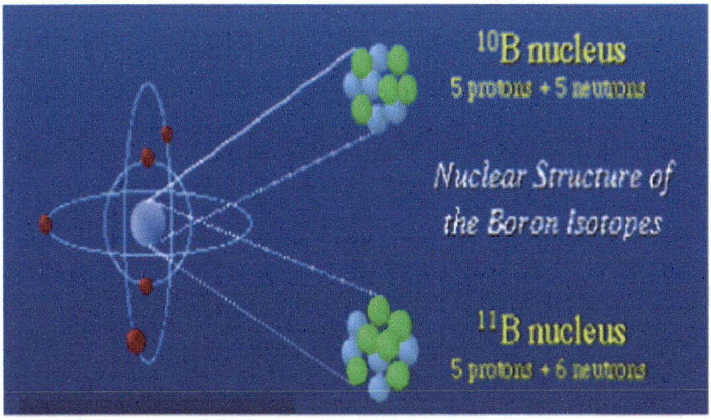

Fig. 4.14 The most stable boron isotopes ^{10}B and ^{11}B [33]

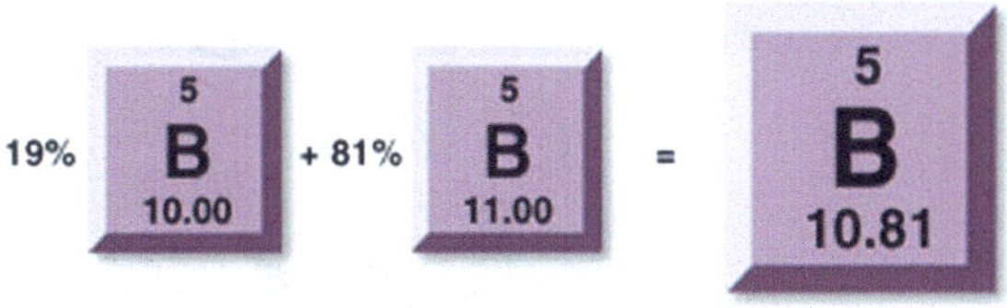

Fig. 4.15 The average atomic mass unit (amu) of boron isotopes is the sum of 0.19×10.00 amu $+0.81 \times 11.00$ amu $=$ 10.81 amu [34]

and semiconductors. The abundance of the both boron isotopes ^{10}B and ^{11}B in the earth's crust is around 19% and ^{11}B of 81%, respectively. Thus the average mass of boron is about 10.81 amu, as shown in Fig. 4.15. Boron is at room temperature non-reactive with oxygen, water, alkalis or acids, and is unaffected by air. The density of amorphous boron is 2.37 g/cc, and that of crystalline boron is about 2.34 g/cc. Boron has a high melting temperature of around 2300 °C (about 1000 °C higher than silicon), and rather higher boiling temperature of 2550 °C. Therefore the preparation of boron is extremely difficult, but it is a very refractory substance. For instance, boron fibers are known with their great strength, which is very important in composite materials [35].

Gaseous Boron: Boron does not only exist in minerals, but also in compounds in gaseous form. Some of them are diborane, the simplest boron hydride, which is in fact colorless, highly toxic, and flammable. Diborane catches fire spontaneously upon exposure to air. It burns in oxygen releasing an enormous amount of energy. It can be produced on an industrial scale by the reaction of BF_3 with sodium hydride. The gas of boron trifluoride BF_3 is toxic and colorless. It forms white fumes in moist air. Manufacturing of BF_3 can be achieved by the reaction of hydrogen fluoride with boron oxides [36, 37].

Boron in Nuclear Industry: Due to the high neutron absorption capability of boron, it is obvious that the nuclear industry serves both isotopes ^{10}B and ^{11}B extensively (Fig. 4.16). On the one side, the ^{10}B isotope is used as a chemical shim in pressurized water reactors, when it is prepared in the form of boric acid H_3BO_3. This ^{10}B isotope

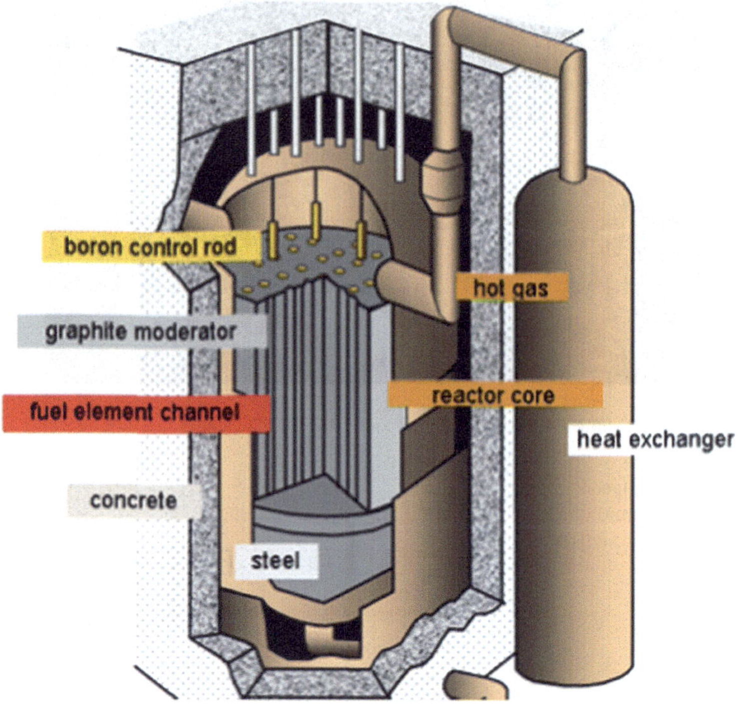

Fig. 4.16 The control rods in nuclear reactor are made with neutron-absorbing materials such as boron [39]

is also used in boiling water reactors as standby liquid control systems, when it is prepared in the form of sodium pentaborate. On the other side, the ^{11}B isotope is deployed as neutron reflector. However, beside the nuclear industry, there is another field where both isotopes are applied, namely in food enabling the research of boron metabolism. Also in the medicine, boron plays a very important role in the abatement of cancer. Therefore, in Boron Neutron Capture Therapy (BNCT) becomes the ^{10}B isotope to be commitment, as already reported in Sect. 4.1. The element boron with its isotopes ^{10}B and ^{11}B can also help in preparing other isotopes, like ^{11}C and ^{13}N [38].

Production of Boron: The fabrication of boron can be attained via different pathways. The most applied methods to produce boron are given as follows: (1) By reduction of hydrogen with boron trichloride BCl_3. (2) By thermal conversion of diboran B_2H_6. (3) By reduction of magnesium with boron trioxide B_2O_3. These methods were used in the 1950s, because the production and repurification of boron on the basis of potassium fluroborate KBF_4 were rather very expensive. However, amorphous boron was reextracted by the US scientists at the beginning of the 1950s. They did achieve boron content of around 90–92% directly, first by a reaction of magnesium with the most common form of boron oxide B_2O_3, then by acid treatment as a following step. At the beginning of the 1960s and due to the great need of

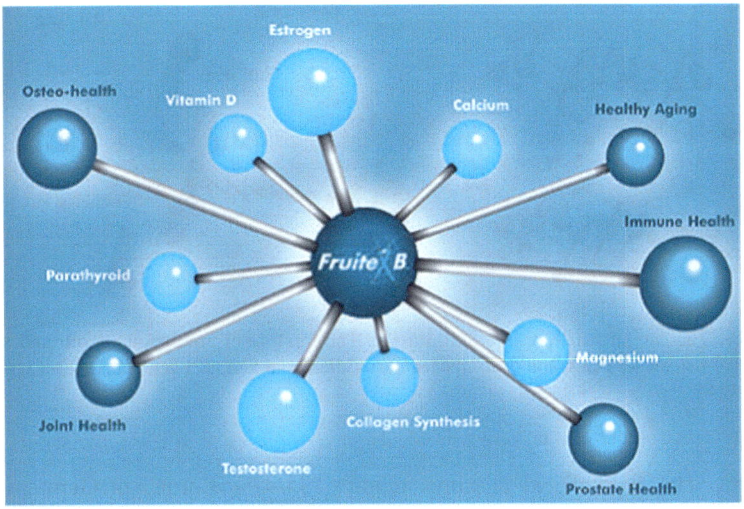

Fig. 4.17 Fruitex-B, otherwise known as calcium Fructoborate, is a special and highly bioavailable form of the trace mineral boron. Boron is essential to healthy bones as it serves as a co-factor in the functions of calcium, magnesium, and vitamin D [42]

improving the propulsion performance of the rocket motors, the US scientists added amorphous boron (90–92% boron) to its solid propellant and hence gaining a wide acceptance. The advantage of amorphous boron is simply its capability to be used as an igniter in rockets as well as in pyrotechnic flares providing prominent green color [40]. In the following years boron compositions were produced using different analytical methods and the main production of "90/92% boron" was manufactured in the United States fulfilling the requirements largely [15].

Boron in Food: Boron is a trace element in human and plant nutrition and very different from other minerals. Boron influences a wide range of bioprocessing enclosing macro-minerals and energy-rich nutrient medium like glucose and triglycerides, proteins and amino acids, prostate health, bone mineralization, free radicals, estrogen metabolism, mental function and numerous body systems. For example, in order to keep cell division and growth of some plants intact, boron is needed to be supplied constantly. Thus, one can conclude that the mineral boron is really crucial to our life and health. Therefore, as a trace element in plant life, boron also plays a great role in our physiological life [36, 41]. Trace amounts of boron seem necessary for good growth of plant life, but large amounts are toxic. Boron accumulated in the body (see Fig. 4.17) through absorption, ingestion and inhalation of its compounds affects the central nervous system. The symptoms are depression of circulation, vomiting and diarrhea, followed by shock, coma and a body rash.

Oxidation of Boron: One of very important applications of amorphous boron in our daily life was emerged at the end of the 1980s. That was the airpag, equipped in lifesaving automobile safety systems and based on the gas generator with ignition

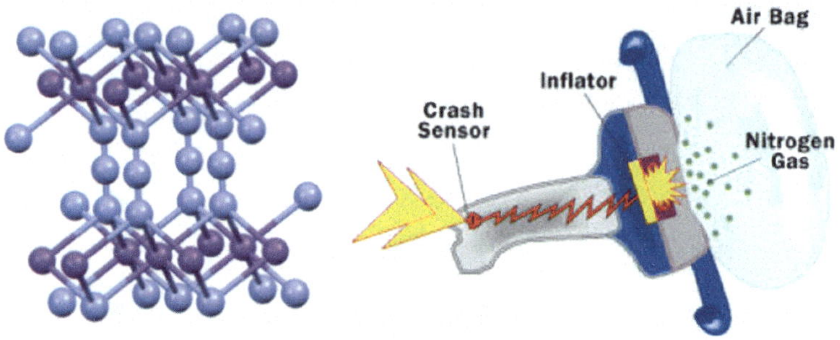

Fig. 4.18 The crystal structure of sodium azide. The mixture of amorphous boron and sodium nitrate detonates after crash and decomposes into sodium salt and nitrogen gas: $2NaN_3 \rightarrow 2Na + 3\ N_2$ [43, 44]

mixtures. The skyrocketed mixtures require "95/97% Boron" in Germany, and "90/92% boron" in the United States. They are nontoxic and absolutely reliable. The mixture of sodium azide (NaN_3) and amorphous boron is a densified powder. An electric impulse in the airbag igniter triggers the powder causing the detonation of the mixture and spreading of gas (Fig. 4.18). In other words, after a primary ignition of NaN_3 and boron mixture, it decomposes in a fraction of a second to nitrogen and sodium in form of pellets in the gas generator. The released nitrogen inflates the airbag, while the sodium is removed with the help of silicates [15].

4.2 Icosahedral-Based Crystalline Boron

4.2.1 First Crystalline

Pure Boron Crystals: As reported in Sect. 4.1.2, the first samples of highly pure crystalline boron were prepared and obtained in 1909 by Weintraub [8], by reduction of BCl_3 with hydrogen in an electric arc. Also, the first crystals of the α-tetragonal boron were obtained in 1943 by Laubengayer et al. [45]. They have undertaken the production of single crystals of boron, the establishment of the high purity of the samples, a comprehensive study of the properties of the element and, in particular, X-ray examination of the material in an effort to ascertain the X-ray powder diffraction pattern characteristic of boron and to obtain data adequate for the determination of its crystal structure. They used the "filament method" which consists essentially in the vapor phase reduction of a volatile compound of boron in the vicinity of a glowing filament, the free boron depositing upon the filament. After considerable preliminary experimentation the reduction of boron tribromide vapor by hydrogen was found to be particularly suitable for the production of pure boron, hydrogen bromide being the other product formed. They observed single crystals of boron of high purity which have been grown in appreciable size.

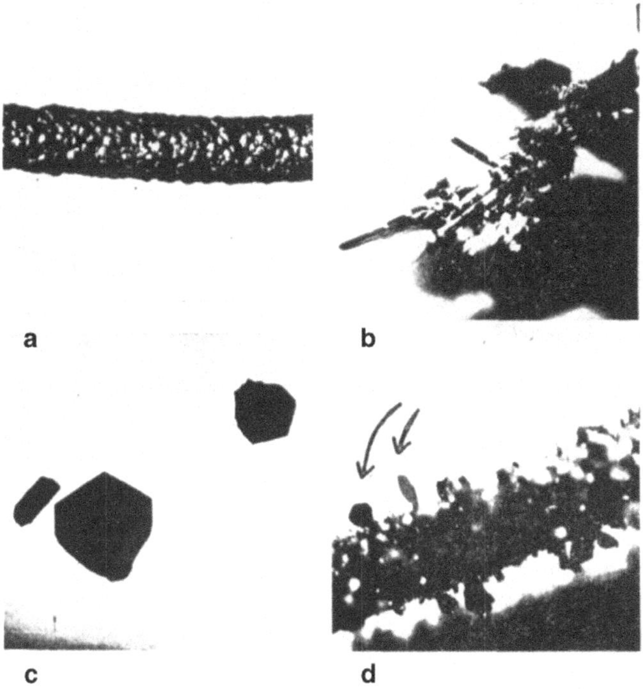

Fig. 4.19 A, vitreous boron; B, needle cluster; C, plate crystals; D, plates growing on wire [45]

Boron Needles and Plates: The most exciting and interesting results of Laubengayer et al. [45] are the large single crystals in the form of *needles* or *hexagonal plates* (Fig. 4.19). Smaller crystals were observed as *hexagonal rods* or massive in form. Optical evidence indicates that both the needle and plate forms of crystalline boron may belong to the monoclinic system. Crystalline boron shows great resistance to attack by various chemical reagents including oxygen and oxidizing agents. Crystalline boron shows an extreme hardness, great opacity, metallic luster, and electrical properties usually associated with semi-metallic solids. The unusual amount of symmetry shown by the needle crystals indicates that the natural system of this form of crystalline boron may be tetragonal under ideal conditions of crystal growth. These structures could be considered as the precursors of modern boron nanostructures, since *needles* could be renamed nowadays to nanowires, *hexagonal plates* to sheets and *hexagonal rods* to nanorods.

Icosahedron as Basic Unit: The structures of crystalline boron were identified in the fifties and sixties. Most of boron crystalline as well as a great number of boron-rich borides are all characterized by the existence of three-dimensional skeletons of boron atoms built from a B_{12} elementary unit (Fig. 4.20), itself consisting of twelve boron atoms at the vertex positions of a quite regular icosahedron. The boron icosahedra can be considered as building blocks for the most of boron-rich materials.

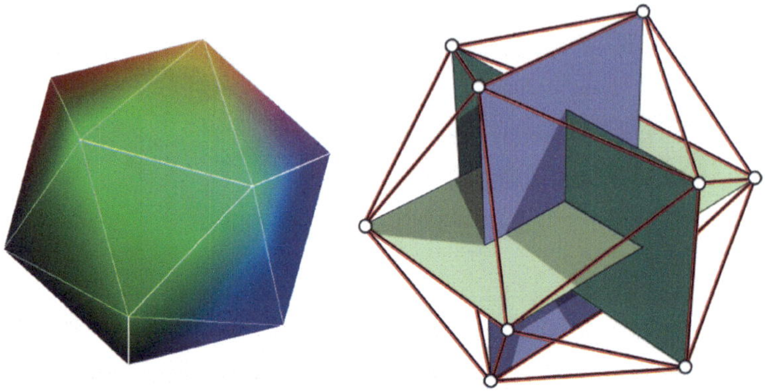

Fig. 4.20 The regular icosahedron is one of the five Platonic solids. The icosahedron vertices form three orthogonal golden rectangular [46, 47]

The presence of icosahedral clusters, though rather unusual in organic structures, is not all that surprising in the case of boron-rich phases with a covalent character, in view of the electron-deficient nature of boron and the efficient manner for distributing electrons provided by the compact grouping of twelve atoms characteristic of an icosahedron. The bonding nature of icosahedral boron solids can be explained by the intra-icosahedral and inter-icosahedral two-center, three-center, and multi-center covalent bonds and can be interpreted by the central and non-central forces on the lattice dynamics [48].

4.2.2 Rhombohedral Boron

α-**Rhombohedral Boron**: The common crystals of boron is the α-rhombohedral boron emerging in four prominent polymorphs. They are α, β, γ and T (tetragonal). The basic unit of the crystal phases of α, β and T is the B_{12} icosahedron. The γ-phase of boron is composed of B_{12} icosahedra and B_2 atomic pairs, similar to a rocksalt-type arrangement. The crystalline α-rhombohedral boron (α-rh.) (Fig. 4.21), called the low-temperature form or red boron, was first determined in 1958 by McCarty et al. [49]. The unit cell of α-rh. boron is composed of of eight B_{12} icosahedra. These icosahedra are distributed at the eight corners of the rhombohedron, at which the centers of the icosahedra are located. The lattice constants of α-rh. boron are (a = 5.075 Å, $\alpha = 58° \ 06'$; space group R3m). The structure was obtained either by precipitation from melts of boron-saturated precious metals, or by low-temperature (800–1200 °C) reduction of BX_3 boron halides for (X = Br or I) [19].

β-**Rhombohedral Boron**: The β-rhombohedral variety of boron, also called the high-temperature form, is prepared either by reduction of a boron halide with hydrogen on a substrate heated to a temperature above approximately 1400 °C [51], or by

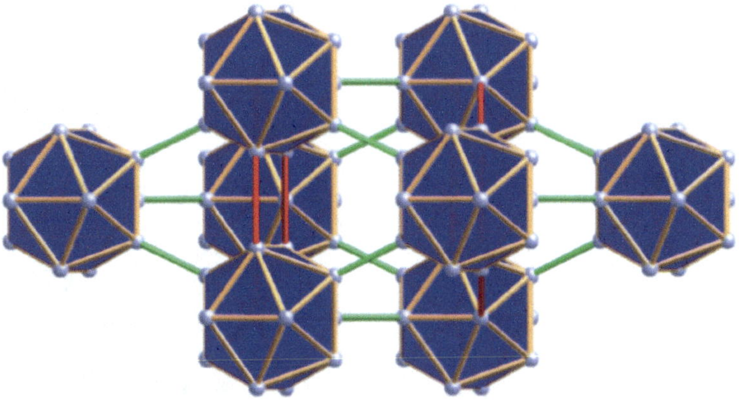

Fig. 4.21 The regular icosahedron is one of the five Platonic solids.The icosahedron vertices form three orthogonal golden rectangular [50]

solidification of liquid boron, or by annealing (α-rh.) boron beyond the (α-rh.) $\rightarrow$ (β-rh.) transformation point. That means, the α-rhombohedral boron [52] transforms by 1200°C into the solid β-boron, the more stable one [53, 54]. The components of β-rh. boron are $B_{12}(B_6)_{12}$ polyhedra exhibiting much complicated structure [50, 55]. However, an example of a β-boron structure is shown in Fig. 4.22. This structure is composed of B_{12} icosahedra as well as of B_{28} subunits [56]. The lattice constants of β-rhombohedral boron are ($a = 10.145$ Å, $\alpha = 65^{\circ}\ 17'$; space group R$\bar{3}$m). It is obviously reasonable clear that the complexity of the β-boron structure is related to its aperiodic character, since the icosahedral unit B_{12} exhibits 5-fold rotational symmetry, which is incompatible with lattice periodicity [57]. The melting point of β-rh. polymorph is around 2450 °C, which is thermodynamically more stable than the α-rh. boron. However, the crystals of α-rh. and β-rh. boron are semi-conductors, and the related experimentally determined band gaps are around 2.0 and 1.6 eV, respectively [58].

4.2.3 Orthorhombic and Tetragonal Boron

γ-**Orthorhombic Boron**: The third crystalline phase of solid boron γ-B_{28} (Fig. 4.23) was recently discovered by Oganov et al. [59] and Zarechnaya et al. [60]. The new solid boron γ-B_{28} was synthesized for the first time under high-temperature (1500–1800 °C) and high-pressure (160 GPa). Both groups found independently this new crystal form of boron. They found that γ-boron is B_{12} icosahedra connected to B_2 pairs (Fig. 4.24). in a orthorhombic NaCl-type arrangement. The unit cell has a space group symmetry Pnnm and contains 28 boron atoms. The structure of this structure is shown in Fig. 4.23. This new phase is rather stable above 9 GPa and up 89 GPa.

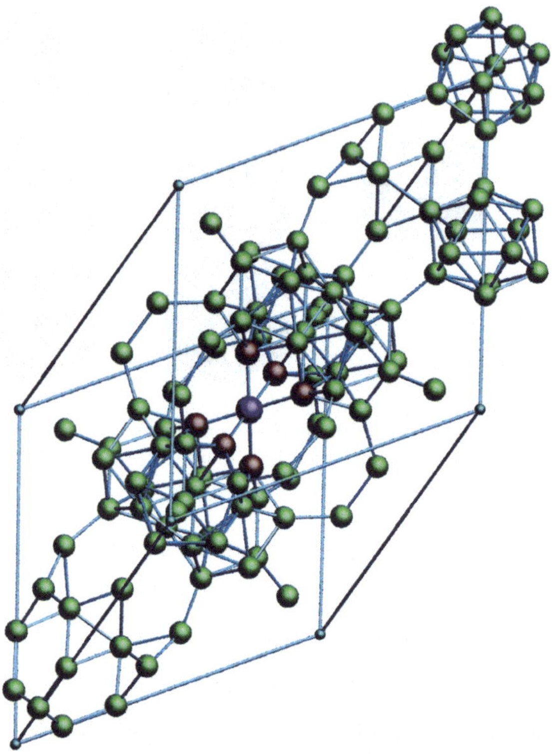

Fig. 4.22 Side view of the unit cell of 105 atom BR boron. Most atoms are part of icosahedra. The central atom (purple) is in the B15 position and connects the two groups of three inter-penetrating icosahedra via atoms at B13 sites (red) [54]

Furthermore, the structure of γ-boron is stable also at room temperature and pressure and can be chilled under ambient conditions. They also found that the calculated diffraction pattern of γ-B_{28} are very similar to the experimental pattern Oganov et al. [59] found that a phase transformation occurs from α-B_{12} into γ-B_{28}, and from γ-B_{28} into α-Ga-type, when the pressure approaches 19 and 89 GPa, respectively. Thus, γ-B_{28} should be much more stable than the common phases of α- and β-boron. This fact can be verified by their hardness. The corresponding Vickers values of the three phases α-rh., β-rh., and γ-B_{28} are 42, 45, and 50–58 GPa, respectively. The structure of other phase at higher pressure, say by compressing boron above 160 GPa is still unknown. However, this phase turns to be a superconductor at low temperatures of around 6–12 K [61].

Tetragonal Boron: Beside the α- and β-rhombohedral boron there are another interesting solid phases of boron in the form of α- and β-tetragonal boron (Fig. 4.25). The unit cell of the lattice of α-Tetragonal boron consists of 4 B_{12} icosahedra and two extra boron atoms, 50 in all. Therefore, this phase is called T-50 and assigned to the $B_{50}N_2$ and $B_{50}C_2$ compounds. The β-tetragonal boron is composed of boron tetrahedral arrangements of B_{12} icosahedra, assigned to T-192. Further appearance of icosahedral solids is the amorphous boron. It was found that even nominally

Fig. 4.23 Structure of γ-B_{28}. For the partially ionic γ-B_{28} structure, two oppositely charged sublattices are marked by different colors (anionic, blue; cationic, orange) [59]

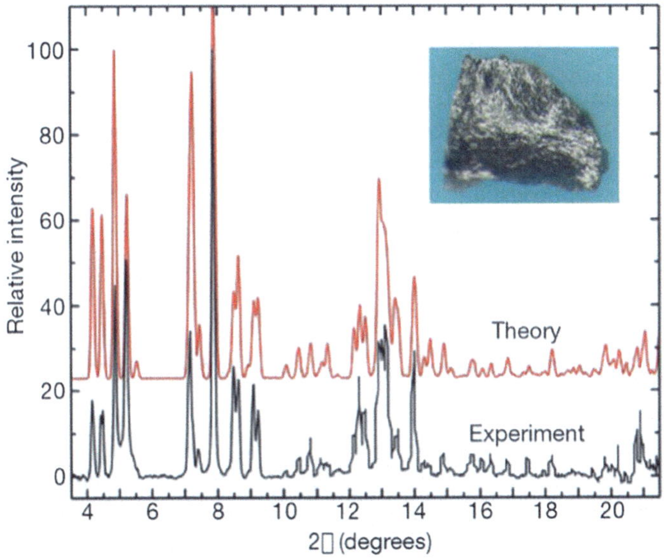

Fig. 4.24 Calculated and measured X-ray diffraction patterns of γ-B_{28} [59]

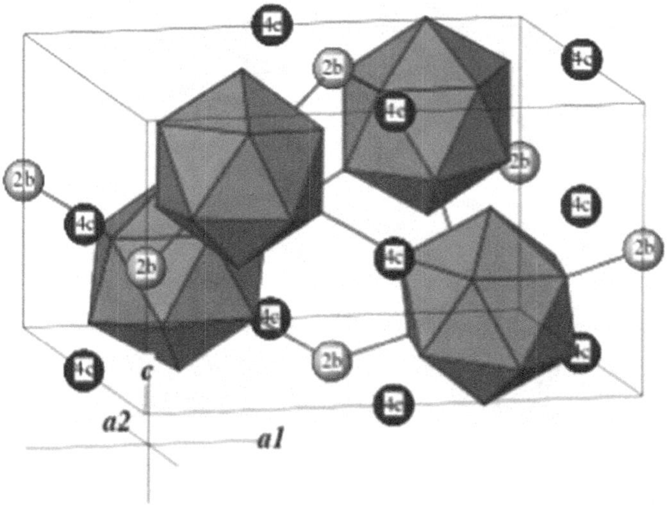

Fig. 4.25 The regular icosahedron is one of the five Platonic solids. The icosahedron vertices form three orthogonal golden rectangular [62]

amorphous boron includes proper boron icosahedra that are, however, connected to each other randomly without any long-range order. The number and bonding state of the icosahedra which constitute to the amorphous boron is around 9–3 B_{12} icosahedra. These icosahedra arrange themselves in so-called out-of-phase in $\sim\pi/5$ rotation to each other through their pseudo-five-fold axes [63].

4.3 Non-icosahedral Boron Clusters B_n ($2 \leq n \leq 14$)

Bottom-Up and Top-Down Approaches: The idea of clusters to form building blocks of matter becomes generally accepted and the bottom-up as well as the top-down approaches, as presented in Fig. 4.26, are nowadays real methods to construct nanostructures and nanodevices. In the eighties started the formation of non-crystalline and of non-icosahedral arrangements boron to be developed repidly. During this decade many theoretical and experimental studies were focused on aggregation and agglomeration of small clusters which represent an intermediate stage of condensed matter and simulate the phase transition from separate atoms or the large single crystals in the form of *needles* or *hexagonal plates* (Fig. 4.19). Smaller crystals were observed molecules to solids. Many groups studied metal and non-metal clusters, alkali-metal and alkali-earth-metal clusters, noble-gas and halogen clusters, transition and lanthanide clusters and many others. Of course boron, carbon and nitrogen clusters were included. In general, clusters were assembled by laser ablation

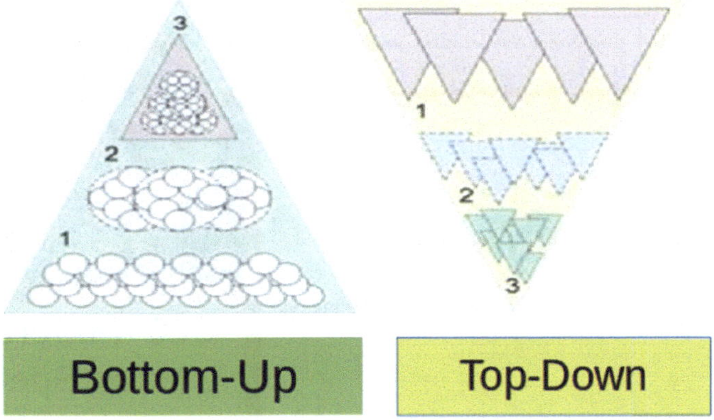

Fig. 4.26 The regular icosahedron is one of the five Platonic solids. The icosahedron vertices form three orthogonal golden rectangular [64]

of refractory materials in a supersonic molecular beam source and then selection of definite sizes of cluster by a mass spectrometer called time-of-flight (TOF).

Why Clusters? It is obviously perceptible that clusters are interesting not only because they form building blocks but also because they exhibit distinguish properties, different than those of solids or of single atoms. These properties can be described as follows. Since the clusters are bridging the domains of atomic and molecular physics on one hand and condensed matter physics on the other, they are dominated by their large surface-to-volume ratio, providing a unique opportunity to study the interplay between surface and volume effects. Clusters exhibit also discrete spectroscopy because of their finite size, as already found in the mass spectroscopy of alkali-metal and carbon clusters in the 80ies. The most prominent results in that decade were the observation of the electronic shells in Sodium clusters in connection with the magic numbers [65] and Buckminster Fullerene C_{60} [66], as shown in Fig. 4.27. Clusters possess electronic shells similar to those in cores or atoms. Therefore shell effects may be pronounced in the ionization potentials of clusters. In addition, the feature and the degree of hybridization particularly between s-type and p-type orbitals, the number of electrons, and the degeneracy of the highest occupied molecular orbitals (HOMOs) influence strongly the cluster geometry. That leads to an interplay between the geometrical and electronic structures and symmetry distortion of the Jahn–Teller-Effect forcing either nearly spherical or nearly planar geometric shapes. All these effects can be seen on examples of Lithium and Sodium clusters [68, 69]. Generally, by small clusters the chemically active valence electrons and the ionic cores are decomposed from each other approaching the microscopic state. Therefore, the ionic configuration as well as the structure of the cluster can be theoretically determined by searching the minimum of the total energy. Thus, different local minima at the potential energy surface correspond to the different isomers of

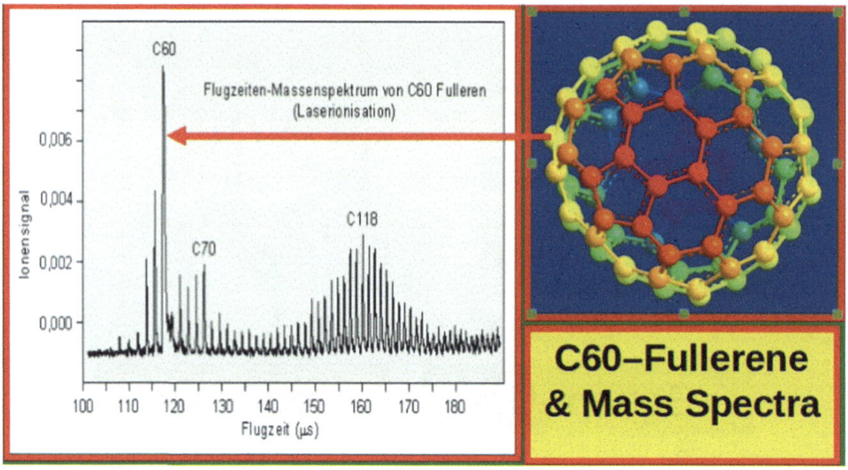

Fig. 4.27 The regular icosahedron is one of the five Platonic solids. The icosahedron vertices form three orthogonal golden rectangular [67]

the cluster and the absolute lowest energy minimum corresponds to the ground state configuration of the most stable cluster. Therefore, the searching for the local minima at the potential energy surface can be achieved by solving the electronic many-body problem [70].

Boron Mass Spectra: As already mentioned in Sect. 4.3, the development of non-icosahedral arrangements as well as of non-crystalline boron started at the end of eighties and the beginning of nineties. Many studies investigated the electronic and geometrical structures of small boron clusters, theoretically as well as experimentally. The determination of mass distribution spectra of boron clusters, up to a clusters size of 24 atoms (Fig. 4.28), was performed by Hanley et al. [71, 72] in a series of experimental studies combined with theoretical calculations. The appearance of these mass spectra was based on the collision-induced dissociation consistent with the intuitive idea that more stable species should be harder to fragment, and based on the oxidation or reaction of small boron cluster ions by oxygen, or with N_2O, respectively. La Placa et al. [73] performed the mass spectra of boron clusters and reported further important experimental study. By means of laser ablation, the authors used hexagonal boron nitride to generate boron clusters. With time-of-flight technique they were able to separated the mass of the charged boron clusters and to produce the related mass selected spectra of boron clusters. All experimental data showed "magic numbers" by boron cations with 5, 7, 10, 11, and with anomalously high experimental stability of the 13 boron cluster. By means of calculations, the stability of these clusters was characterized with two-, three-, and multi-centered σ-type and π-type bonds.

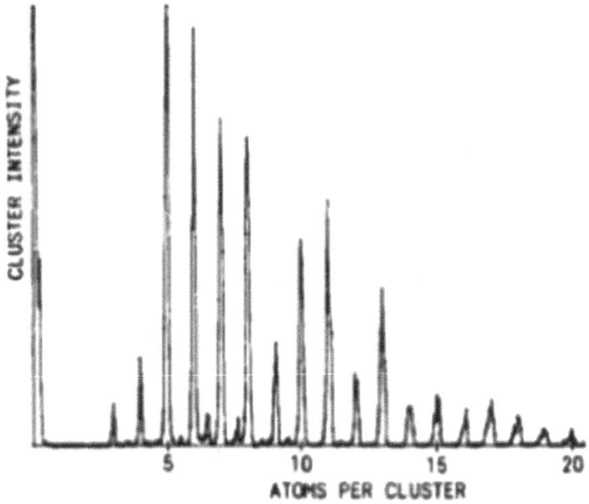

Fig. 4.28 Typical mass distribution of B_n^+ from the laser ablation source [71, 72]

Oxidation of Boron Clusters: Absolute cross sections for all ionic products formed by the reaction $B_n^+ + O_2$ for $n = 2-13$ of boron clusters with oxygen have been measured under single collision condition and at encounter energies from 0.25 to 10 eV [74]. The reaction mechanisms appear to be threefold importance: collision induced dissociation (CID), boron atom abstraction, and oxidative fragmentation. The central oxidation process is exoergic for all small sizes of boron clusters, but for larger cluster ions seems to have bottlenecks or activation barriers. Figure 4.29 shows a plot of the measured boron ion stabilities from the CID experiments. The binding energies of $[B_n\text{-}O]^+$ deduced from the B_nO^+ thresholds, as plotted with an inverse scale. As can be seen, there is a strong correlation between increasing cluster stability and decreasing $[B_n\text{-}O]^+$ binding energies. Perhaps the molecular orbital hybridization that strengthens the B-O bonds in solid B_2O_3 and $[B_2O_3]^+$ clusters, weakens as the B:O ratio increases in the B_nO^+ clusters.

Beyond Icosahedra : As already adverted, the most important theoretical studies and the interpretation of the experimental features of small boron clusters beyond the non-crystalline and non-icosahedral arrangements were started at the late eighties and the beginning of nineties. These studies were carried out by Hanley et al. [74], Gadiyak et al. [75], Kawai et al. [76] and Ray et al. [77]. They predicted 3-dimensional (3D) structures for boron clusters B_n with $n \leq 7$. In contrast, Kato et al. [78, 79] found that the most stable boron clusters are 2-dimensional (2D) with cyclic or butterfly structures, as shown in Fig. 4.30, considering the 3D structures to be less stable. The results of these theoretical studies contradict each others and give no insight to the experimental features. These documented studies were not able to reproduce neither model interpretations nor any trend in cluster formation.

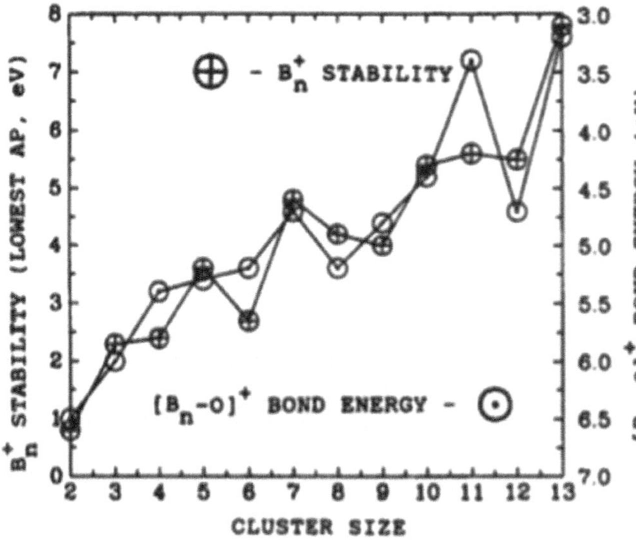

Fig. 4.29 B_n^+ stabilities compared with estimated $[B_n - O]^+$ bond energies [74]

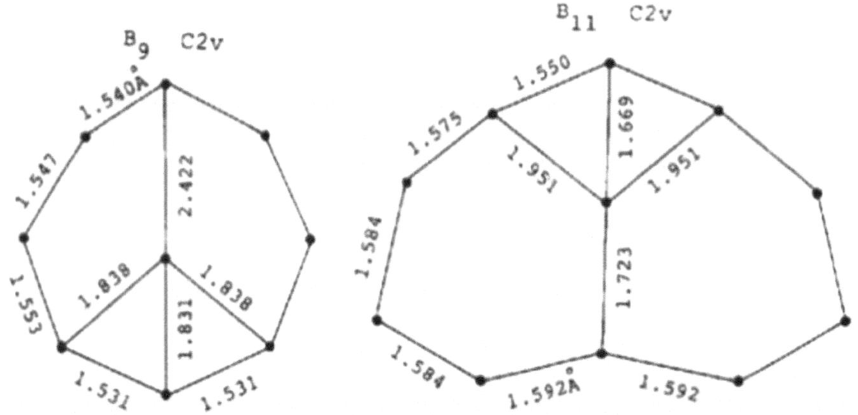

Fig. 4.30 Geometries of the most stable B_9^+ and B_{10}^+ boron clusters calculated at the HF/3-21G level of theory. These structure are marked by cyclic and butterfly characters [79]

4.3.1 Small Boron Clusters

Our First Attempt : The first attempt to investigate boron clusters was carried out in 1990 by Boustani et al. [80]. It was a systematic ab initio study on neutral, singly and doubly ionized boron clusters B_n for (n = 2–8), carried out in the framework of the Hartree-Fock (HF) self-consistent-field (SCF) and configuration interaction (CI) theory, using a double-ζ basis set with polarization function (DZ+P) [80]. We

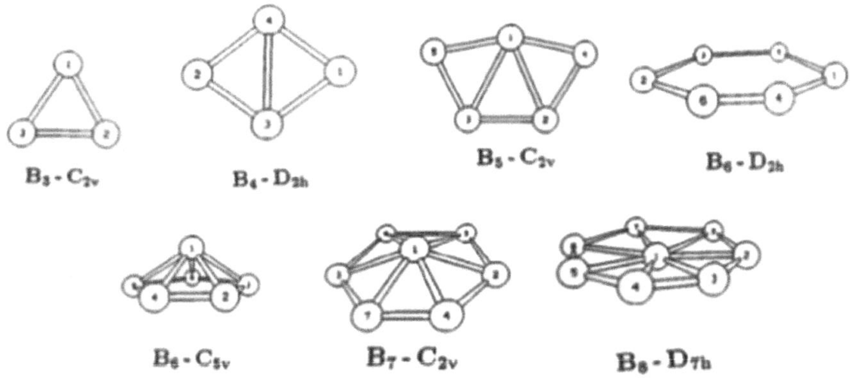

Fig. 4.31 The CI binding energy per atom E_b/n as a function of cluster size n for the most stable neutral, singly and doubly ionized boron clusters B_n, B_n^+ and B_n^{++}, for n = 2 − −8. The topologies of the optimized geometries are drawn [80, 81]

have found that the optimized structures of the neutral and charged boron clusters possess planar geometries except of B_7, which has a quasi-planar structure in form of distorted hexagonal pyramid, whose apical atom lies slightly above the plane.

The Elemental Boron Structures B_n **for n** ≤ 8: These small boron clusters (Fig. 4.31) are optimized at the SCF-HF and DZ+P basis set. They are considered as the elemental clusters from which larger cluster sizes can be developed and built after the "Aufbau Principle", proposed by Boustani [85], as will be explained in the following sections. The ground state of the neutral dimer B_2 is $D_{\infty h}(^3\Sigma_g^-)$ with a bond-length of 1.66 Å, while for the cationic dimer B_2^+ is $D_{\infty h}(^2\Sigma_g^+)$ with a larger bond-length of 2.32 Å. The cation B_3^+ has a symmetric triangular geometry $D_{3h}(^1A_1^{'})$ having a degenerate HOMO and LUMO with a bond-length of 1.56 Å. As expected, by adding an electron to the LUMO to determine the neutral B_3 cluster, the degeneracy breaks due to the Jahn–Teller-Effect getting symmetry lowering $C_{2v}(^2A_1)$ and leading to a distorted isosceles triangle with the corresponding bond-lengths of 1.5 and 1.6 Å. The structure of the neutral tetramer B_4 is a distorted square $D_{2h}(^1A_g)$ with an average bond-length of 1.52 Å. The structure of the cationic tetramer B_4^+ is a regular square $D_{4h}(^2A_{1g})$ with a bond-length of 1.55 Å. The structure of the cation B_5^+ is a regular pentagon $D_{5h}(^1A_1^{'})$ with a bond-length of 1.55 Å. The neutral and doubly charged B_5 cluster have a distorted pentagon forming a shape of three triangles. The geometry of the neutral and singly charged B_6 clusters is a distorted planar hexagon B_6–$D_{2h}(^3B_{2u})$ and B_6^+–$D_{2h}(^2B_{1u})$. They can be considered as two bonded isosceles triangles. The structure of the doubly charged B_6^{++} cluster is a regular hexagon. The cationic B_7^+ cluster has a non-distorted hexagonal pyramid $C_{6v}(^1A_1)$. The bond-length of the hexagon is 1.58 Å and the distance to the apical atom is about 1.75 Å. The structure of the neutral B_7 clusters becomes the Jahn–Teller distortion

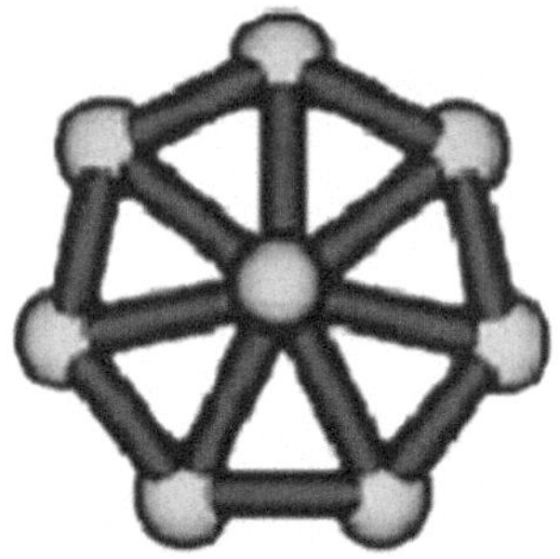

Fig. 4.32 The molecular wheel structure of the highly stable B_8 (D_{7h}–$^3A'_2$) boron cluster, calculated at the HF/3-21G level of theory [80]

which leads to lowing the point group symmetry from $C_{6v}(^1A_1)$ to $C_{2v}(^2B_2)$. The octamer B_8 is planar and forms a wheel-like structure, as it will be discussed in the next paragraph.

The First Boron Molecular Wheel: The most interesting structure is that of the so-called molecular wheel B_8 [80], which is composed of a central boron atoms surrounded by a regular boron heptagon (Fig. 4.32). The ground state of the B_8 molecular wheel is $D_{7h}(^3A'_2)$. The bond length of the heptagon is 1.516 Å, and the distance to the central atom is about 1.747 Å. The geometry of the singly ionized (cationic) B_n^+ cluster becomes a slight Jahn–Teller distortion of C_s-$^2A'$ symmetry and state. In contrast, the doubly ionized boron clusters B_8^{++} is of $D_{7h}(^1A'_1)$ symmetry and state, similar to the neutral cluster.

The Stability of Neutral and Ionized Clusters: The binding energy per atom E_b/n as a function of cluster size n of the neutral, singly and doubly ionized boron clusters is illustrated in Fig. 4.33. As can be seen, The functions are increasing with increasing the cluster size. The geometries of Fig. 4.31, and those of singly ionized (cations) and doubly ionized clusters are shown in Fig. 4.33. Most of the singly and doubly ionized clusters B_n^+ and B_n^{++}, are symmetric [80]. The calculations of the elemental boron clusters, described above, show that the average binding energy per atom E_b/n of the neutral and singly charged boron clusters varies from 1.2 eV for the dimers up to 3.25 eV for the wheels. The E_b/n of the doubly charged boron clusters decreases and lies between 0.25 eV for the tetramer and 2.5 eV for the wheels, while the doubly charged boron dimers and trimers do not exist because of the Coulomb explosion. However, one year later Bonaĉić-Koutecký et al. [81] published in part the Boustani's results, quoted as 211 in [81].

4.3.2 *Prediction of Planarity and Aufbau Principle*

Small Boron clusters B_n ($n \leq 14$): The breakthrough in the search of structural formations of small boron cluster occurred in 1994, namely by our second attempt in the theoretical study of small boron clusters B_n for ($n \leq 14$) [82] By means of

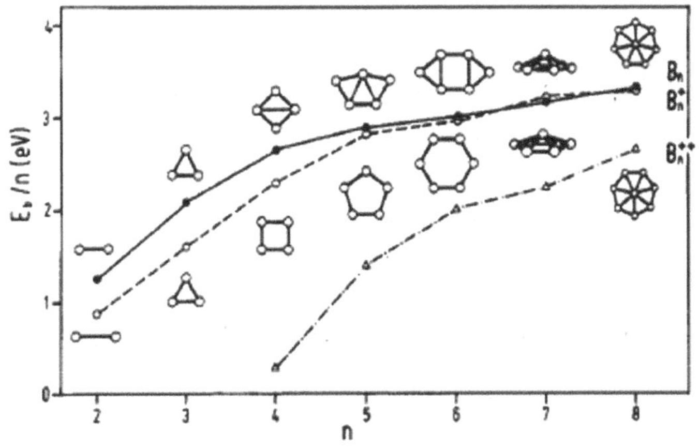

Fig. 4.33 The CI binding energy per atom E_b/n as a function of cluster size n for the most stable neutral, singly and doubly ionized boron clusters B_n, B_n^+ and B_n^{++}, for n = 2–8. The topologies of the optimized geometries are drawn [80, 81]

the density functional theory and the non-local corrections we have studied systematically a lot of geometrical configurations of boron clusters exploring the absolute minima of the potential energy surfaces and the corresponding ground state configurations. However, boron clusters associated to the absolute energy minimum can be distributed into two groups: the first group contains 2D clusters with convex, planar or quasi-planar structures and the second group contains 3D clusters with open spherical structures, as shown in further studies [83, 84].

The 2D clusters, presented in Figs. 4.34 and 4.35, exhibit the lowest energy minima and consequently are energetically favored over the 3D clusters. The preference of 2D over 3D in boron clusters can be coupled with the π character of molecular orbitals. The boron atoms undergo the sp^2 hybridization. These hybridized atomic orbitals form strongly directed chemical bond in form of in-plane σ bonds inward-pointing hybrids and out-of-plane π bonds outward-pointing hybrid orbitals [86], as shown for B_{13} in Fig. 4.36. These π orbitals substantiates the planarity (chemically aromaticity) of the boron clusters.

Extraction the Building Block Elements of Boron Clusters: Figures 4.34 and 4.35 exhibit two kinds of boron clusters. They are called convex and quasi-planar clusters. By a careful inspection of the structures of this cluster group, which are mostly assigned by triangular faces, one can find hexagonal pyramidal subunit, as shown in Fig. 4.37. The hexagonal pyramidal subunits dovetail each other by axial bonds and the out-of-plane apices of these subunits are either lying above the plane of the peripheral atoms, hereafter referred to as the convex clusters, or lying in an alternating pattern above or below the plane forming top and bottom atoms, respectively, hereafter referred to as the quasi-planar clusters. The convex or quasi-planar structures of the clusters can easily be constructed starting from the hexagonal pyramid

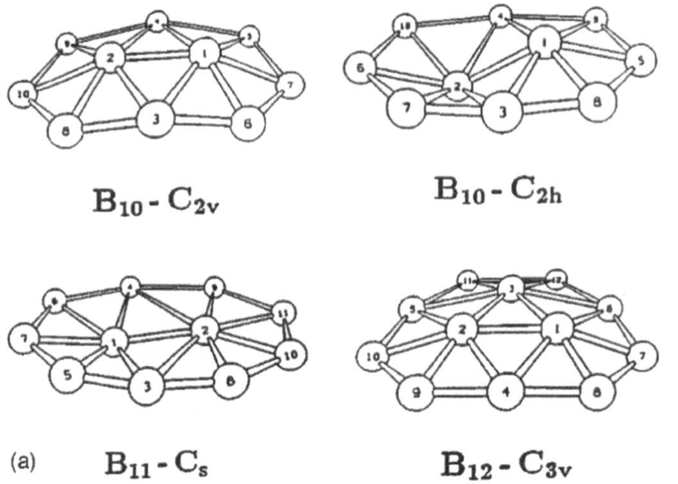

Fig. 4.34 Geometries of the convex and quasi-planar neutral B_{10}, B_{11} and B_{12} boron clusters, obtained at the SCF-HF/3-21G level of theory. They are built from the hexagonal pyramidal subunits B_7, which are dovetailed into each other [85]

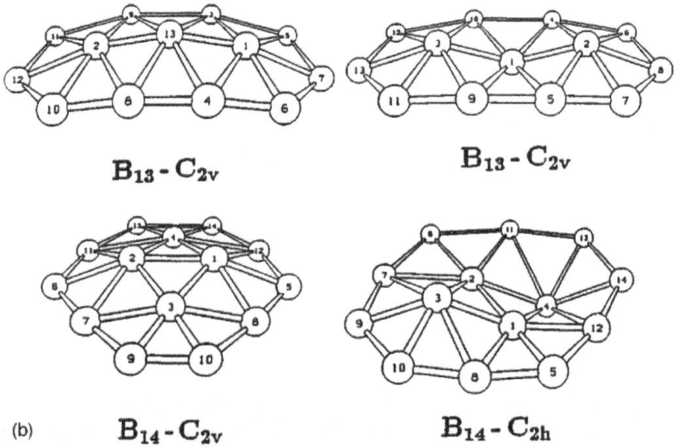

Fig. 4.35 Geometries of the convex (left) and quasi-planar (right) neutral B_{13} and B_{14} boron clusters, obtained at the SCF-HF/3-21G level of theory. They are built from the hexagonal pyramidal subunits B_7, which are dovetailed into each other [85]

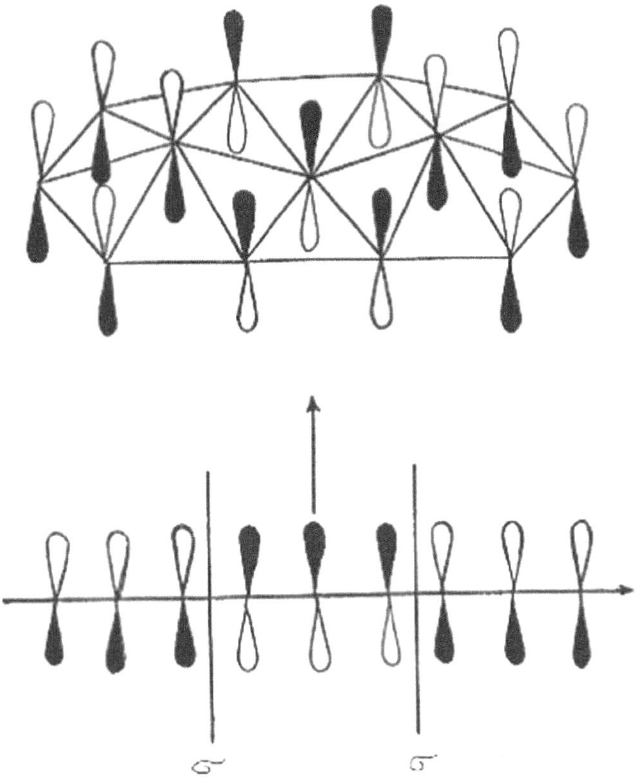

Fig. 4.36 Graphical description of the HOMOs of the quasi-planar B_{13}. The shape (top) is lying in xy-plane and showing the π-orbitals. The side perspective (bottom) shows two nodal planes, indicated by σ vertical to the xy-plane and parallel to the xz-plane [86]

and adding atoms consecutively to form a new hexagonal pyramid. These convex and quasi-planar boron clusters can be considered as building block clusters constructed from the pentagonal and hexagonal pyramidal subunits B_6 and B_7, presented in Fig. 4.37.

Disposition the "Aufbau Principle" for Boron Clusters: Based on the idea that the obtained convex, planar, or quasi-planar boron clusters can be considered as segments of a surface or fragments of a sphere, Boustani foresaw in a continuing study [85], that these small boron clusters can operate as building blocks of larger boron 2D surfaces and 3D shapes and spheres. Therefore Boustani formulated the so-called "Aufbau Principle" (AP) for boron clusters. The AP reveals that highly stable boron clusters, sheets, nanotubes and spheres of convex, planar or quasi-planar surfaces can be constructed from these two basic units B_6 and B_7 by adding atoms consecutively forming new pentagonal or hexagonal pyramids, as shown in Fig. 4.38. The B_6 unit is characteristic for spherical or open-spherical clusters, whereas the hexagonal

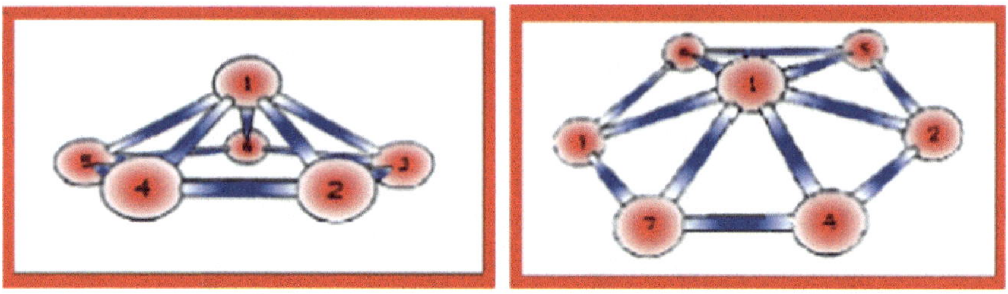

Fig. 4.37 The pentagonal and hexagonal pyramidal subunits B_6 and B_7, which are the building block uniforms for boron clusters, surfaces, sheets, nanotubes and spheres [85]

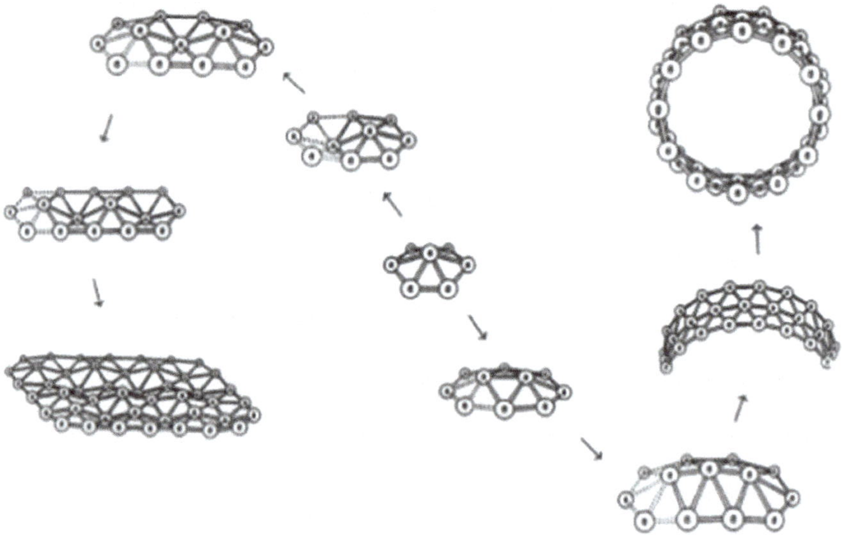

Fig. 4.38 Schematic diagram of the boron cluster growth according to the "Aufbau principle". Starting from the basic unit, the hexagonal pyramid B_7 and by adding atoms consecutively, forming new hexagonal pyramids to obtain either quasi-planar or convex structures generating buckled triangular sheets or rings and nanotubes [85]

B_7 cluster is the characteristic unit for surfaces of quasi-planar or convex sheets, of nanotubes or spheres. The resulting finite spherical surfaces can be considered as dovetailed units of pentagonal and hexagonal pyramids, while the larger boron clusters or infinite surfaces of quasi-planar sheets and nanotubes can be considered as dovetailed units of hexagonal pyramids only. Consequently, the highly symmetric icosahedron B_{12} is no more the basic unit to build up new boron formations, as known in crystalline boron, but rather cluster units of pentagonal or/and hexagonal pyramids are. This fact was the break of this dogma in the "Boron Solid-State Chemistry". Herewith Boustani [85] presented novel models for small boron clusters and predicted, for the first time, the existence of highly stable planar, quasi-planar, or convex boron formations, unknown in nature by then, but in the following years experimentally confirmed for many and larger cluster sizes, as will be explained later.

4.3.3 Theoretical Confirmations

The first theoretical confirmation for the predicted quasi-planarity of boron clusters of [82] followed two years later by an ab initio quantum chemical study by Ricca et al. [87]. They studied boron clusters B_n for $n = 2$–14 and established the same trend of planarity for B_n for $10 < n < 14$ possessing structures with quasi-planar formations. In the following ten years many investigations on boron clusters were carried out using different theoretical methods and various experimental techniques. For example, the neutral and charged small boron clusters B_n for $n = 4$–7 were investigated by Li et al. [88–91]. They studied their electronic and geometrical structures applying similar theoretical approaches. They confirmed that their results are in a good agreement with the results of Boustani [82] verifying similar configurations for this kind of boron clusters. Cao et al. [92] investigated the structures of boron clusters B_7, B_{10} and B_{13}. These authors applied a molecular-dynamics method using linear-muffin-tin-orbitals including full-potentials. They generated respectively 7, 15 and 3 stable configurations of these boron clusters. They verified that boron clusters with the highest stability exhibit 2D structures with quasi-planar formations, confirming the results of Boustani and obeying the "Aufbau Principle" to configure highly stable boron clusters. Further theoretical study on boron clusters B_n for $n = 2$–20 was carried out by Atiş et al. [93, 94] using accurate ab initio quantum chemical methods and density functional theory. They confirmed the quasi-planarity of boron clusters and verified the same trend of cluster expansion. Based on the density functional theory, Pandey et al. [95] reported the electronic properties and the vibrational frequencies for different isomers of cationic and neutral B_{12} clusters. The calculated vibrational frequencies of these clusters are laying in the region of 200–1328 cm^{-1}. They also confirmed that the convex and quasi-planar structures of these boron clusters correspond solely to their ground state, meaning the lowest-energy state. They explained that the high stability of these clusters is connected with the delocalization of the charge density as well as with the multi-centered bond character. These two dominant factors favor higher stability for the 2D over the 3D configurations.

4.3.4 Experimental Confirmations

Boron Clusters B_3 and B_3^-: Zhai et al. [96] studied theoretically and experimentally the neutral and negatively charged boron clusters B_3, B_3^-, B_4 and B_4^-. Theoretically, they used ab initio quantum chemical methods at the B3LYP/RCCSD(T)/ 6-311+G* level of theory. Experimentally, the authors applied photoelectron spectroscopy (PES) and measured the spectra of boron clusters. The electronic

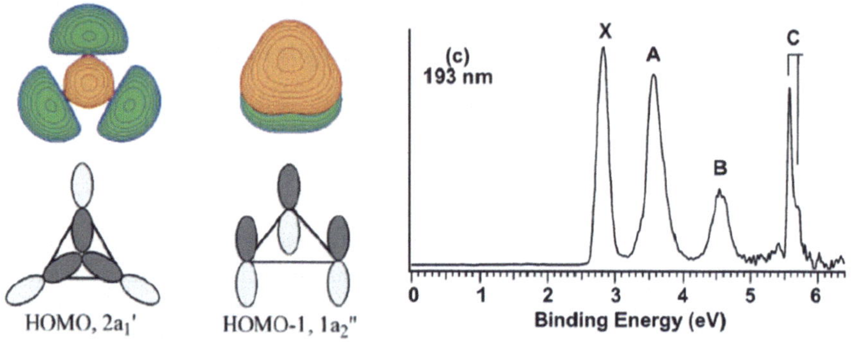

Fig. 4.39 Occupied valence molecular orbitals of B_3^- (D_{3h}, $^1A_1'$) and the related photoelectron spectra observed at 193 nm (6.424 eV). The experimental value of the X band 2.82 eV is in a good agreement with the calculated VDE value 2.88 eV, obtained at RCCSD(T) level [96]

transitions between the anionic ground states and the ground and excited states of the neutral clusters are represented by the PES spectra revealing well-resolved detachment characteristics for boron clusters. These characteristics can be considered as 'fingerprints' for the anionic clusters. These experimental 'fingerprints' of the anionic clusters can be compared with the theoretically determined detachment energies from the anion ground state, as we will see for B_3 and B_3^- and for larger boron clusters in the following sections. They confirmed that the neutral and anionic clusters, B_3 and B_3^-, have triangular shape and are σ and π aromatic with D_{3h} symmetry. The photoelectron spectra of B_3^-, obtained at 193 nm (6.424 eV), show four observed bands X, A, B, and C, as shown in Fig. 4.39. The electron affinity[4] of B_3, in corresponding to the measured adiabatic electron detachment (ADE),[5] was 2.82 $\pm$ 0.02 eV. The measured vertical electron detachment (VDE)[6] was also 2.82 $\pm$ 0.02 eV. The calculated VDE, obtained at the RCCSD(T)/6-311+G(2df), is 2.88 eV and in excellent agreement with the experimental value. The calculated totally symmetric stretching frequencies for the anionic cluster B_3^- [$\omega_1(a_1) = 1239\,\text{cm}^{-1}$] and for the neutral cluster B_3 [$\omega_1(a_1)$ $1223\,\text{cm}^{-1}$], obtained at the B3LYP level of theory, are also in good agreement with the vibrational frequencies determined from the spectrum of the X band for the anion B_3^- ($1230 \pm 40\,\text{cm}^{-1}$) and for the neutral B_3 ($1020 \pm 50\,\text{cm}^{-1}$). As shown in Fig. 4.39, the bonding nature of the $2a_1'$ HOMO is presented in connection with the reduced vibrational frequency upon electron detachment.

---

[4]Electron affinity (EA) is defined as the change in energy (ΔE) of a neutral atom (in the gaseous phase) when an electron is added to the atom to form a negative ion: $X + e^- = X^-$, or simply the binding energy of an electron to the molecule.

[5]Adiabatic detachment energy is the energy ($h\nu$) needed to eject or separate an electron from the anion ground state to the neutral ground state.

[6]Vertical detachment energy is the energy ($h\nu$) needed to eject or separate an electron from the anion in its ground electronic state: $X^- + h\nu = X + e^-$.

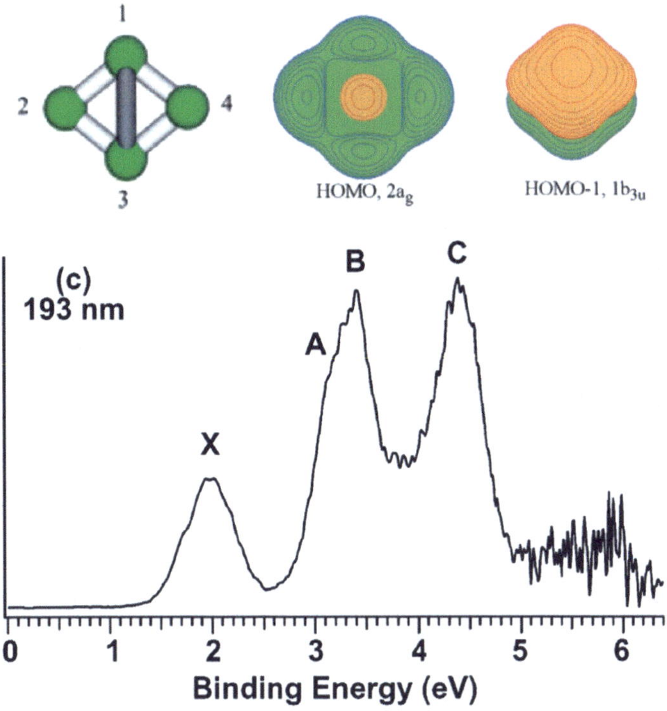

Fig. 4.40 The rhombus structure, the molecular orbitals of B_4 (D_{2h}, ${}^1A_g$), and the photoelectron spectra of B_4^-, observed at 193 nm. A good agreement between the experimental value of the X band 1.99 ± 0.05 eV and the calculated VDE value 1.85 eV, obtained at the RCCSD(T)/6-311+G(2df) level [96]

Boron Clusters B_4 **and** B_4^-: Zhai et al. [96] also confirmed that the neutral and anionic clusters, B_4 and B_4^-. Both structures exhibit rhombus form with the point group symmetry D_{2h} and with the spectroscopic states ${}^1A_g$ and ${}^2B_{1u}$, respectively. As shown in Fig. 4.40, the pattern of the molecular orbital HOMO ($2a_g$) is a of $2p_z$ atomic orbitals, where the charge density is distributed above and below the rhombus plane, similar to benzene. Thus the system can be characterized as σ- and π-aromatic as well, and B_4 cluster is also doubly aromatic. The electron affinity was measured to be 2.33 ± 0.02 eV, which is in fact the adiabatic electron detachment ADE. However, the electron detachment leads to a transition between the ground state ${}^2B_{1u}$ of B_4^- to the ground state ${}^1A_g$ of B_4. This transition leads to a large change in the rhombus structure in consistent agreement with the broad X band of B_4^-, observed in the photoelectron spectrum as shown in Fig. 4.40.

Boron Clusters B_5 **and** B_5^-: Zhai et al. [97] proceeded their experimental and theoretical investigations to study the neutral and anionic boron clusters of B_5 and B_5^-. They used also the same photoelectron spectroscopy and ab initio methods as they have done by the B_3 and B_4 clusters. The structures of the global minimum for the

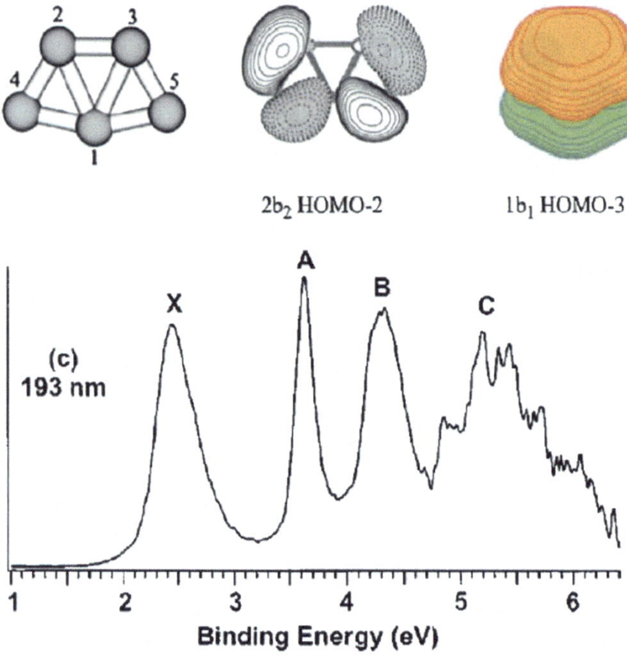

Fig. 4.41 The structure, the occupied valence molecular orbitals of B_5^- (C_{2v}, $^1A_1$), and the related photoelectron spectra, observed at 193 nm (6.424 eV). The π-bonding orbital ($1b_1$) of HOMO-3 is completely delocalized over all five atoms. An excellent agreement was obtained between the experimental value of the X band at 2.40 ± 0.05 eV of the binding energy and the calculated VDE value 2.45 eV, achieved at the RCCSD(T)/6-311+G(2df) level [97]

neutral B_5 and anionic B_5^- are planar with C_{2v} symmetry. The corresponding ground states respectively are $^2B_2$ and $^1A_1$. These results were obtained at the B3LYP/6-311+G* level of theory. The B_5 structure, as shown in Fig. 4.41, can be deduced from the counterpart the anionic cluster B_5^- by detaching the outmost valence electron exactly from its HOMO. The experimentally measured vibrational frequency 550 cm^{-1} of the X band is in excellent agreement with the calculated deformational normal mode ω_4 (a_1) 596 cm^{-1}. Thus, the identification of the ground state structures for both B_5^- and B_5 clusters is now firmly established after the excellent agreement for the ground-state transition between the theoretical and experimental results.

Boron Clusters B_6 and B_6^-: Alexandrova et al. [98] continued their theoretical and experimental investigations to study the neutral and anionic boron clusters of B_6 and B_6^-. They used once again the same photoelectron spectroscopy and ab initio methods as they have done by the B_5 and B_4 clusters, mentioned in the previous section. As presented in Fig. 4.42, the photoelectron spectra of B_6^- cluster indicate the transition X with an intense threshold, followed by band A with closely space. However, this texture exhibits that the ground-state of the neutral cluster according to this detachment transition belongs definitely to open shell system. Thus, the global minimum

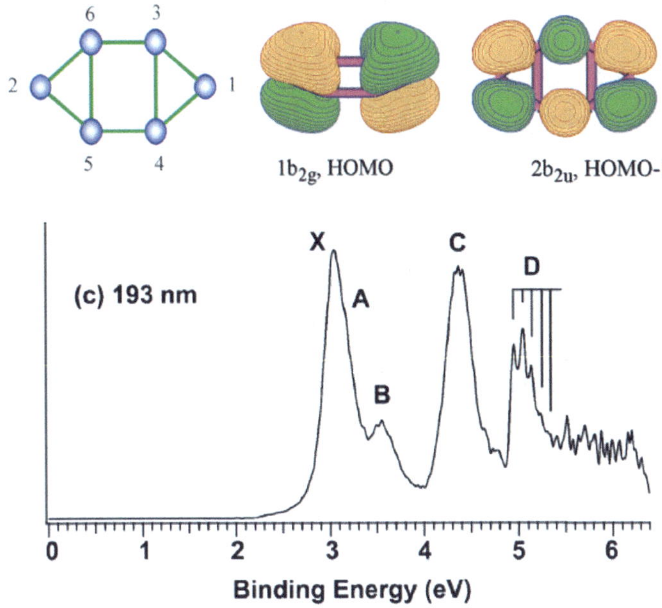

Fig. 4.42 The structure, the molecular orbitals of B_6^- (D_{2h}, $^2B_{2g}$), and the related photoelectron spectra, observed at 193 nm. The HOMOs show anti-bonding four σ- and four π-electrons. Thus the B_6^- system is ani-aromatic. An excellent agreement was obtained between the experimental value of the X band at 3.01 ± 0.04 eV of the binding energy and the calculated VDE value 2.90 eV of $1b_{2g}$-HOMO, achieved at the RCCSD(T)/6-311+G(2df) level [98]

for the neutral B_6 and anionic B_6^- was found to be a C_{2h}, and D_{2h} planar structures with the ground states ^3A_u and $^2B_{2g}$, respectively, obtained at the B3LYP level using 6-311+G* basis set. They also asserted that the next stable neutral B_6 cluster has a pentagonal pyramidal structure C_{5v} with an energy difference of 1.2 kcal/mol only. The recorded vibrational frequency (510 cm^{-1}) can be ascribed to the elongation of the B5-B6 and B3-B4 distances of the structure represented in Fig. 4.42 upon detachment according to the $2b_{2u}$ orbital. Finally one can conclude that the observed vibrational spacing is in a good agreement with the calculated vibrational frequency for ω_4 (517 cm^{-1}).

Boron Clusters B_7 **and** B_7^-: Alexandrova et al. [99] investigated the electronic structures, isomerism, as well as chemical bonding of the B_7 and B_7^- clusters using ab initio quantum chemical methods accompanying with photoelectron spectroscopy. The obtained photoelectron spectra for the B_7^- anion, preserved at several photon energies, are fairly complicated but however well-resolved. These spectra show distinct contributions for diverse isomers of the B_7^- anionic cluster. The global minimum of the low-lying isomer, yielded at the B3LYP/6-311+G* level of theory, corresponds to a relative flat hexagonal pyramidal structure. It possesses a C_{6v} symmetry and shown in Fig. 4.43. The neutral structure of B_7 is the slightly distorted hexagonal pyramid

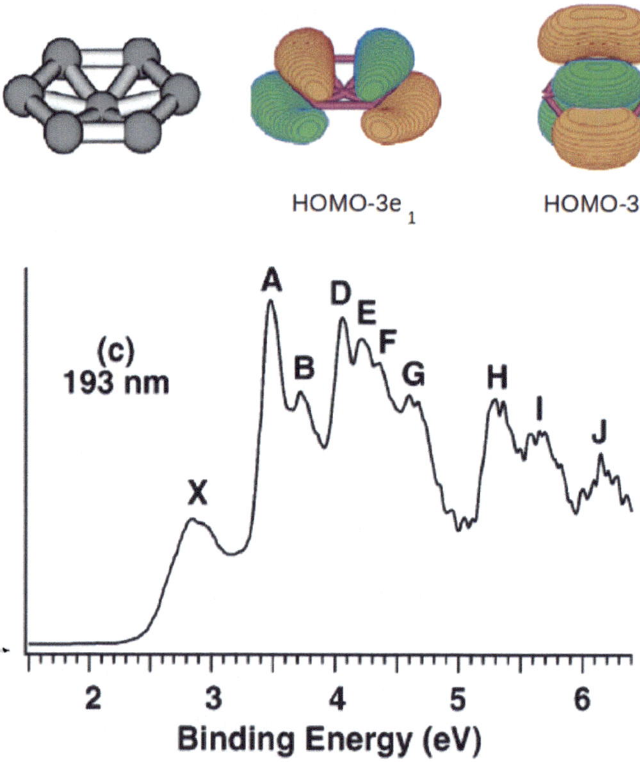

Fig. 4.43 The structure, the molecular orbitals of B_7^-, and the related photoelectron spectra, observed at 193 nm. The triplet structure of the anion B_7^- (C_{6v}, $^3A_1$) can be considered as being doubly (σ and π) aromatic. The HOMO' ($3e_1$) and HOMO ($3e_1$) are a set of π orbitals, which give the triplet C_{6v} structure π-aromaticity [99]

with a C_{2v} symmetry. An excellent agreement was achieved between the experimental value of the X band at 2.85 eV of the binding energy and the calculated VDE value 2.89 eV, obtained at the RCCSD(T)/6-311+G(2df) level. The calculated vibrational frequency for $\omega_4(a_1)$ (440.5 cm^{-1}), obtained at the B3LYP/6-311+G* level of theory, is in a good agreement with the observed vibrational frequency (480 cm^{-1}).

The Molecular Wheels of B_8 **and** B_9 : Zhai et al. [100] investigated experimentally and theoretically the electronic and geometric structures of the hepta- and octa-coordinated boron clusters. The most stable geometrical structures of these clusters were found to have centered cyclic rings. The authors used a hybrid (density functional and Hartree'Fock') of computationally extensive method to search for the global minimum structures of the neutral and charged clusters B_8, B_8^-, B_8^{2-}, B_9, B_9^-. All the five species of these clusters were found to exhibit planar structures in a wheel form, as can be seen in Fig. 4.44. The molecular wheels of B_8 and B_8^{2-} clusters are perfect heptagons with D_{7h} symmetry, while the B_9^- cluster is a perfect octagon with

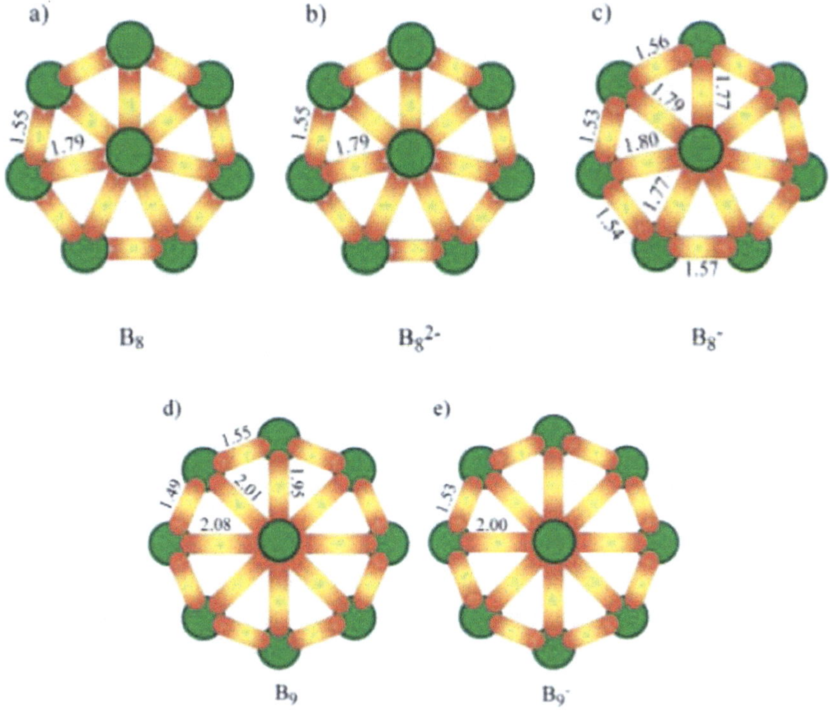

Fig. 4.44 The molecular wheel structures of neutral and anionic B_8 and B_9 clusters. The octamers B_8 (D_{7h}–$^3A'_2$) (**a**) and B_8^{2-} (D_{7h}–$^1A'_1$) (**b**) are perfect heptagons (closed shells), while B_8^- (C_{2v}–$^2B_1$) (**c**) is a distorted heptagon. B_9^- (D_{8h} - $^1A_{1g}$) (**d**) is a perfect octagon, whereas B_9 (D_{2h}–$^2B_{1g}$) (**e**) has a slight in-plane distortion within the octagonal structure [100]

D_{8h} symmetry. The B_8^- and B_9 clusters possess respectively planar heptagonal and octagonal structures with slight in-plane distortions. In analogous to their previous studies on the smaller B_5 rings [97] and B_6 rings clusters [98], the authors were able to reproduce the B_8^- and B_9^- wheels experimentally, using laser vaporization cluster source and probing their electronic structures via photoelectron spectroscopy (PES).

The Photoelectron Spectra of B_8^- **and** B_9^- are presented in Fig. 4.45. An excellent agreement between experiment and theory was documented by the observed and calculated vertical detachment energies VDE of both species B_8^- and B_9^-. The measured and calculated VDE values of the X band for B_8^- (C_{2v}, $^2B_1$) were 3.02 and 2.99 eV, whereas the measured and calculated VDE values of the X band for B_9^- (D_{8h}, $^1A_{1g}$) were 3.46 and 3.47 eV, respectively. Upon one-electron detachment of B_8^{2-}–D_{7h}, the Jahn–Teller Effect distorts slightly the structure of the obtained doublet B_8^- and lowers the symmetry to C_{2v}. To C_{2v} symmetry. However, the resulting electron configuration for the B_8^- cluster is settled as: $1a_1^2 2a_1^2 1b_2^2 2b_2^2 3a_1^2 4a_1^2 3b_2^2 5a_1^2 1b_1^2 4b_2^2 6a_1^2 1a_2^2 2b_1^1$. Thus, the doubly degenerate $1e_1''$-HOMO of B_8^{2-} (D_{7h}) is now splitting into $1a_2$ (HOMO-1)

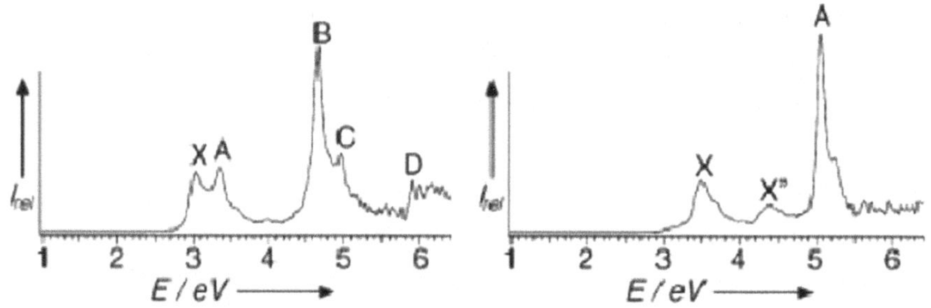

Fig. 4.45 The photoelectron spectra, of B_8^- (left) and B_9^-, observed at 193 nm (6.424 eV). The measured and calculated VDE values of the X band, obtained for B_8^- (C_{2v}, $^2B_1$) with 3.02 and 2.99 eV and for B_9^- (D_{8h}, $^1A_{1g}$) with 3.46 and 3.47 eV, respectively, are in excellent agreement [100]

and $2b_1$ (HOMO) in C_{2v} structure. When a further electron is removed from B_8^- to gain B_8, the final state can be either a singlet (detachment from the singly occupied $2b_1$ MO) or a triplet (detachment from the doubly occupied $1a_2$ MO). When two electrons are removed from B_8^{2-}, then the degenerate $1e_1''$ HOMO is partially occupied or half-filled leading to a structural change and resulting a triplet ground state ^3A_g of B_8. Due to the alteration of the Hückel rule for triplet states, this state is highly π aromatic. Concerning some calculations, the 481 and 1243 cm^{-1} vibrational modes can be accountable for the transformation from C_{2v} into D_{7h}. These theoretically determined frequencies are in a good agreement with the experimentally measured frequencies at 510 ± 50 and $1160 \pm 60\,cm^{-1}$.

The Molecular Orbitals of B_9^-: Concerning the molecular bonding, now arises the question: (concept) the reason behind the adoption of these extraordinary and admirable structures? The cognition can be obtained by their molecular orbitals and their occupation by the valence electrons. Figure 4.46 shows the molecular orbitals for the closed shell B_9^- cluster. One can acknowledge two kinds of bonding characters, namely σ-, and π-bonding characters, which are responsible for the high stability of the cluster. Three σ-bonds are localized between the peripheral atoms, assigned by the orbitals HOMO-1, $2e_{1u}$, and HOMO-4, $2a_{1g}$. The other π-bonds are localized between the central atom and the peripheral atoms, assigned by the three π orbitals of HOMO, $1e_{1g}$ and HOMO-2, $1a_{2u}$. These HOMOs show that the B_9^- cluster is doubly σ and π aromatic.

The Planar Boron Clusters B_n **for** ($10 \leq n \leq 15$): Zhai et al. [101] reported once again a combined theoretical and photoelectron spectroscopy (PES) investigation studying larger boron species for cluster sizes between ten and fifteen. The theoretical and experimental results confirm that the all structures of the boron clusters are dominated by planar or quasi-planar formations. The authors discovered rather surprisingly that these boron clusters possess aromatic and anti-aromatic characters in analogous to hydrocarbons following in fact the well-known Hückel rules. The structural searches for B_{10}^-, B_{11}^-, B_{14}^- and B_{15}^- clusters started from the structures of B_{12}^- and B_{13}^- isomers of the lowest-energy. For example, they built from the ground-state

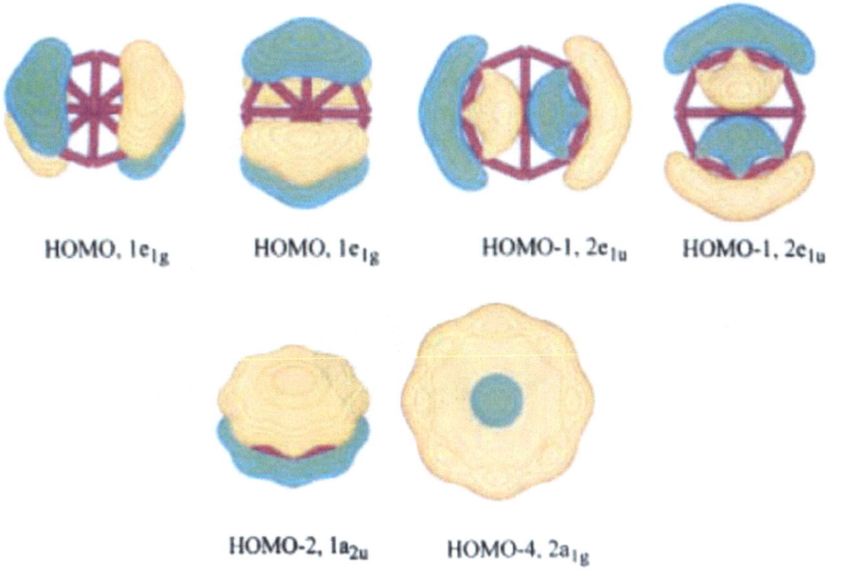

Fig. 4.46 The highest molecular orbital pictures of B_9^-, showing three σ-type orbitals (HOMO-1, $2e_{1u}$ and HOMO-4, $2a_{1g}$), and three π-type orbitals (HOMO, $1e_{1g}$, and HOMO-2, $1a_{2u}$), being doubly σ and π aromatic [100]

structure of B_{12} the structures of the 11-atom clusters, and from the ground-state structure of B_{11}^- they built the 10-atom cluster. They have rebuilt these structures either by eliminating a peripheral atom or an inner atom from the initial geometry.

The Boron Clusters B_{10}^- **and** B_{10}: Zhai et al. [101] found that the most stable B_{10}^- and B_{10} clusters have convex and Fig. 4.47. Quasi-planar structures respectively with C_s–$^2A''$ and C_{2h}–^1A_g symmetry and ground state, as can be seen in The PES spectra of B_{10}^- (Fig. 4.47) revealed well-resolved detachment features, representing electronic transformations from the ground state of the anionic cluster B_{10}^- to the ground as well as excited states of B_{10} the neutral boron cluster. The measured ADE and VDE values of the X band, 2.88 ± 0.09 and 3.06 ± 0.03 eV, obtained for B_{10}^-, are respectively in excellent agreement with the calculated values 2.84 and 3.10 eV. The molecular orbitals of the B_{10} cluster possess three π-type ones. These π-orbitals are almost identical with those of benzene. The π bonds of the quasi-planar clusters are apparently delocalized over the and below the plane of the clusters following the aromatic $(4n + 2)$ rules of Hückel.

The Boron Clusters B_{11}^- **and** B_{11}: Zhai et al. [101] found that the most stable B_{11}^- and B_{11} clusters have similar circular perfect planar structures. The structures of B_{11}^- and B_{11} clusters have C_{2v} symmetry and respectively $^1A_1$ and $^2B_2$ ground state, as shown in Fig. 4.48. The well-resolved detachment features of the PES spectra of the B_{11}^- anionic cluster represent distinctly electronic transitions from the ground state of the anionic B_{11}^- cluster to the ground and excited states of B_{11} the neutral boron cluster.

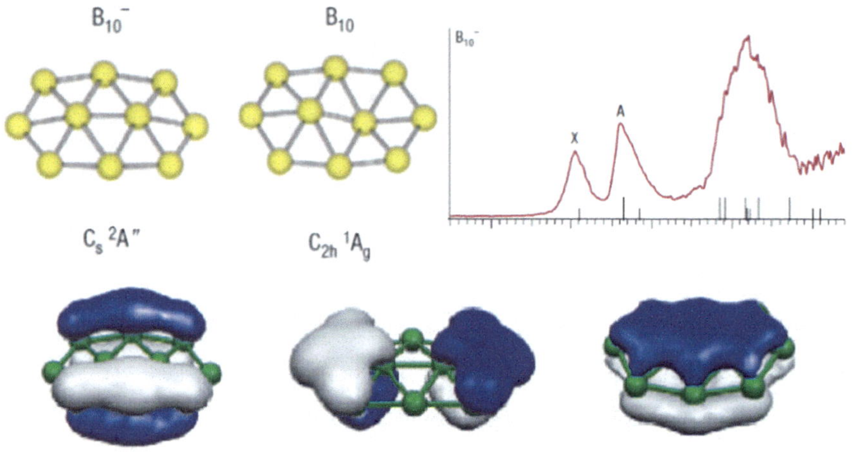

Fig. 4.47 The convex and quasi-planar structures of B_{10}^- (C_s–$^2A''$) and B_{10} (C_{2h}–^1A_g), the PES of B_{10}^-, observed at 193 nm (6.424 eV). The HOMO pictures of B_{10} possess six π electrons and thus B_{10} is aromatic [101]

The measured ADE and VDE values of the X band, 3.426±0.01 and 3.426±0.01 eV, obtained for B_{11}^-, are respectively in good agreement with the calculated values 3.27 and 3.38 eV. The B_{11}^- cluster has three delocalized occupied π-type molecular orbitals. As in the case of the decagonal clusters, mentioned in the above section, the π-bonds in the MOs of the B_{11}^- cluster follow similarly the aromatic $(4n+2)$ rules of Hückel. After the electron detachment, the structure of the ground state of the anionic B_{11}^- cluster transforms into the structure of the ground state of the neutral B_{11} cluster but with relatively minor change. Nonetheless, both structures maintain the same C_{2v} symmetry [101]. The measured high-frequency mode 1,040 ± 50 cm^{-1} corresponds mainly to the B'B stretching of the inner two atoms but it is in good agreement with the calculated frequency 1,092 cm^{-1}). However, the measured frequency around 480 ± 40 cm^{-1} and the pretty identical calculated frequency at 481 cm^{-1} belong together to the lower-frequency mode, which correlates to the stretching of the peripheral atoms surrounding the central dimer of the cluster.

The Boron Clusters B_{12}^- and B_{12}: Zhai et al. [101] investigated furthermore B_{12}^- and B_{12} clusters and found that the structure of the neutral cluster at the ground-state is similar to the anion ground-state structure however raising to higher symmetry. The B_{12}^- and B_{12} clusters exhibit convex or quasi-planar structures with respectively C_s and C_{3v} symmetry corresponding to the $^2A'$ and $^1A_1$ ground stats, as shown in Fig. 4.49. The photoelectron spectra of the anion B_{12}^- as the most outstanding observation of this kind of clusters, indicate for the molecular orbitals of the neutral B_{12} cluster extremely large HOMO-LUMO gap of around 2.0 eV, observed from the segmentation between the X and A bands. This large energy gap was induced by the aromatic stabilization due to the six delocalized π electrons, analogous to benzene. These six π electrons are distributed in three delocalized π molecular orbitals or π

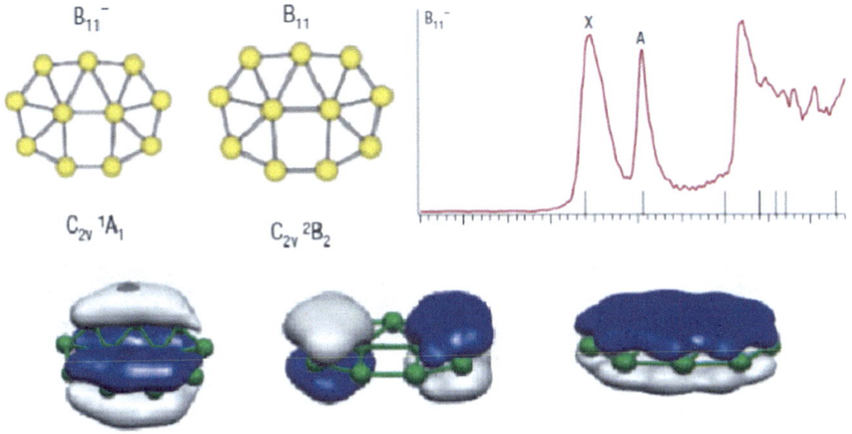

Fig. 4.48 The circular structures of B_{11}^- $(C_{2v}{-}^1A_1)$ and B_{11} $(C_{2v}{-}^2B_2)$, the PES of B_{11}^-, observed at 193 nm (6.424 eV). The HOMO pictures of the closed shell B_{11}^- cluster possess six π electrons and thus B_{11} is aromatic [101]

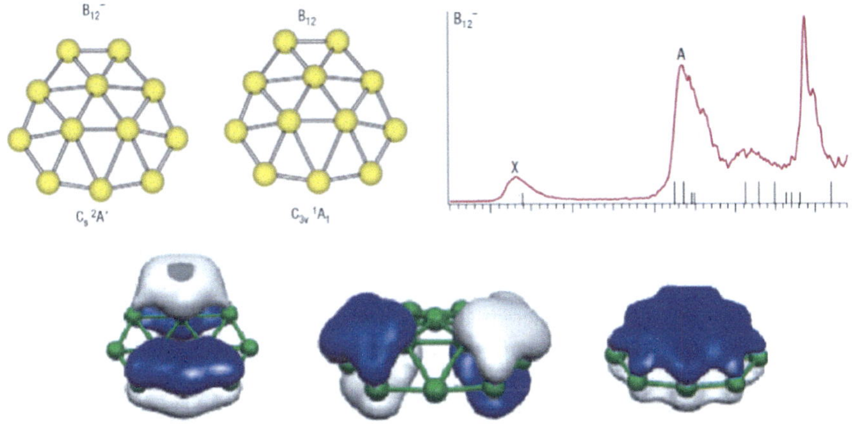

Fig. 4.49 The convex structures of B_{12}^- $(C_s{-}^2A')$ and B_{12} $(C_{3v}{-}^1A_1)$, the PES of B_{12}^-, observed at 193 nm (6.424 eV). The HOMO pictures of the closed shell B_{12} cluster possess six π electrons and thus B_{12} is aromatic [101]

bonds and follow the aromatic $(4n+2)$ rules of Hückel. The well-resolved detachment features, revealed by the PES spectrum of the anion B_{12}^- (Fig. 4.49), represent in fact the electronic transition from B_{12}^- the anion ground-state to B_{12} the neutral ground and excited states. The measured ADE and VDE values of the X band, 2.21 ± 0.04 and 2.26 ± 0.04 eV, obtained for B_{12}^-, are respectively in a good agreement with the calculated values 2.25 and 2.38 eV.

The Boron Clusters B_{13}^- **and** B_{13}: Zhai et al. [101] continued their experimental and theoretical investigations on B_{13}^- and B_{13} clusters and found different anionic and

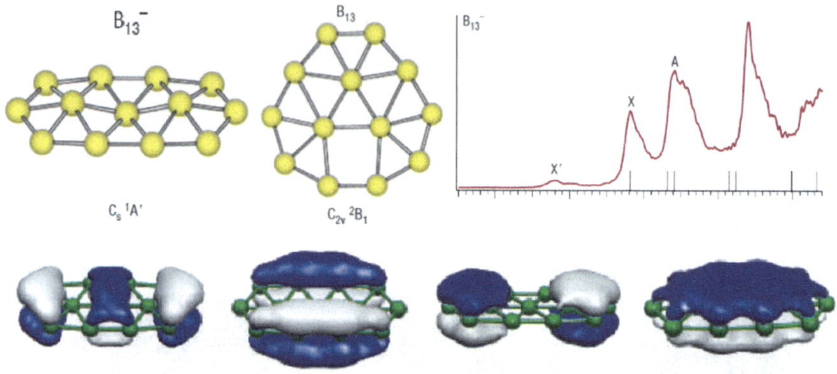

Fig. 4.50 The elongated B_{13}^- (C_s–$^1A'$) and circular B_{13} (C_{2v}–$^2B_1$) structures, the PES of B_{13}^-, observed at 193 nm (6.424 eV). The HOMOs of the closed shell B_{13}^- cluster possess eight π electrons and thus B_{13}^- is anti-aromatic [101]

neutral ground-state structures. The B_{13} was the favorite cluster in many experimental and theoretical studies. For example, Hanley and Anderson [72] produced small boron clusters via collision-induced dissociation and laser vaporization and observed in their mass spectra the prominent peak of the B_{13}^+ cluster. Fowler and Ugalde [104] studied the neutral and ionic B_{13} clusters and found that the charge of the cluster affects strongly the ground-state structure and an icosahedral configuration of B_{13} with an atom at the center is totally unstable. The authors studied different isomers of B_{13}^- and B_{13} and the low-lying structures are presented in Fig. 4.50. The structure of B_{13}^- has an elongated quasi-planar form with C_s symmetry and singlet ground state. It is interesting to learn that two structures for the B_{13}^- cluster were found to be degenerate low-lying isomers both with triplet ground state exhibiting circular shapes in which the boron atoms are connected together in slightly different distances. For neutral B_{13}, they calculated different structures with spin doublet and found the circular structure, with C_{2v} - $^2B_2$ symmetry and ground state, was slightly more stable than the elongated one, as shown in Fig. 4.50. The measured ADE and VDE values of the X band, 3.78 ± 0.02 and 3.78 ± 0.02 eV, obtained for B_{13}^-, are respectively in excellent agreement with the calculated values 3.69 and 3.81 eV. The authors found that beside the multi-center σ bonds, the planar B_{13}^- clusters possess also delocalized π bonds which obey the anti-aromatic 4n rules of Hückel. Figure 4.50 shows the structures of B_{13} and B_{13}^- clusters, the corresponding molecular orbitals HOMOs as well as the photoelectron spectrum of the B_{13}^- cluster. As can be seen, the B_{13}^- cluster is anti-aromatic, because the π bonds possess eight π electrons.

The Boron Clusters B_{14}^- and B_{14}: Zhai et al. [101] studied the B_{14}^- and B_{14} clusters using the same experimental and theoretical methods as was the case of the B_{13}^- and B_{13} clusters. They found different anionic and neutral ground-state structures. The most stable anionic cluster B_{14}^- has a quasi-planar structure with an elongated form and with C_{2v}–$^2A''$ symmetry and ground state. In contrary, the most stable

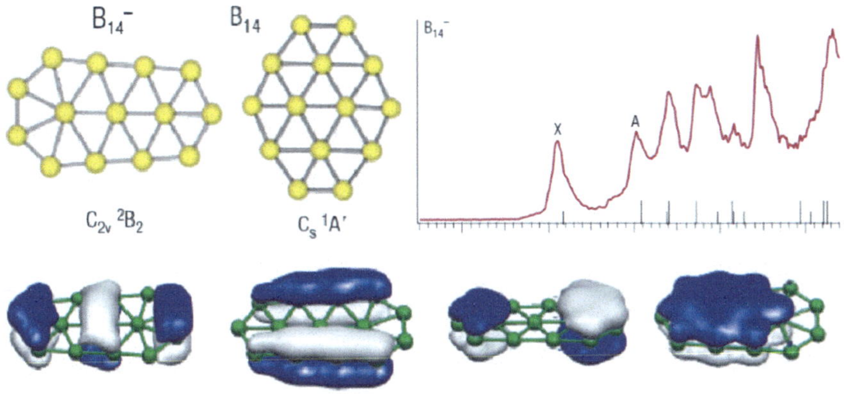

Fig. 4.51 The elongated planar B_{14}^- (C_{2v}–$^2B_2$) and circular B_{14} (C_s–$^1A'$) structures, the PES of B_{11}^-, observed at 193 nm (6.424 eV). The HOMO pictures of the closed shell B_{14} cluster possess eight π electrons and thus B_{14}^- is anti-aromatic [101]

neutral cluster B_{14} has a circular structure with C_s–$^1A'$ symmetry and ground state, as can be seen in Fig. 4.51. Both structures of B_{14}^- and B_{14} have degenerate low-laying isomers with similar circular and elongated forms. The ground state of B_{14}^- is the doublet $^2B_2$ exhibiting quasi-planar structure with an elongated form and of C_{2v} symmetry. Another boron isomer with degenerate low-lying state was found for B_{14}^-, which has a shape similar to the neutral B_{14}. The latter cluster has a circular structure with C_s–$^1A'$ symmetry and ground state, as shown in Fig. 4.51. Also B_{14} has one degenerate low-lying elongated isomer with a similar structure of B_{14}^-, which is slightly less stable than the circular one. The measured ADE and VDE values of the X band, 3.102 ± 0.01 and 3.102 ± 0.01 eV, obtained for B_{14}^-, are respectively in a good agreement with the calculated values 3.14 and 3.17 eV. Concerning the nature of the bonding in the circular planar boron cluster B_{14}, the authors found that in addition to the multi-center σ bonding, the circular planar boron cluster of B_{14} possess eight π electrons and thus it is anti-aromatic, following the anti-aromatic 4n rules of Hückel. In summary, the anionic and neutral boron clusters B_n^- and B_n for ($10 \leq n \leq 14$) exhibit either planar or quasi-planar structures.

The Dicationic Boron Cluster B_{14}^{2+}: Yuan and Cheng [102] investigated different isomers of B_{14}^{2+} boron clusters using the density functional theory (DFT) within TPSSh functional (see Sect. 1.6). They first adopted a small basis set (3-21G) for the global optimization procedure then they relaxed the low-lying TPSSh/3-21G geometries at the TPSSh/6-311+G* level of theory after global optimization. They studied 7 isomers of B_{14}^{2+} and found that the dicationic B_{14}^{2+} cluster has the lowest energy and exhibits a double-ring structure, as shown in Fig. 4.52. The authors found also that the second low-lying isomer of B_{14}^{2+} possesses a quasi-planar structure with 1.2 eV above the double-ring cluster.

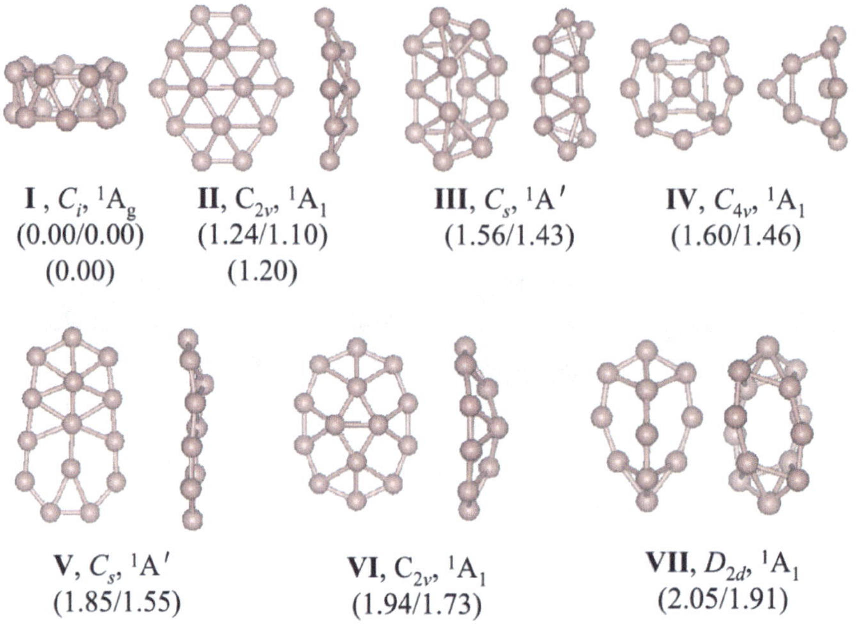

Fig. 4.52 Optimized geometry of the low-energy isomers of B_{14}^{2+} at TPSSh/6-311+G*. Symmetry and electronic state of each isomer are labeled, and enclosed are energies/free energies in eV relative to the first isomer I. For the first two isomers, relative energies at CCSD(T)/aug-cc-pVTZ are also labeled. This figure was reproduced with kind permission of the Journal of Chemical Physics of the American Institute of Physics [102]

The Chemical Bonding of B_{14}^{2+}: Yuan and Cheng [102] carried out canonical molecular orbital analysis, in order to investigate the nature of the chemical bonding of the double-ring cluster B_{14}^{+2}. The dicationic cluster possesses 40 valence electrons, which are distributed pairwise in 20 molecular orbitals (MOs). As shown in Fig. 4.53, the valence electrons are occupied by the orbital number 15 up to the orbital number 34. As can be seen in Fig. 4.53a, 14 pairs of valence electrons are localized on the MOs 15–24 and 26–29 with B-B σ-bond character, determined with the AdNDP method, which was developed by Zubarev and Boldyrev [103]. Figure 4.53b shows the six molecular orbitals 25, 30–34 with delocalized feature. These orbitals are composed of three radial and three tangential characters. The double-ring B_{14}^{2+} with 3 radial and 3 tangentail electron pairs is double-aromatic consisted of both radial and tangetial Hückel aromaticity. According to the AdNDP analysis, the B-B bonding frameworks exhibit two pattern, as can be recognized in Fig. 4.53c. The first pattern includes 14 2c-2e B-B σ-bonds with 1.73 |e| occupation number (ON). The second pattern includes 14 3c-2e σ-bonds with 1.97 |e| ON. By the first pattern, each atom in the 2c-2e σ-bonds contributes equally of 0.87 |e| and therefore these bonds are equivalent. In contrast, by the second pattern the 3c-2e σ-bonds are extremely

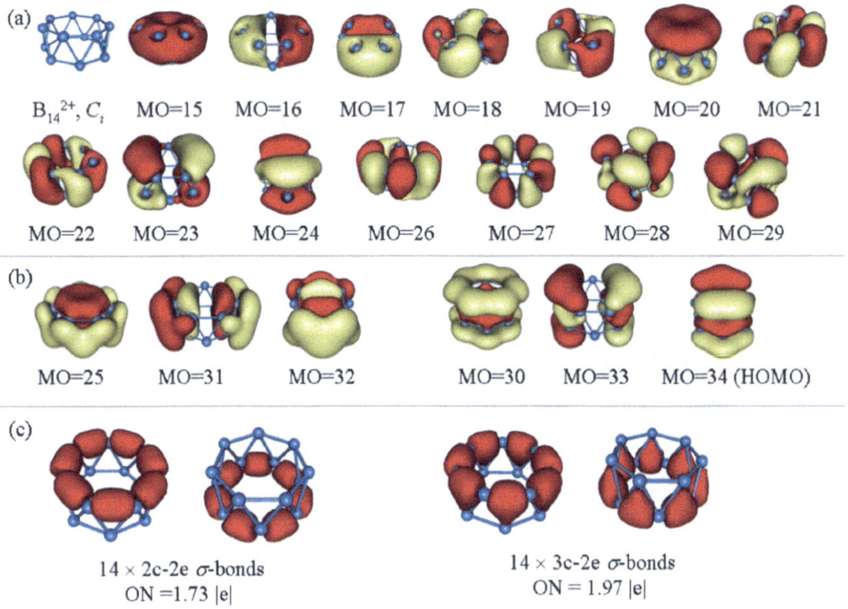

Fig. 4.53 The Structure and canonical MOs of the B_{14}^{2+} (C_i) cluster (**a**) and (**b**); Results of the AdNDP localization (**c**). This figure was reproduced with kind permission of the Journal of Chemical Physics of the American Institute of Physics [102]

inequivalent. Accordingly, one can conclude that the B-B bonding of the B_{14}^{2+} cluster is laying between the Lewis bonding and delocalized bonding.

4.3.5 Hydrocarbon and Boron Clusters

Analogy to Hydrocarbon Benzene: The three occupied π MOs of B_{12} cluster in Fig. 4.49 are nearly identical to those of benzene, as presented in Fig. 4.54. The neutral B_{12} cluster in Fig. 4.49 exhibits a structure composed of a central B_3 triangle surrounded by 12 triangles with a C_{3v} symmetry. From another point of view, this structure can be considered as a composition of six BBB triangles without edge-sharing, as shown in Fig. 4.55. Suppose each single triangle of the B_{12}–C_{3v} cluster provides a π electron that altogether converting the π framework over the whole structure, then appears a ring system composed of six connected B_3 triangles, similar to the cyclic π system of benzene. Accordingly, one can assert that the B_{12}–C_{3v} cluster exhibits π framework similar to benzene. Thus, the B_{12} and benzene systems are isomorphic. Therefore the analogous of the both π systems is a natural consequence for the similarity between the π MOs of B_{12} and benzene. One can also ascertain that the B_{12} cluster represents the most prototypical aromatic boron system similar

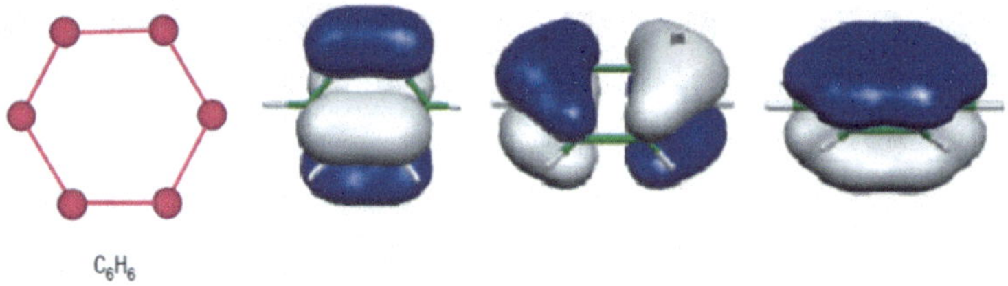

Fig. 4.54 The highest molecular orbital pictures of C_6H_6, showing three π-type orbitals [101]

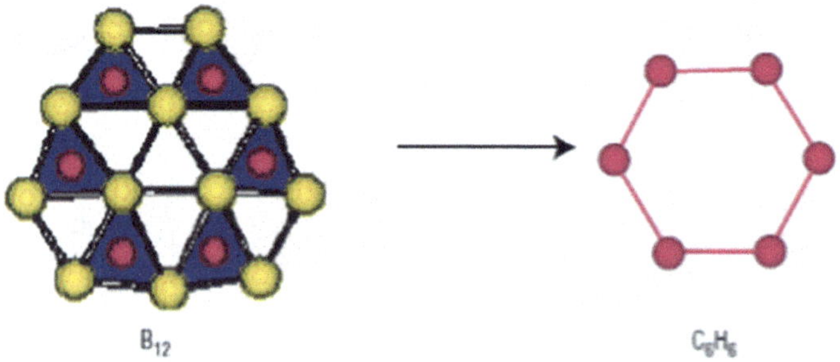

Fig. 4.55 Analogy of the aromaticity between boron clusters and hydrocarbons. The analogy between the π MOs framework of B_{12} and that of benzene [101]

to the system of the aromatic hydrocarbons particularly of benzene. The unusual high stability of the B_{12} cluster can be verified by the observed PES spectra of B_{12}^-, indicating extremely large energy gap.

Analogy to Hydrocarbon Cyclobutadiene: The planarity of boron clusters is reminiscent of the unsaturated polycyclic hydrocarbons. Hence, the discovered aromaticity and anti-aromaticity in boron clusters are unique and thus the reason for their flatness, as predicted for the flat boron cluster B_{13}^+ [104, 105]. However, we have already seen that the two anti-aromatic clusters, B_{13}^- and B_{14}, are prolated. As can be seen in Fig. 4.56, the square-to-rectangular structural transformation of C_4H_4, the anti-aromatic cyclobutadiene, which possesses two localized π bonds, can correspond to the structural transformation between the both isomers of B_{13}^-. The nodal properties of the π MOs in the B_{10}, B_{11}^-, B_{12}, B_{13}^-, and B_{14} boron clusters are similar to those in hydrocarbons, as already shown in the Figs. 4.47, 4.48, 4.49, 4.50 and 4.51. One can also recognize that the first π-MO ((π_0) exhibits a zero node. The following π-MOs with degenerate or near-degenerate states possess an increasing number of nodes. For example, the π_1 pair possesses one node, the π_2 pair two nodes, the π_3 pair three nodes, and so on.

Hückel Rules for Small Boron Clusters: The chemical bonding of boron clusters B_n for $3 \leq n \leq 14$ were found to possess multi-center σ bonds and π bonds with delocalized character. These σ and π bonds are following unanticipated the $4n + 2$ and $4n$ Hückel rules respectively determined for aromaticity and anti-aromaticity.

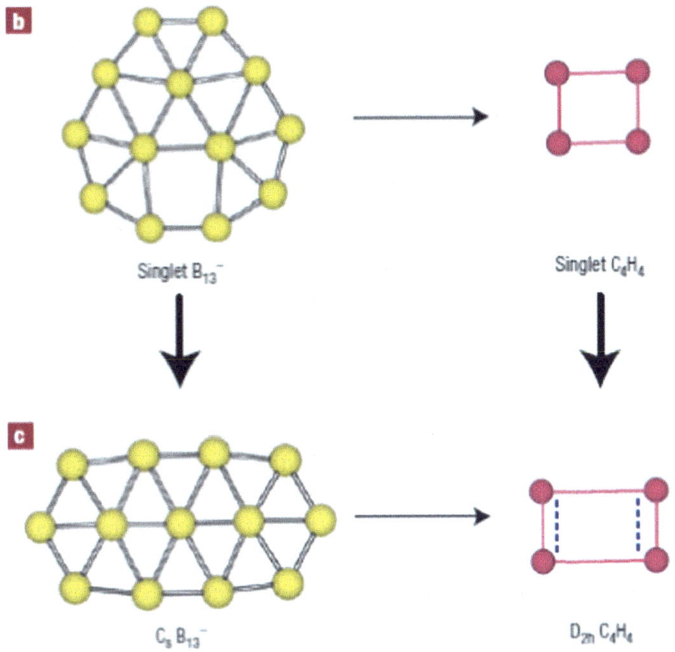

Fig. 4.56 Analogy of the (anti)-aromaticity between B_{13}^- and hydrocarbons. **b** The unstable B_{13}^- and the unstable square C_4H_4. **c** The analogy between elongation of B_{13}^- and that of C_4H_4 due to anti-aromaticity [101]

All the ground states of the boron clusters B_{10}, B_{11}^-, B_{12}, B_{13}^-, B_{14} and B_{15}^- have low spins with closed-shell electronic configurations and a singlet-state. Each of the B_{10} and B_{11}^- clusters possesses six π electrons exhibiting aromatic character. B_{12} also possesses six π electrons and is aromatic. The aromaticity of B_{12} provokes an extremely large energy gap between the (HOMO) and (LUMO), respectively the highest occupied molecular orbital and the lowest unoccupied molecular orbital, causing unusual stability in its quasi-planar structure. In contrast, each of the B_{13}^- and B_{14} clusters possesses eight π electrons with an anti-aromatic character, exhibiting elongated structure and as expected, induced by its anti-aromaticity, in a similar feature to cyclobutadiene and its square-to-rectangular structural distortion.

Planarity versus Aromaticity and Hückel Rules: It's really worth to mention that Fowler and Ugalde [104] were the first researchers who connected the planarity of the B_{13}^+ cluster with the delocalized π-electrons. They found that the stability and the low reactivity of the B_{13}^+ cluster are related to the aromaticity, which in turn is caused by the delocalization of the π-electrons distributed over the HOMOs. Both authors also asserted that the high stability of the B_{13}^+ cluster and its quasi-planar structure are referred to the its aromatic character in analogous to benzene. Also Kiran et al. [106] studied the B_{12} and B_{13}^+ clusters and found that these clusters possess unusual high stability which in turn arose by the existence of the multiple aromatic $4n + 2$ sextets. They confirmed that the energy gaps between the (HOMO) and (LUMO) of the B_{12} and B_{13}^{+} boron clusters are mainly based on the nature of the units and their interaction but not only on the numbers of the π-electrons. Aihara et al. [107] established in terms of topological resonance energy the aromaticity of planar boron clusters. From the latter two papers one can ascertain that in similarity to the general organic chemistry the concept of the aromaticity seems to work and to be useful in boron chemistry. They confirmed furthermore that “Aufbau Principle” is very useful draft for constructing highly stable structures of boron clusters.

4.3.6 Coulomb Explosion

Multi-charged Boron Clusters: Using ab initio quantum chemical methods and density functional theory Akman et al. [108] carried out a further study on neutral and multi-charged boron clusters B_n for ($n = 2$–13). They determined the stability, electronic structures, fragmentation channels, ionization energies, and Coulomb explosion of these clusters. The authors confirmed in their calculations that when the electron-deficient boron clusters become negatively multi-charged (see Fig. 4.57), then they preserve more stability and toughness and are hardly to fragment. The stability of boron clusters, labeled as the binding energy per atom (E_B), declines by wasting the valence electrons. Therefore, when the ionization increases decreases the stability. By a single ionization of the boron clusters, the odd-size clusters are more stable than the even-size ones. Multiple ionization of the clusters prompts the repulsive Coulomb forces to getting stronger and the chemical binding between the constituent atoms of the clusters becoming weaker leading to meta-stable state and consequently to the Coulomb explosion of the boron clusters.

None of the cations studied remain stable after six times ionization destabilization of clusters (Fig. 4.58). Stability of boron clusters increases by increasing the clusters size (Fig. 4.59), and the BE decreases as the Q value increases (Fig. 4.60). In Fig. 4.61 they observed that B_3^{+2}, B_4^{+3}, B_5^{+3}, B_6^{+4}, B_7^{+4}, B_8^{+5}, B_9^{+5}, B_{10}^{+5}, B_{11}^{+6}, B_{12}^{+6}, and B_{13}^{+6}, have all negative BE (i.e., they are unstable). In other words, multi-ionization leads to The excess charges on the ionized clusters causes the repulsive Coulomb potential energy to exceed their binding energy. Now, how does the binding energy per atom (BE) of the anionic and cationic boron clusters change as its charge increases?. To answer this question, the authors [108] plotted the BE of several clusters as a function of increasing the cluster size and of increasing the electron charge Q, as presented in Fig. 4.57. The BE of multi-charged give their energy away and explode. Thus, the critical charge for meta-stability (Q_m) is estimated as $Q_m \leqslant \frac{n}{2}$ for even-size clusters, and $Q_m \leqslant \frac{1}{2}(n-1)$ for odd-size clusters. In addition, the critical charge for the Coulomb explosion (Q_c) is found to be $Q_c = \frac{n}{2} + 1$ for even-size clusters, and $Q_c = \frac{1}{2}(n+1)$ for odd-size clusters.

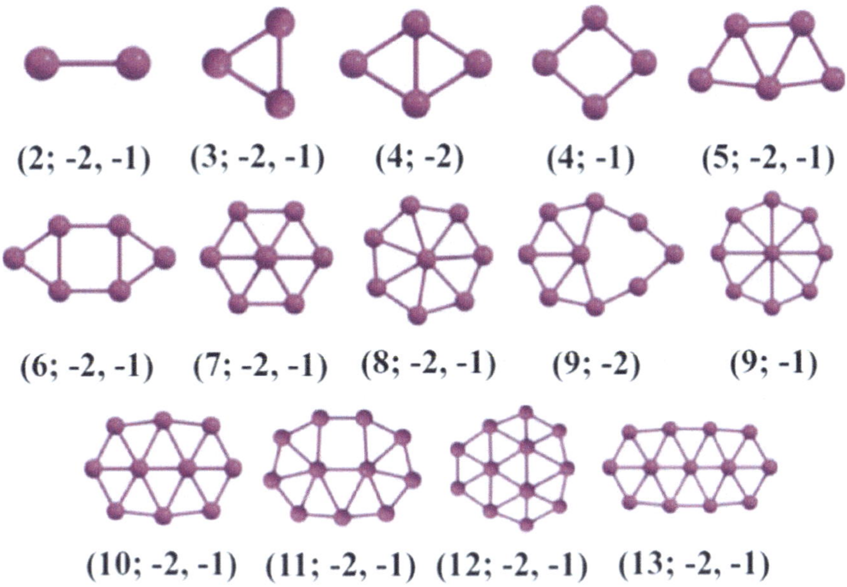

Fig. 4.57 Top views of anionic boron clusters. Numbers in parentheses denote, respectively, the cluster size and charge states [108]

Agreement with Boustani's Predictions: All boron clusters exhibit convex, planar or quasi-planar structures with circular or elongated shapes. The observed bonding energies of the photoelectron spectroscopy in the form of VDE are in excellent agreement with calculated VDE values for the proposed convex or quasi-planar structures. The finding of Zhai et al. [101] confirms the convex or quasi-planar structures in Figs. 4.33, 4.34, and 4.35, respectively, which were predicted by Boustani [80–82, 85] for the first time.

4.4 Neutral and Charged Boron Clusters B_n ($n \leq 40$)

4.4.1 Boron Clusters B_n ($15 \leq n \leq 25$)

The Boron Clusters B_{15}^- **and** B_{15}: Zhai et al. [101] investigated the B_{15}^- and B_{15} clusters using the same experimental and theoretical methods as was the case of B_{14}^- and B_{14}. They found that the ground state of the B_{15}^- and B_{15} boron clusters exhibit circular shapes possessing no symmetry C_1, as presented in Fig. 4.62. The structures of the next low-lying isomers of the B_{15}^- and B_{15} clusters have elongated forms with C_s symmetry. Concerning the photoelectron spectra of B_{15}^- cluster, for the X band there were no resolved vibrational structures in evidence and the related ADEs were

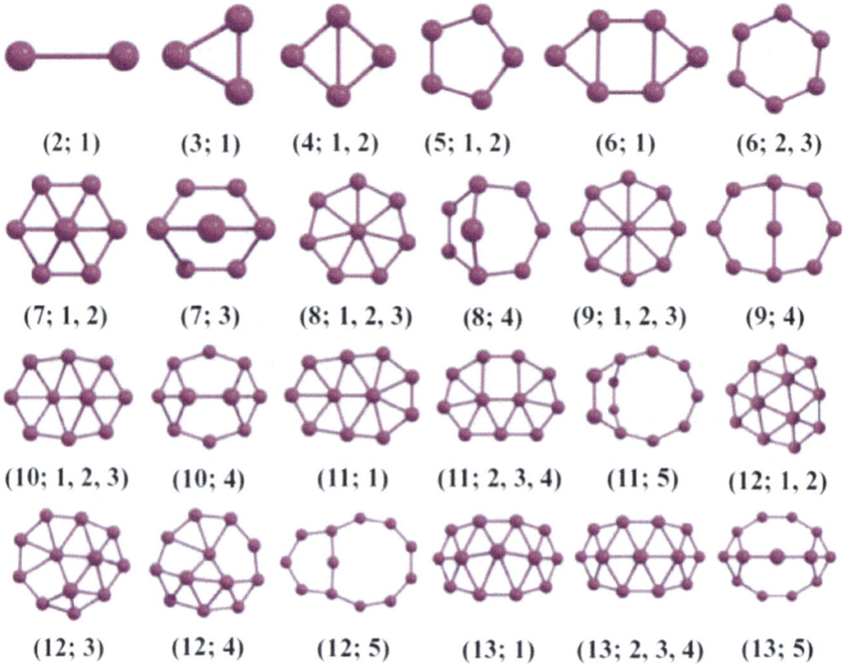

Fig. 4.58 Top views of cationic boron clusters. Numbers in parentheses denote, respectively, the cluster size and the charge states [108]

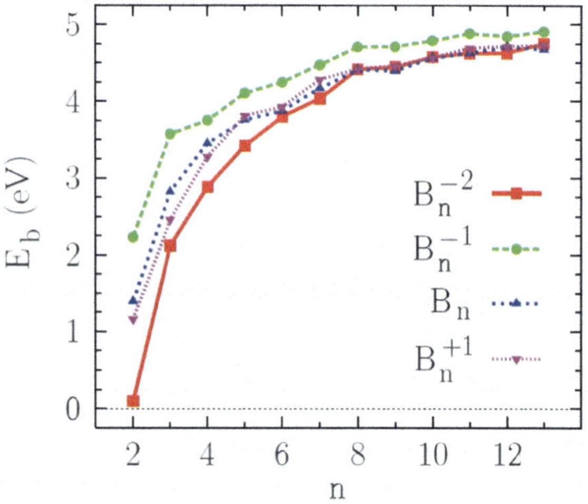

Fig. 4.59 The binding energy per atom of neutral and multi-charged boron clusters as a function of cluster size [108]

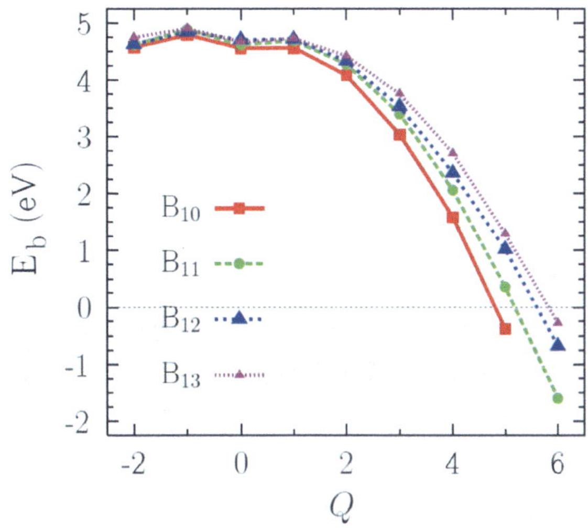

Fig. 4.60 The binding energy per atom of boron clusters B_n ($n = 10$–13) with respect to the cluster charge [108]

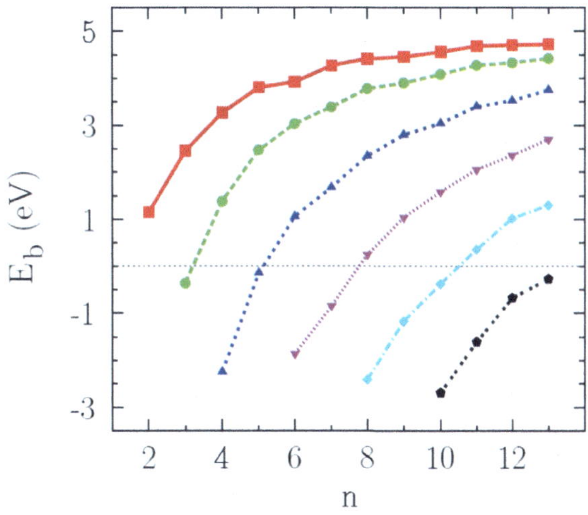

Fig. 4.61 The binding energy per atom of multi-ionized cationic boron clusters. Red curve is for $Q = 1$. The lower curves stand respectively for $Q = 2, 3, 4, 5$, and 6 [108]

estimated directly from the threshold. For the B_{15}^- cluster, the measured ADE and VDE values of the X band 3.34 ± 0.04 and 3.43 ± 0.04 eV, and the calculated values 3.24 and 3.37 eV are in a good agreement, respectively. For monocyclic systems, it is known that they obey the Hückel rules. The authors found surprisingly that the bonding nature of the circular planar B_{15} cluster exhibits not only multi-center σ-bonds but also ten π electrons and thus it is aromatic, following the $4n + 2$ Hückel rules for aromaticity. However, Fig. 4.62 shows that parts of the π MOs of B_{15}^- are distributed around the cluster. Therefore, due to these fragmented π MOs the Hückel rules seem to be not applicable for larger clusters.

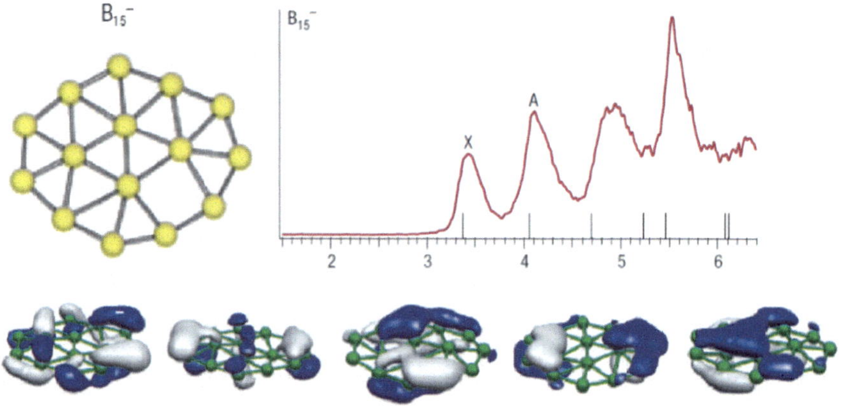

Fig. 4.62 The circular planar structure of both B_{15}^- (C_1–^1A) and B_{15} (C_1–^2A), the PES of B_{15}^-, observed at 193 nm (6.424 eV). The HOMO pictures of the closed shell B_{15}^- cluster possess ten π electrons and is aromatic [101]

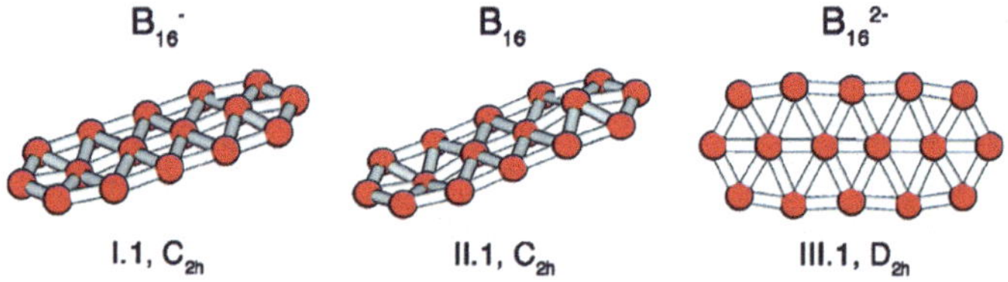

Fig. 4.63 The optimized global minimum structures of B_{16}^- (^2A_u), B_{16} (^1A_g) and B_{16}^{2-} (^1A_g) at B3LYP/6-311+G*. This figure was reproduced with kind permission of the Journal of the American Chemical Society [109]

The Anionic and Neutral Boron Clusters B_{16}^- **and** B_{16}: Sergeeva et al. [109] investigated the electronic and geometrical structures as well as the bonding character of B_{16}^- and B_{16}^{2-} clusters applying photoelectron spectroscopy in combination with theoretical calculations. They found that B_{16}^- and B_{16} possess quasi-planar structures (C_{2h}), as shown in Fig. 4.63. The authors found that the elongated B_{16}^- cluster becomes a perfectly planar B_{16}^{2-} (D_{2h}) structure, when an extra electron is added. The patterns of σ bonds in the B_{16}^{2-} cluster appear to be very important because they could play a very important role in the formation the planar structures of boron clusters. In order to verify the feasibility of using the B_{16}^{2-} cluster as building blocks in cluster-assembling of novel nanomaterials, the authors added two Li^+ cations to the B_{16}^{2-} cluster then optimized the new structure of Li_2B_{16} cluster, as shown in Fig. S3 of [109]. The addition of the Li^+ cations enabled the planar B_{16}^{2-} structural unit to be quite elastic and to be crooked upon coordination by Li^+. The reason behind this flexibility can be related to the in-plane B-B σ bonds, which are relatively weak and also delocalized.

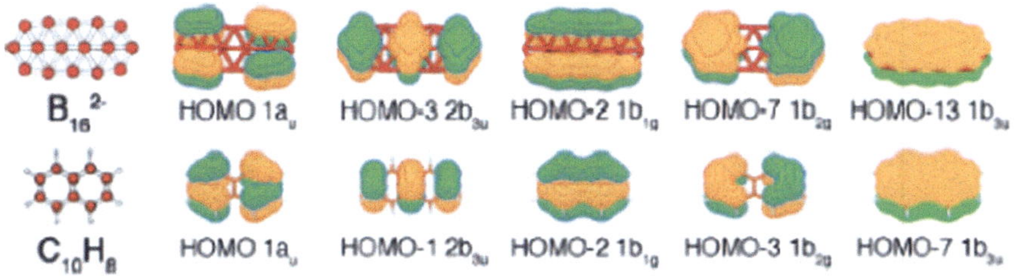

Fig. 4.64 Comparison between the structures and π bonding of B_{16}^{2-} (D_{2h}–^1A_g) and naphthalene. This figure was reproduced with kind permission of the Journal of the American Chemical Society [109]

Similarity of B_{16}^{2-} to Naphthalene: More importantly, Sergeeva et al. [109] found, as mentioned above, that adding of a single electron to B_{16}^- produces a perfectly planar B_{16}^{2-}(D_{2h}) structure, that possesses 10 π electrons exhibiting a π bonding pattern in analogous to naphthalene, as presented in Fig. 4.64. Thus, the B_{16}^{2-} cluster can be acknowledged as an "all-boron-naphthalene". It is exciting to ascertain that when the B_{16}^{2-} cluster loses two π electrons from the HOMO, the neutral B_{16} cluster undergoes out-of-plane deformation and becomes π-anti-aromatic, similar to the out-of-plane structural distortion of the common anti-aromatic cyclooctatetrene (C_8H_8). As can be seen in Fig. 4.63, the structure of the B_{16} cluster has two atoms at the center of two hexagons located at the both ends. These two atoms are bending or hiking above and below the molecular plane by 0.24 Å, inducing a quasi-planar structure. The anionic B_{16}^- cluster with only one electron in the open shell molecular orbital is laying between the π-aromatic B_{16}^{2-} and the π-anti-aromatic B_{16} structures. Therefore, this structure suffers fewer out-of-plane distortion and the both central atoms emerging only by 0.08 Å.

Photoelectron Spectra of B_{16}^-: As mentioned above, Sergeeva et al. [109] investigated the electronic and geometrical structures of B_{16}^- cluster applying photoelectron spectroscopy in combination with theoretical calculations. They used two photon energies to obtain photoelectron spectra (PES) spectra. They applied laser vaporization technique in order to produce the B_{16}^- cluster. The target was ^{10}B-enriched material, and the boron clusters were mass-selected by time-of-flight mass spectrometry. As shown in Fig. 4.65, the observed VDE values 3.39 and 3.78 eV of the X and A bands in the 193 nm (PES) spectrum of B_{16}^-, obtained for the ground state minimum structure of B_{16}^- (C_{2h}, ^2A_u), are in a good agreement with calculated values 3.37 and 3.70 eV, respectively. The calculations were carried out at the TDDFT and B3LYP levels using the 6-311+G(2df) basis sets.

The Neutral Boron Clusters B_{16} and B_{17}: In combination of vibrational spectroscopy techniques and ab initio quantum chemical calculations, Romanesco et al. [110] investigated the geometrical structures and the electronic properties of the neutral B_{11}, B_{16} and B_{17} boron clusters. Experimentally, they produced infrared absorption spectra in the wavelength range of 650–1550 cm^{-1} for the three low-

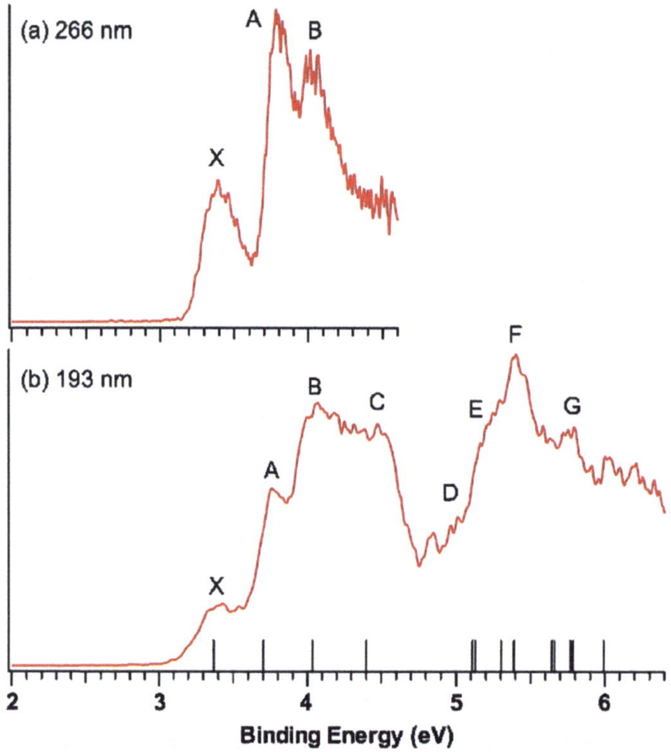

Fig. 4.65 Photoelectron Spectra of B_{16}^-, observed at 266 nm (4.661 eV) and 193 nm (6.424 eV). The vertical bars represent the calculated VDEs at the TD-B3LYP level. This figure was reproduced with kind permission of the Journal of the American Chemical Society [109]

energy neutral boron clusters. Figure 4.66 shows that the structures of the B_{11}, B_{16} and B_{17} boron clusters exhibit planar or quasi-planar formations, similar to those of the anionic counterparts B_n^-. The latter were also found to possess global minima in form of planar structures confirmed by theoretical calculations and photoelectron spectroscopy. The calculations were carried out using B3LYP, PBE0, and TPSS functionals as well as 6-311+G(d) basis set. After generation of roughly 300 random structures for each of the B_{11} and B_{16} clusters, the authors carried out an optimization process at the PBE0/3-21G levels of theory. These at the PBE0/3-21G level obtained isomers were reoptimized once again using the larger 6-311+G(d) basis set. The initial structure for B_{17} was taken from [111].

The IR-Spectra of B_{16} **and** B_{17}: Romanesco et al. [110] determined the IR frequencies of B_{16} and B_{17} experimentally via photoelectron spectroscopy and theoretically at the B3LYP/6-311+G(d) level of theory. The IR spectrum of the B_{16} cluster, as presented in Fig. 4.67, exhibits a strong band around 850 cm^{-1} and 1120 cm^{-1}, while the IR spectrum of the B_{17} cluster shows a broad band around 860 cm^{-1}, and other

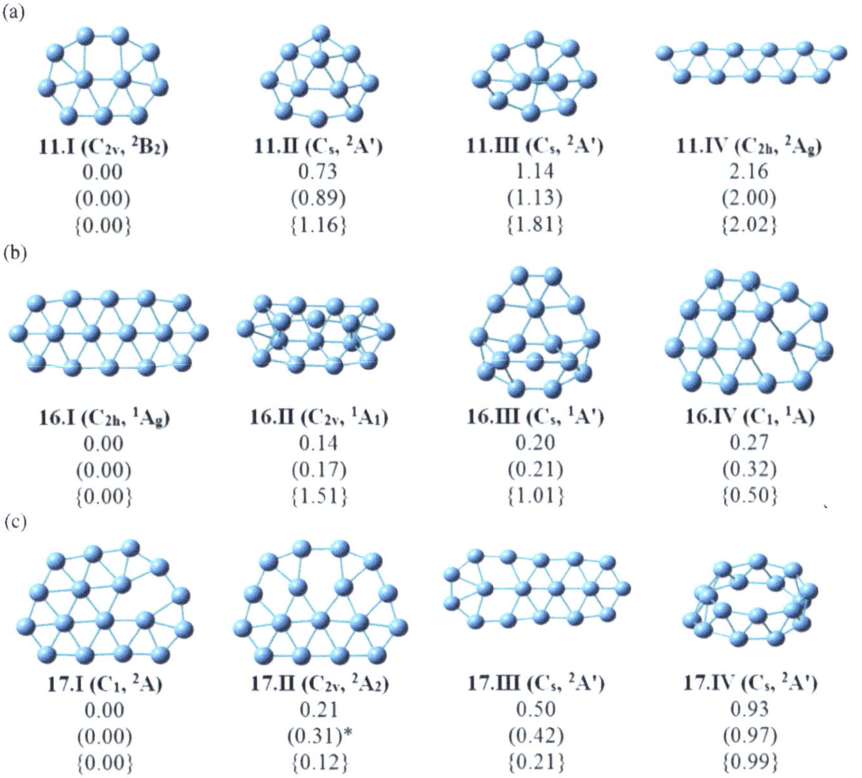

Fig. 4.66 The global minima and low-lying isomers of B_{11} (**a**), B_{16} (**b**), and B_{17} (**c**). Relative energies in eV are given at the PBE0/6-311+G(d), TPSS/6-311+G(d) (in parenthesis), and B3LYP/6-311+G(d) (curly brackets). This figure was reproduced with kind permission of the Journal of Chemical Physics of the American Institute of Physics [110]

three nearly sharper bands at $1150\,cm^{-1}$, $1260\,cm^{-1}$, and $1450\,cm^{-1}$. However, the excellent agreement between the calculated spectrum and the experimental IR spectrum for 16-I in Fig. 4.67, confirms unambiguously that the isomer 16.I of the neutral B_{16} cluster is the global minimum, as already shown in Fig. 4.63. The isomer 17.I is also the global minimum for B_{17} as confirmed by the excellent agreement of the experimental and the calculated IR spectra. In summary, the neutral clusters B_{16} and B_{17} exhibit structures with planar and quasi-planar shapes, in similarity with their related anionic counterparts.

Confirmation of Neutral Planar B_{18} **and Tubular** B_{20}: Hsing et al. [112] used the density functional theory (DFT) as well as the diffusion quantum Monte Carlo (DMC) method to calculate the relative energies of different isomers of covalent and metallic clusters. The investigated isomers of B_{18} have double ring tube, hollow cage with O_h symmetry, and quasi-planar structures, while the isomers of B_{20} have a tube and three planar structures, as can be seen in Figs. 4.68 and 4.69. Different

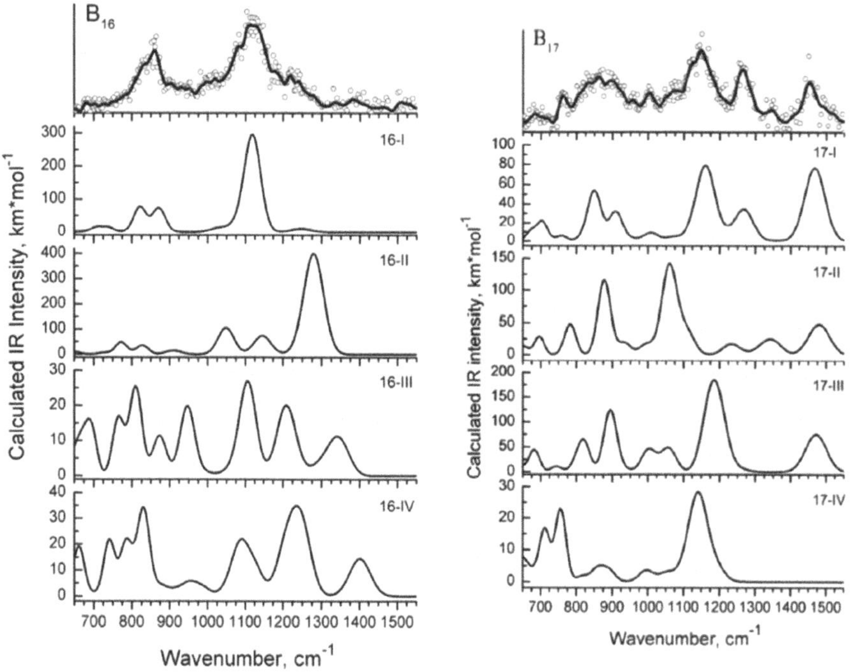

Fig. 4.67 IR spectrum of B_{16} and B_{17} (top) compared with the IR spectra of low lying isomers from DFT calculation (PBE0/6-311+G(d)). The open circles in the top spectrum correspond to the original data points folded with a 7-point binomial weighted average (solid line). The calculated spectra were generated using Gaussian line widths of 2% of the central wave number. This figure was reproduced with kind permission of the Journal of Chemical Physics of the American Institute of Physics [110]

exchange-correlation functionals (XC) within the DFT were applied, like the local density functional approximation (LDA), and the non-local generalized gradient approximations (GGA) such as Perdew-Wang 1991 (PW91) and Perdew-Burke-Ernzerhof (PBE). They compared the DMC results with the DFT ones obtained by the LDA, PW91 and PBE methods. The LDA, PBE, and DMC calculations all predict that the planar-2 isomer of B_{18} is the most stable structure. The DFT-LDA, DFT-PBE, and DMC calculations all predict that the double-ring tubular isomer of B_{20} is the most stable structure. The next stable structure is the convex planar-2 structure in Fig. 4.69. The double-ring isomer B_{20} is identical to the structure 5 in Fig. 4.76 found by Kiran et al. [115], confirming the tubular character of the neutral boron cluster B_{20}.

The Boron Clusters B_{19}^- and B_{19}: Huang et al. [113] extended their investigations to study the electronic and geometrical structures of the B_{19}^- and B_{19} boron clusters, using ab initio quantum chemical calculations and photoelectron spectroscopy (PES) techniques. Experimentally, they applied magnetic bottle photoelectron spectroscopy

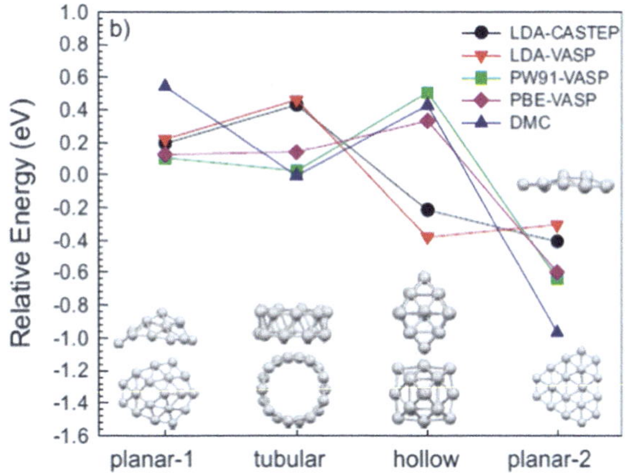

Fig. 4.68 Energy relative to the average for isomers of B_{18}, obtained using the DFT-LDA, DFT-PBE, DFT-PW91, and DMC methods [112]

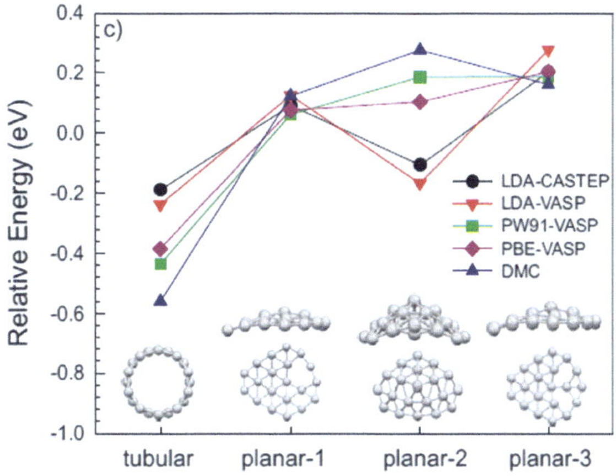

Fig. 4.69 Energy relative to the average for isomers of B_{20}, obtained using the DFT-LDA, DFT-PBE, DFT-PW91, and DMC methods [112]

apparatus equipped with a laser vaporization source. The authors used two programs for the search of the global minimum of B_{17}^- and B_{18}^- boron clusters. The programs are the Coalescence Kick and the Basin Hopping search programs,[7] in order to search for the global minimum of B_{19}^-, both initially at the B3LYP/3-21G level of theory. They found that the structure of the global minimum of B_{19}^- cluster is planar, composed of two cyclic rings with a central atom, as can be seen in Fig. 4.70a. The interior ring is a pentagon and the exterior ring is of 13 atoms. The average distance between the central atom and the exterior ring atoms is 3.20 Å. As can be seen in Fig. 4.70b, the first low-lying isomer, assigned by II (C_{2v}, $^1A_1$), is about 1.75 kcal/mol (CCSD(T)/6-311+G*) higher in energy than the the global minimum. The second low-lying isomer B_{19}^-, assigned by III (C_{2v}, $^1A_1$), as shown in Fig. 4.70c, has an elongated structure

[7]The Coalescence Kick program is written by by B. B. Averkiev and the Basin Hopping search program is written by W. Huang (see [113] and the references therein).

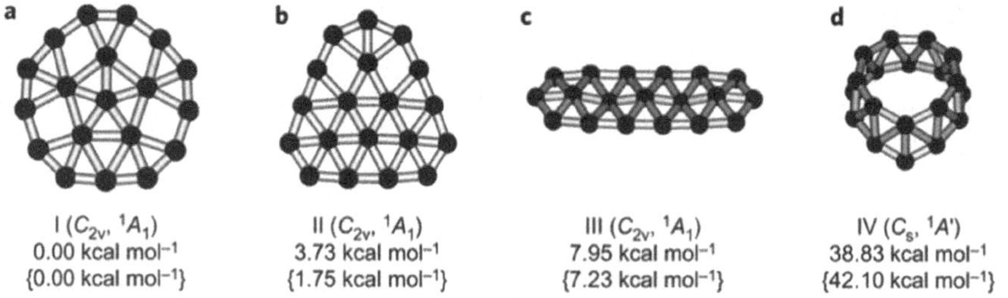

Fig. 4.70 Four representative optimized isomers of B_{19}^-. **a–d** The relative energies are given at B3LYP/6-311+G* and at CCSD(T)/6-311+G*, with the optimized B3LYP/6-311+G* geometry given in curly brackets, both corrected for zero-point energies at B3LYP/6-311+G*. This figure was reproduced with kind permission of the Nature Publishing Group [113]

with about 7.23 kcal/mol (CCSD(T)/6-311+G∗) above the global minimum. The tubular-like isomer, presented in Fig. 4.70d and assigned by IV (C_s, $^1A'$), is about 42.1 kcal/mol higher than the global minimum.

Photoelectron spectra of B_{19}^-: The ground state and fairly broad band (X), observed at the 266 nm of the PES spectra of B_{19}^-, shown in Fig. 4.71a and observed at 4.4 eV, suggests a structural transition between the anion and the neutral ground state. The next band at 4.5 eV is resolved into two sharp peaks (A and B). A very high adiabatic detachment energy of B_{19}^- was estimated to be 4.2 ± 0.1 eV. The PES spectrum, observed at 193 nm photon energy and presented in Fig. 4.71b, shows five resolved bands. The calculated vertical electron-detachment VDE 4.37 and 4.48 eV, obtained at the TD-B3LYP/6-311+G(2df) level of theory, are in an excellent agreement with the observed VDE values of the X and A bands 4.34 ± 0.05 and 4.48 ± 0.02 eV.

Similarity to Annulene $C_{10}H_{10}$: As known, benzene or annulene are the representatives of aromatic molecules with six π-electrons which satisfy the $4n + 2$ Hückel rule with ($n = 1$). The planar mono-cyclic cyclodeca-1,3,5,7,9-pentaene hydrocarbon molecule annulene ($C_{10}H_{10}$) with $n = 2$ is presented in Fig. 4.72a, i. The concepts of of aromaticity and anti-aromaticity, as mentioned in previous sections can be used to rationalize the Chemical bonding, structure, reactivity and stability of clusters with planar and quasi-planar structures. In particular, all B_8^{2-} and B_9^-, B_{10}, B_{11}^-, B_{12}, and B_{13}^+ boron clusters can be interpreted in analogues to benzene based on their six π-bonds exactly similar to those in benzene. Thus, B_{19}^- was found to possess a structure with almost cyclic planar ring, as shown in Fig. 4.72b, with a central pentagon B_6 inside a B_{13} ring exhibiting delocalized π-bonds in analogous to those π-bonds assigned in the planar aromatic annulene ($C_{10}H_{10}$), as presented in Fig. 4.72a, ii, and in circulene ($C_{24}H_{12}$), as shown in Fig. 4.72c.

Chemical Bonding of B_{19}^-: Molecular orbital analysis of the anionic cluster B_{19}^- and the hydrocarbon molecule annulene ($C_{10}H_{10}$) shows that the B_{19}^- isomer I in Fig. 4.70 possesses a unique double aromaticity in two concentric π-systems. Two of the 12 π-electrons are delocalized over the inner ring B_6, while the other ten π-electrons are

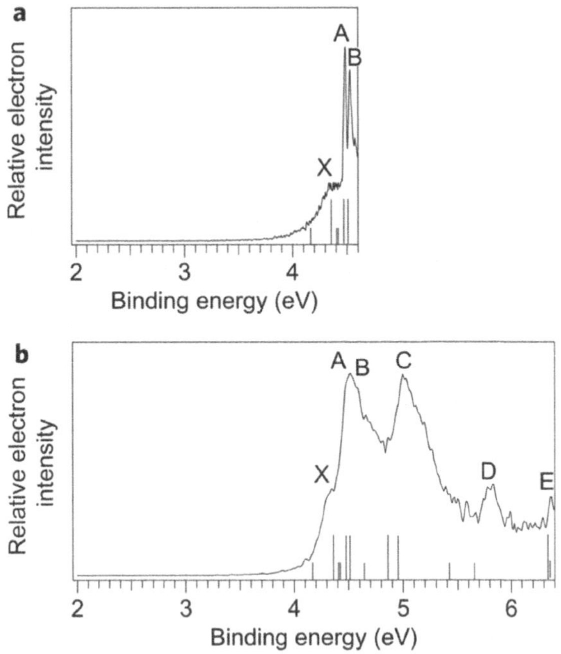

Fig. 4.71 Photoelectron Spectra of B_{19}^-, observed at the 266 nm (4.661 eV) (**a**) and 193 nm (6.424 eV) (**b**). The calculated VDEs for two isomers of B_{19}^- are represented as the vertical bars. The long bars correspond to the VDEs calculated for the global minimum isomer I and the short bars for the low-lying isomer II. This figure was reproduced with kind permission of the Nature Publishing Group [113]

delocalized over the outer ring B_{13}. The corresponding HOMOs of B_{19}^- are presented in Fig. 4.73d. In comparison with the annulene molecule, the HOMOs of ($C_{10}H_{10}$) in Fig. 4.73c show similar orbital structures.

Theoretical Confirmation of Planar B_{19}^-: Boustani et al. [114] interpreted theoretically the experimental data of Huang et al. [113] using density functional theory (DFT) to determine the geometrical and electronic structures, vertical electron-detachment energies and the vibrational spectra of the neutral, anionic and cationic boron clusters B_{19}, B_{19}^- and B_{19}^+. They carried out the local density approximation (LDA) and Ceperley-Alder exchange-correlation functional as parameterized by Perdew and Zunger. They also used norm-conserving pseudopotentials for the interactions between the ions and the valence electrons, but however with separable nonlocal operators. The electron wave functions were expand by atomic orbitals with a double-ζ polarization and with an energy cutoff of 100 Ry for the real-space mesh. In order to determine the cutoff radii of the atomic orbitals, the authors used a low value of around 3 mRy as a confinement energy shift. Using the conjugate gradient method, the authors performed geometry optimization for a large number of plausible candidate structures without any symmetry constraints. Furthermore, they used

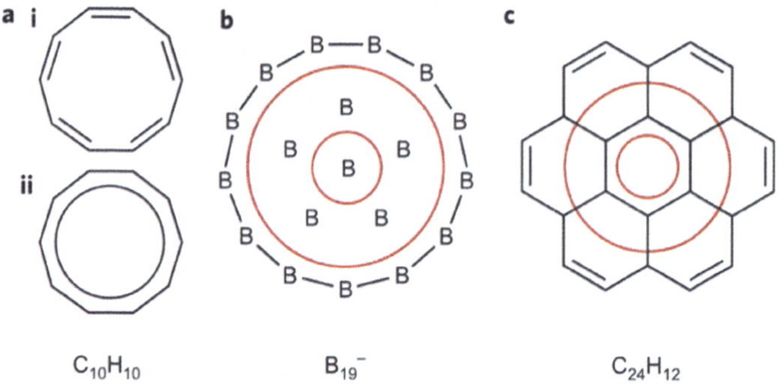

Fig. 4.72 Analogy of the chemical bonds and structures between the anionic B_{19}^- cluster and the hydrocarbon molecules ($C_{10}H_{10}$) and ($C_{24}H_{12}$). The localized C-C and B-B σ-bonds are represented by single lines, while the delocalized π-bonds are represented as circles. **a** The classic annulene ($C_{10}H_{10}$) molecule exists in the non-planar form (i) is non-aromatic. The model planar ($C_{10}H_{10}$) molecule has a cyclic form (ii) and exhibits delocalized π-bonds and is aromatic. **b** The anion B_{19}^- is composed of two cyclic π-electrons. The first one is delocalized over the central pentagon B_6, while the second π-system is delocalized over the peripheral B_{13} ring. Both delocalized π-systems are aromatic. **c** As by B_{19}^- the coronene ($C_{24}H_{12}$) molecule is consisting two carbon rings. The first one is the inner ring exhibiting a delocalized π-system with an aromatic character. The second one is the outer ring exhibiting also a delocalized π-system with an aromatic character. The outer π-system is responsible for the bonding between the both C_6 and C_{18} rings. This figure was reproduced with kind permission of the Nature Publishing Group [113]

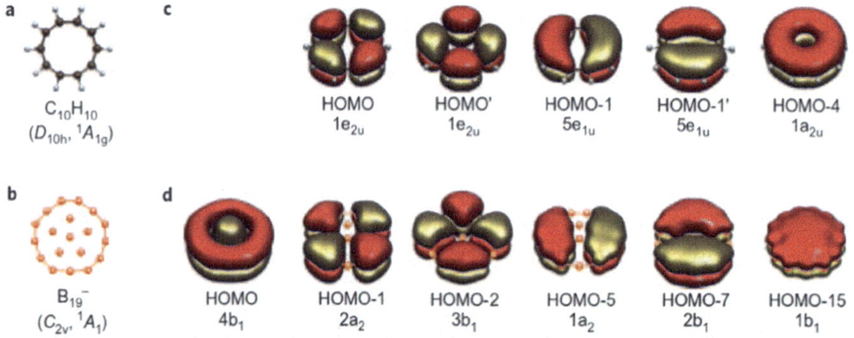

Fig. 4.73 Comparison between the molecular orbitals of the global minimum of B_{19}^- b-d, and of the hydrocarbon molecule annulene ($C_{10}H_{10}$) a-b. The similarity of the orbital characters is obvious [113]

the B3LYP functional as well as the 6-311+G* and 6-311+G (2df) basis sets, applied previously to metallic boron structures, in order to compare the results with those obtained at the empirical hybrid exchange-correlation functionals for the 19-atomic systems [114].

Relative Energies of B_{19} and $B_{19}^{\pm}$ Isomers: Boustani et al. [114] determined the neutral and cationic structures of the B_{19} and B_{19}^+ clusters by reoptimization the

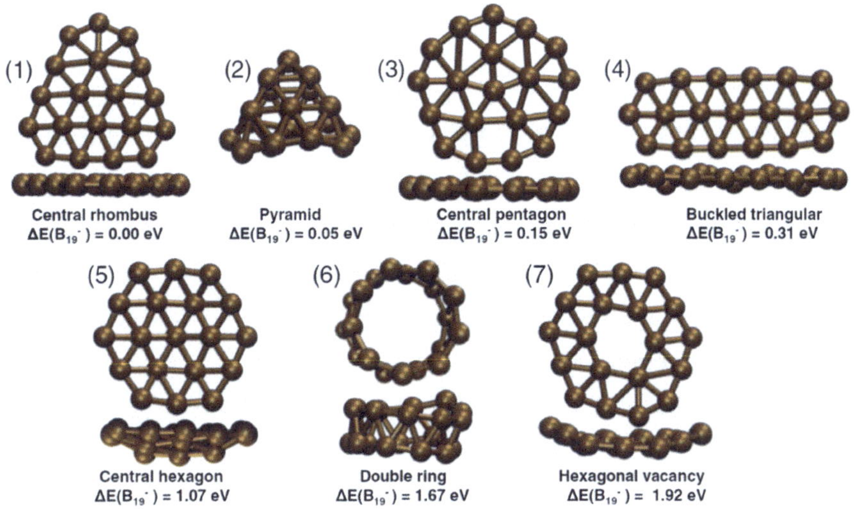

Fig. 4.74 Equilibrium geometry and total energy differences, corrected for zero-point energies, between different structural isomers of B_{19}^- anion clusters. For selected structures, the varying degree of planarity is depicted in top and side view [114]

anionic structures of B_{19}^- in Fig. 4.74. Then they compared the obtained energies of the three sets: neutral, anionic and cationic clusters, relative to the most stable isomer of each set, assigned by 0.00 eV. The calculated relative energies are given in Table 4.1. This table shows that the most stable neutral B_{19} and cationic B_{19}^+ clusters have a 3D structure, assigned by (2) Pyramid, while the most stable anionic B_{19}^- cluster exhibits a structure with a central rhombus, assigned by (1) in the Table 4.1 and in the Fig. 4.74. Thus, one can conclude that adding or removing an electron to a small boron cluster causes a structural transformation. The electronic closed shell of anionic B_{19}^- cluster has 2D structure, while the electronic open shells of B_{19}^+ and B_{19} clusters have 3D structures. This fact was verified by the anionic B_{20}^- cluster with 2D planar structure, while the neutral B_{20} has 3D double ring structure, as we will see in the next section.

The Photoelectron Spectra of B_{20}^-: Kiran et al. [115] pursued their experimental and theoretical studies to investigate the neutral boron cluster B_{20} and the negatively charged boron clusters B_{20}^-, in the same manner of investigations done on previous clusters B_3^- to B_{19}^-. Theoretically, *ab intio* density functional methods, coupled with basin-hopping algorithm were applied to search for the global minimum. Moreover, the sampling method of Monte Carlo (MC) was used in combination with the potential energy transformation. In addition, an optimization procedure was carried after each accepted Monta Cralo move. Also, gradient corrected functionals within the density functional theory including plane-wave pseudopotentials were applied to carry out the optimization procedure. Experimentally, on the other side, an apparatus composed of laser vaporization supersonic cluster source, a magnetic-bottle time-of-

Table 4.1 Relative energies of B_{19} neutral, B_{19}^- anion, and B_{19}^+ cation clusters. Total energy differences contain zero-point motion corrections. The labeling of the structural isomers follows Fig. 4.74 [114]

Isomer identification	$\Delta E(B_{19}^-)$ (eV)	$\Delta E(B_{19})$ (eV)	$\Delta E(B_{19}^+)$ (eV)
(1) Central rhombus	0.00	0.21	0.88
(2) Pyramid	0.05	0.00	0.00
(3) Central pentagon	0.15	0.76	1.99
(4) Buckled triangular	0.31	0.62	1.54
(5) Central hexagon	1.07	0.61	0.89
(6) Double ring	1.67	0.85	0.95
(7) Hexagonal vacancy	1.92	1.80	2.47

flight measurement technique, as well as photoelectron spectroscopy were deployed to generate and identify the boron clusters. These B_n, generated by laser vaporization of enriched ^{10}B isotope of around (99.75%), were negatively charged by crossing with an electron beam. The produced anionic clusters B_n^- were in turn analyzed with the time-of-flight mass spectrometer. Before an irradiation by a photodetachment laser beam, the anionic B_{20}^- clusters were mass-selected and decelerated. By the magnetic bottle and analyzed in a 3.5-m-long electron flight tube, photoelectrons were collected at nearly 100% efficiency. Figure 4.75 presents the PES spectra of B_{20}^- at two photon energies. The first numerous observed electronic transitions (X and A-H) are well resolved. The peak maximum of the X band being measured by 3.11 eV, which corresponds to the vertical detachment energy. At the well defined sharp onset of the band X, the electron affinity of the neutral B_{20} cluster is evaluated to be 3.02 eV. The weak intensity of feature A in Fig. 4.75 is implying an indication of one B_{20}^- isomer, which seems to be weakly populated. In addition, between the features X and B a large energy separation estimated by $\approx$1.3 eV was observed. This separation in the spectral pattern can be connected with the energy difference between the highest occupied molecular orbital (HOMO) and the lowest unoccupied molecular orbital (LUMO) of the neutral B_{20} cluster indicating that this closed-shell molecule exhibits a large HOMO-LUMO gap and meaning high stability. Now arises the question: (concept) behind the large HOMO-LUMO energy gap?

The Planar B_{20}^- and Tubular B_{20}: As already mentioned above, the ab initio density functional theory (DFT) technique coupled with the basin-hopping global optimization method were used to search for the global minima of the B_{20}^- and B_{20} clusters [115]. Furthermore, the authors identified more than 200 low energy minima, considering an energy difference of about $\leq$0.1 Hartree from the global minimum. They

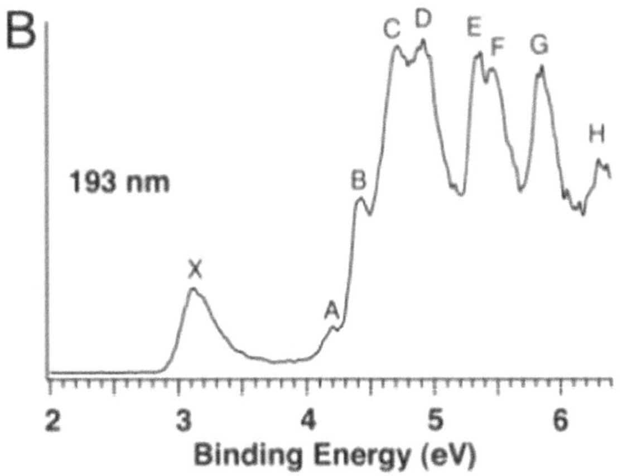

Fig. 4.75 Photoelectron Spectra of B_{20}^- at 193 nm (6.424 eV) [115]

also used all-electron DFT approximations to perform optimization and vibration frequency calculations for the top 10 lowest energy isomers of the B_{20}^- and B_{20} clusters. The structures of the four low-lying isomers of the B_{20}^- and B_{20} clusters are presented in Fig. 4.76. This figure shows that the most stable B_{20}^- and B_{20} clusters exhibit a double-ring tubular structure with a diameter of 5.2 Å. In addition, Kiran et al. [115] calculated the VDEs and EA (Table 4.2) in order to confirm the computational results and facilitate comparison with the experimental data. From their density functional theory computations they also simulated the PES spectra of the double-ring, elongated and bowl structures, and presented the PES of the bowl in Fig. 4.77. The vertical detachment energy can be determined by the energy difference between the ground state energies of the anion and neutral boron clusters at the geometry of the anion. The electron affinity (AE) can be achieved after a geometry relaxation of the neutral cluster back to its equilibrium. As can be seen in the Table 4.2, theoretical VDE and EA values (2.32, and 2.17 eV) of the 3D double-ring isomer are not in a good agreement with the experimental values. In contrast, the computed values of the 2D isomers correspond favorably with the experimental data. For example, the calculated VDE and EA values (3.15, and 3.03 eV) of the 2D elongated structure (2) and (2.79, and 2.88 eV) of the 2D bowl structure (3) agree well with the experimental data (3.11, and 3.02 eV), respectively. Therefore, a closer comparison between the theoretical and experimental values as given in Table 4.2, as well as in Figs. 4.75 and 4.77 shows obviously that the simulated spectrum of the third isomer with the bowl structure exhibits one-to-one correspondence to the experimental spectrum.

The Aromatic Tubular B_{20}: The tubular structure of the B_{20} boron cluster can be considered as dovetailed of two ten atomic rings. On the one side, the strong sp^2 hybridized σ bonds within the wall provide the structure high stability, while the delocalized π bonds, on the other side, enhance the stability by covering the inner and outer surfaces of the wall. Therefore is the tubular structure highly aromatic. These delocalized π bonds can be seen in Fig. 4.78, as Kohn-Sham orbitals, determined by Botti et al. [116]. The latter authors investigated the B_{20} family using time-dependent

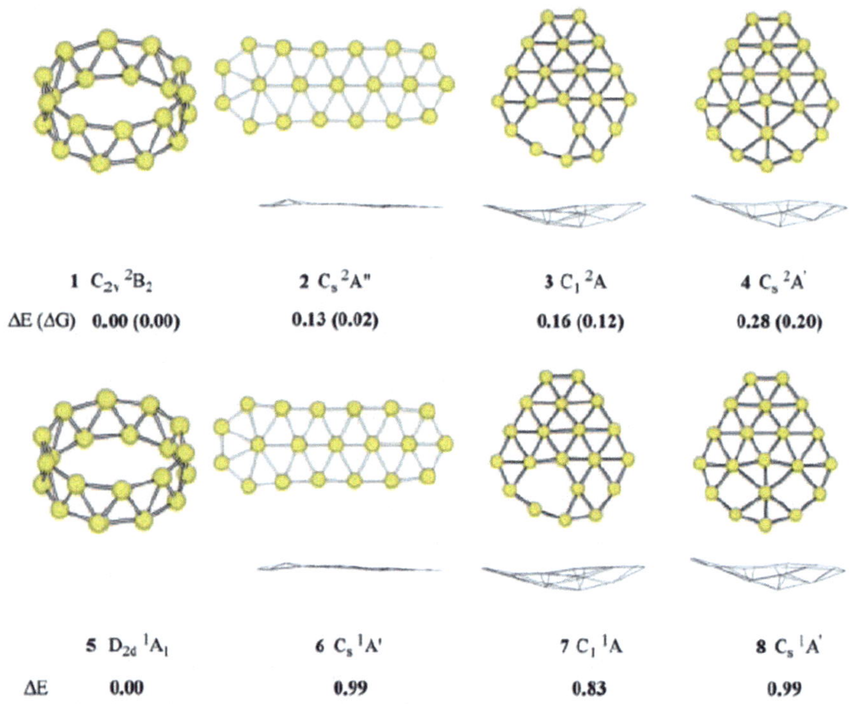

Fig. 4.76 Low-lying structures of B_{20}^- (**1** – **4**), and B_{20} (**5** – **8**) along with their relative energies (in eV) at B3LYP/6-311+G* level [115]

Table 4.2 Observed VDEs of B_{20}^- in comparison with calculated VDEs/EAs, obtained at the B3LYP/6-311+G* level for structures (1,2,3) in Fig. 4.76 [115]

Observed feature	VDE (experimental), eV	VDE (theoretical), eV		
		Structure 1	Structure 2	Structure 3
X†	3.11(2)/3.02(2)	2.32/2.17	3.15/3.03	2.97/2.88
A‡	4.20(3)	3.49/*3.52*	4.04/*4.14*	
B	4.40(3)	4.52/*4.81*	4.49/*4.54*	4.22/*4.37*
C	4.71(4)		4.66/*4.94*	4.59/*4.71*
D	4.92(4)		5.03/*5.17*	4.70/*4.87*
E	5.35(4)			5.21/*5.27*
F	5.46(4)			5.23/*5.49*
G	5.84(3)	5.65/*5.70*	5.79/*5.87*	5.66/*5.75*
H	6.33(3)	6.35/*6.42*	6.12/*6.33*	6.17/*6.21*

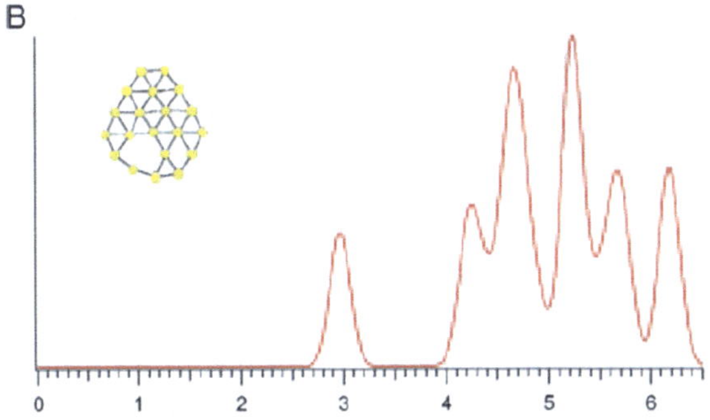

Fig. 4.77 Simulated PES for the bowl low-lying isomer B_{20}^-. The spectra were constructed by fitting the distribution of the calculated VDEs with unit-area Gaussian functions of 0.2 eV full width at half maximum [115]

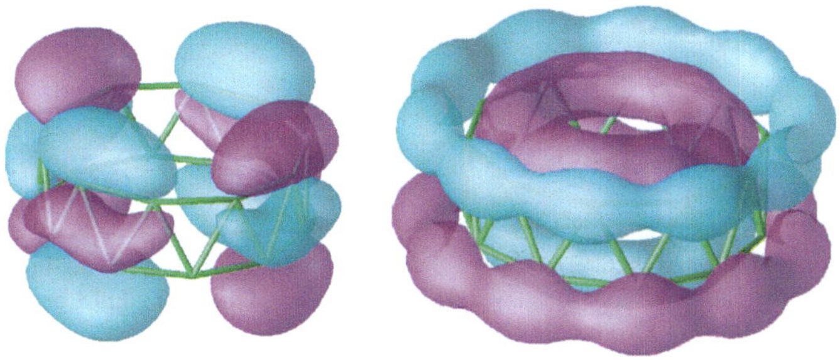

Fig. 4.78 The predominant π character of Kohn-Sham orbitals of the tubular double ring B_{20} [116]

density-functional theory within a real-space, real-time scheme studying the lowest-energy isomers of the B_{20} clusters and their optical response. They found that the simulated spectra of the B_{20} of the isomers are very sensitive to their shapes.

Theoretical Confirmation of Planar B_{20}^- **and Tubular** B_{20}: Wei An et al. [117] employed high-level ab initio molecular orbital methods to study the geometrical and electronic structures of the neutral and anionic B_{20} boron clusters and their relative stability. They investigated different low-lying planar isomers versus the double-ring tubular isomer (see Fig. 4.79) using two accurate methods. The first one is the coupled-cluster method [CCSD0T] including single, double, and non-iteratively perturbative triple excitation and the 6-311G(d) basis set. The second method is the fourth order Möller-Plesset perturbation theory [MP4(SDQ)] and Dunning's correlation consistent polarized valence triple zeta basis set. The calculations of both methods MP40SDQ and CCSD0T produce similar energy sequence for the neutral clusters, as was found by Kiran et al. [115]. Both computations show that the most stable isomer is the tubular cluster with double-ring structure 1. For the anionic clusters B_{20}^-, the MP40SDQ calculations show that the planar structure 4 presented in Fig. 4.79 exhibits the lowest-energy, while the double-ring isomer pos-

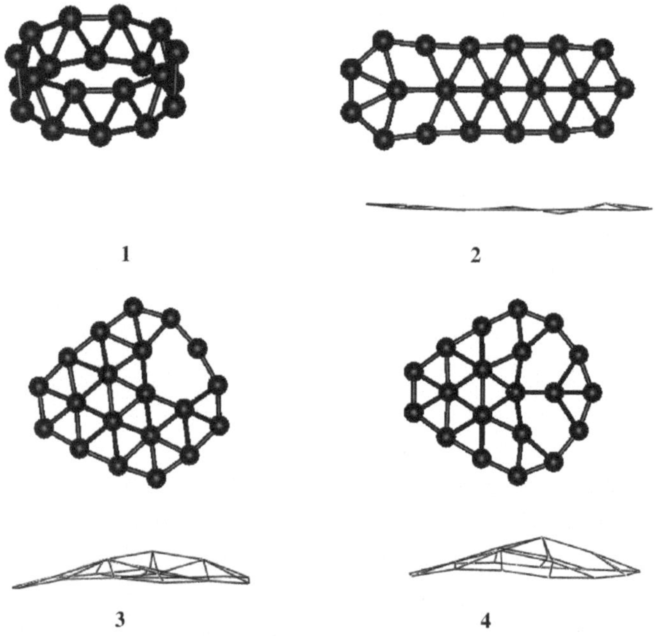

Fig. 4.79 The double-ring tubular 1 and three planar low-lying isomers 2, 3, and 4 of the neutral B_{20} clusters. This figure was reproduced with kind permission of the Journal of Chemical Physics of the American Institute of Physics [117]

sesses the highest-energy structure between the anionic clusters. The authors found that the tubular B_{20} is strongly aromatic due to the large negative NICS value of (40) calculated at the DFT level of theory. Therefore, they confirmed that the structural transition from planar to tubular can start at B_{20} for neutral clusters, but for the anionic clusters it should occur beyond the size of B_{20}^{-}.

The Planarity Confirmation of Neutral B_n ($14 \leq n \leq 20$): Tai et al. [118] reexamined the structures of the neutral boron clusters B_n with $n = 14$–20, using the stochastic search method along with high accuracy quantum chemical calculations. The geometries and harmonic vibrational frequencies of the lower-lying isomers are determined using density functional theory (DFT) at the popular hybrid density functional B3LYP and the 6-311+G* basis set. The low-lying isomers and corresponding global minima of the B_n clusters for $n = 14$–18, are shown in Fig. 4.80, while those of the B_n clusters for $n = 19$–20 are presented in Fig. 4.81. However, the structure 14.1 seems to possess the global minimum with an energy gap of 3.6 kcal/mol below 14.2, as can be seen in Fig. 4.80. The structures 15.1 (C_1, ^2A) and 15.2 (C_s, $^2A'$) are the two lowest-lying isomers of B_{15} with similar energy content at the B3LYP/6-311 + G(d) level. At G3B3 approach, 15.1 is confirmed as the global minimum being 4.5 kcal/mol more stable than 15.2. The structure 16.1 (C_{2h}, ^1A_g) is the most stable isomer being at least 7.5 kcal/mol lower in energy than the remaining isomers. The

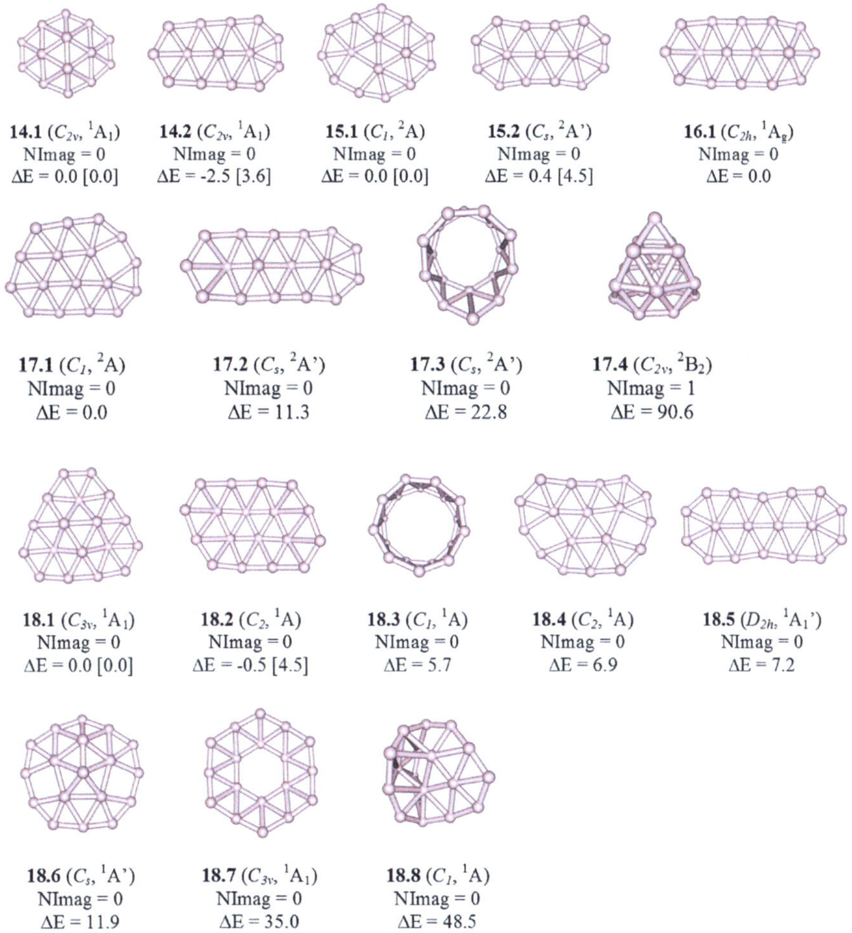

Fig. 4.80 Structures, relative energies (kcal/mol), number of imaginary frequencies and symmetry point groups of lower-lying structures of B_n ($n = 14$–18), using B3LYP/6- 311 + G(d) and G3B3 (values in square bracket) methods. This figure was reproduced from the Chemical Physics Letters with kind permission of the Elsevier Ltd [118]

quasi-planar structure 17.1 (C_1, ^2A), in which a five-membered shape, surrounded by a 12-membered ring, is the most stable isomer for the B_{17} cluster. The following isomer 17.2 (C_s, $^2A'$) is lying at 11.3 kcal/mol higher in energy. The same figure shows that both structures 18.1 (C_{2v}, $^1A_1$) and 18.2 (C_2, ^1A) are the lowest-lying isomers with comparable energies. In each of these, a six-membered ring is located inside a 12-membered ring. The both structures 19.1 (C_s, $^2A'$) and 19.2 (C_s, $^2A'$) in Fig. 4.81 were found to be almost degenerate in energy, and constitute the lowest-lying isomers for B_{19}, in opposite to that 3D structure found by Boustani, as shown in Fig. 4.74. However, the calculations pointed out that that 19.1 is the global minimum with a

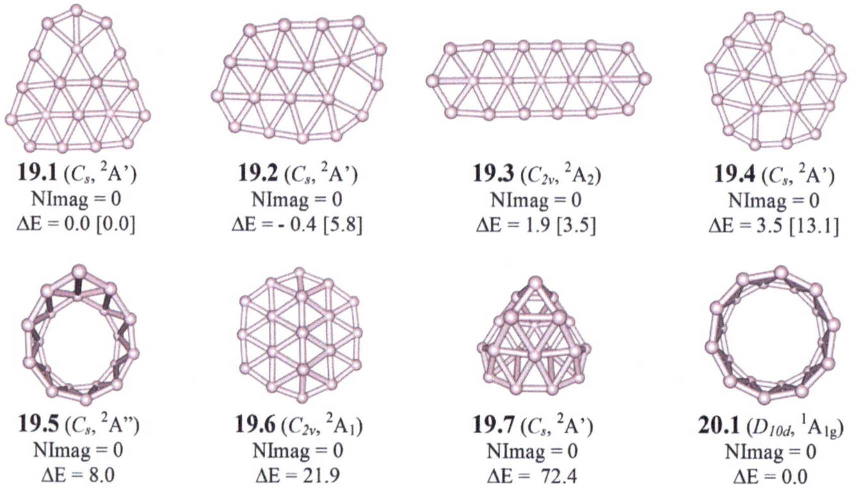

Fig. 4.81 Structures, relative energies (kcal/mol), number of imaginary frequencies and symmetry point groups of lower-lying structures of B_n (n = 19–20), using B3LYP/6- 311 + G(d) and G3B3 (values in square bracket) methods. This figure was reproduced from the Chemical Physics Letters with kind permission of the Elsevier Ltd [118]

relative energy of 5.8 kcal/mol with respect to 19.2. The next isomer for this size is the elongated form 19.3 (C_{2v}, $^2A_2$, being only 1.9 kcal/mol less stable than structure 19.1. Finally, the tubular form 20.1 (D_{10d}, $^1A_1$), shown in Figure 4.81, is the global minimum for B_{20}. Other isomers are much less stable, being at least 18 kcal/mol higher in energy than the global minimum. Tai et al. [118] confirmed that the neutral boron clusters up to a cluster size of $n = 19$ possess planar or the quasi-planar structures and confirmed that the transition from 2D to 3D structure occurs by the size of B_{20}. Nevertheless, Kiran et al. [115], found that B_{20}^- still exhibits planar structure.

The Photoelectron spectrum of B_{21}^-: Piazza et al. [119] extended further investigations on boron clusters and studied experimentally and theoretically the isomers of B_{21}^-. Theoretically, the authors used ab initio quantum chemical RCCSD(T) methods and density functional B3LYP as well as the 6-311+G* basis set. Experimentally, they applied photoelectron spectroscopy (PES) and shown in Fig. 4.82 the photoelectron spectrum of B_{21}^-, measured at the 193 nm. The spectrum exhibits at around 4.6 eV a broad detachment band X, while the following bands, of an almost continuous spectral region, are laying in the binding energy range from 5 to 6 eV and exhibiting discernible fine features. As can be seen in Fig. 4.82, the broad and high intense band at 4.6 eV is composed of more than one detachment channel. The authors measured a VDE value of 4.58 eV for the X band at the leading edge of the peak to the X band and a shoulder on the higher energy side to the A band. The authors obtained 4.38 ± 0.05 eV for the adiabatic detachment energy (ADE), which in turn depicts the electron affinity (EA) of the neutral cluster B_{21}. Furthermore, they estimated a VDE value of 4.79 eV for the A band.

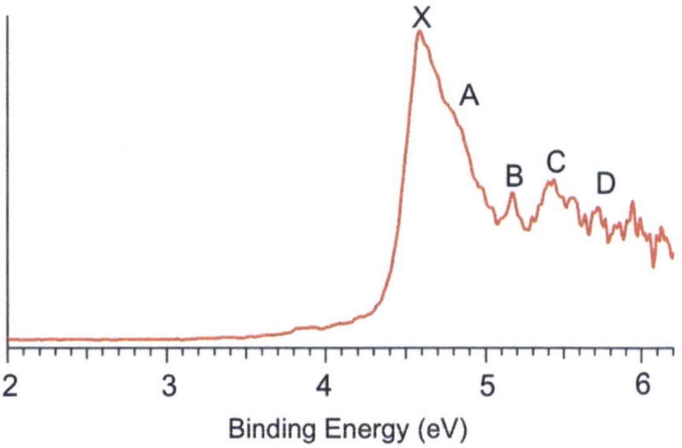

Fig. 4.82 PES spectra of B_{21}^-, observed at 193 nm (6.424 eV). This figure was reproduced with kind permission of the Journal of Chemical Physics of the American Institute of Physics [119]

The Global Minimum Structure of B_{21}^-: Piazza et al. [119] used two methods to search extensively for the global minimum structure of B_{21}^-. The first one is called the Coalescence Kick method and the second one is the recently developed Cartesian Walking method. The authors obtained seven low lying isomers for the B_{21}^- clusters and all of these clusters are exhibiting planar or quasi-planar structures, as shown in Fig. 4.83. At the B3LYP/6-311+G* level of theory, the five low lying isomers of the upper row structures in Fig. 4.83 differ from each other by 6.7 kcal/mol only. However, the choice of the exchange-correlation potential seems to affect the order of these isomers. Thus, at the highest level of theory, CCSD(T)/6-311+G* (using geometries optimized at the B3LYP/6-311+G*) the lowest five structures were found to be within 4.2 kcal/mol, particularly being isomer II above the global minimum isomer I by just 1.9 kcal/mol. The results obtained at the MP2/6-311+G* level differ significantly from those obtained by the most accurate CCSD(T)/6-311+G* method. The MP2/6-311+G* results agree in fact more closely with those results obtained at the DFT (B3LYP/6-311+G*) level of theory, as the former observation of boron clusters with smaller sizes. However, the isomers I and II in Fig. 4.83 were found nearly isoenergetic. As already shown in Sect. 4.4, the two low-lying isomers of B_{19}^- are very similar to the two corresponding low-lying isomers of B_{21}^-. Thus, the B_{19}^- cluster exhibits a perfect planar structure at its the global minimum and belongs to doubly π aromatic system. This structure is composed of centered pentagon B_5 and surrounded by 13-atom outer ring. By adding two B atoms, one to an apex of the interior central pentagonal ring, and one to the 13-atom outer ring of the global minimum of B_{19}^-, one obtains the structure of the B_{21}^- global minimum. In similar way, by adding one B atom to the interior and one B atom to the outer ring of the second lowest-lying planar structure of B_{19}^-, one receives the global minimum structure of the second low-lying isomer B_{21}^-.

Chemical Bonding of B_{21}^-: As expected, Piazza et al. [119] reported that the peripheral 14 atomic ring of the B_{21}^- isomer I are bonded together by localized B-B σ-bonds, in a similar manner to the bonds found by previous investigations of smaller planar or quasi-planar boron clusters. As shown in Fig. 4.84, there is a 2c-2e B-B σ-bond found between the apex of the interior central pentagonal ring and the direct neighboring

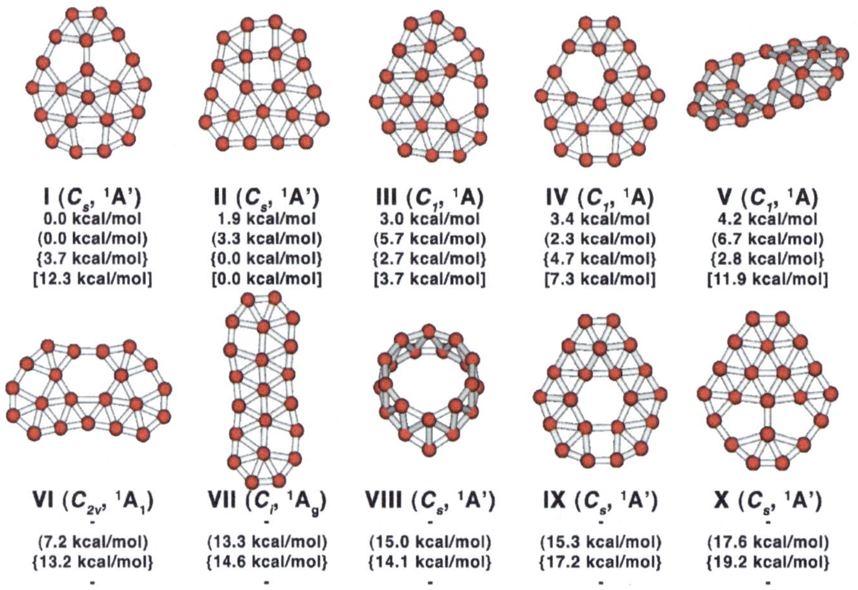

Fig. 4.83 The low-lying isomers of B_{21}^-. Relative energies are given at the CCSD(T)/6-311+G*//B3LYP/6-311+G* level (I-V), at the B3LYP/6-311+G* level (in parenthesis), at the PBE1PBE/6-311+G* level (in curly brackets), and at MP2/6-311+G* level (in square brackets for I-V). All values are corrected for zero-point vibrational energies. Note that lines are drawn for B-B distances of $\leq$2.0 Å to help the visualization of the structures and they do not necessarily represent single B-B bonds. This figure was reproduced with kind permission of the Journal of Chemical Physics of the American Institute of Physics [119]

atom, exhibiting an internal localized σ-bond in the B_{21}^- cluster.Furthermore, besides the 14 peripheral and the single internal 2c-2e B-B σ-bonds, there are further nine 3c-2e and two 4c-2e localized σ-bonds, and four 4c-2e, one 5c-2e, and one 6c-2e delocalized π-bonds.

Boron clusters B_{22}^- **and** B_{23}^-: By a joint study of ab initio quantum chemical methods and photoelectron spectroscopy, Sergeeva et al. [120] continued their investigations on larger boron cluster anions like B_{22}^-, B_{23}^-. They used two different global minimum search methods, the coalescence kick (CK) and the Cartesian walking (CW). The CW is designed for more efficient searches for planar structures. As presented in Fig. 4.85, there are two isomers of the B_{22}^- cluster exhibiting nearly degenerate energies. The first isomer I possesses a double-ring structure with tubular form, while the second isomer II has a quasi-planar structure. However, only the quasi-planar isomer was observed experimentally. This slightly buckled triangular structure of isomer II, composed of 8 interior and 14 peripheral boron atoms, is considered to be the global minimum of B_{22}^-.

Photoelectron spectrum of B_{22}^-: Figure 4.86 shows the band structures of the photoelectron spectrum of the anionic B_{22}^- cluster, spanning the binding energy range from

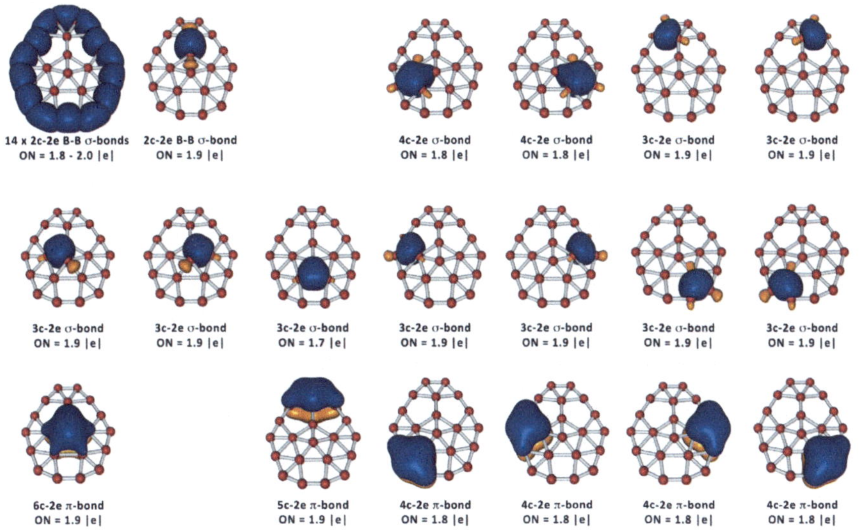

Fig. 4.84 AdNDP analysis for isomer I (C_s, $^1A'$) of the anion B_{21}^-. ON stands for occupation number and is equal to 2.0 |e| in an ideal case. This figure was reproduced with kind permission of the Journal of Chemical Physics of the American Institute of Physics [119]

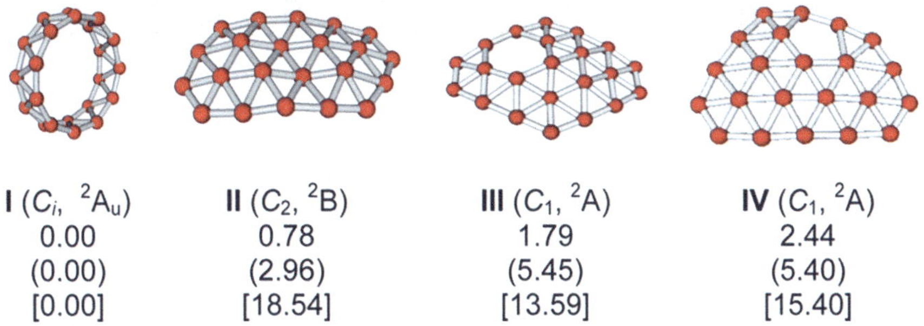

Fig. 4.85 Low-lying isomers of B_{22}^-. Relative energies are given in kcal/mol at the CCSD(T)/6-311+G*//PBE1PBE/6-311+G*, PBE1PBE/6-311+G* (in parentheses), and B3LYP/6-311+G* (in square brackets) levels of theory. This figure was reproduced with kind permission of the Journal of the American Chemical Society [120]

3.3 eV to beyond 6 eV, assigned by quite congested with photodetachment features. The measured adiabatic detachment energy (ADE) is about 3.34 eV and the vertical detachment energy (VDE) is about 3.48 eV for the ground-state band X. From the comparison between theory and experiment one can conclude that the structures of negatively charged boron clusters can maintain their planar or quasi-planar pattern even up to 23 atoms. Since the experimental spectrum of the tubular structure does not agree with the calculated VDE value 3.1 eV, therefore the tubular structure of the B_{22}^- cluster can be ruled out being as a contributor to the observed spectrum. The

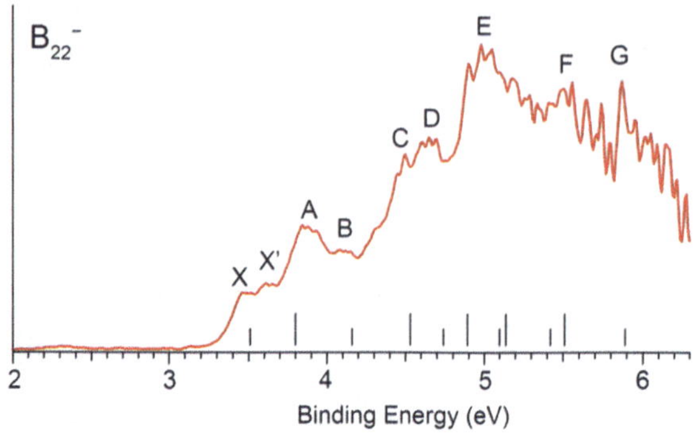

Fig. 4.86 Photoelectron spectrum of B_{22}^- at 193 nm. Vertical bars represent calculated VDEs at the TD-PBE0/6-311+G(2df)//PBE0/6-311+G* level of theory. The short and long bars represent singlet and triplet final states, respectively. This figure was reproduced with kind permission of the Journal of the American Chemical Society [120]

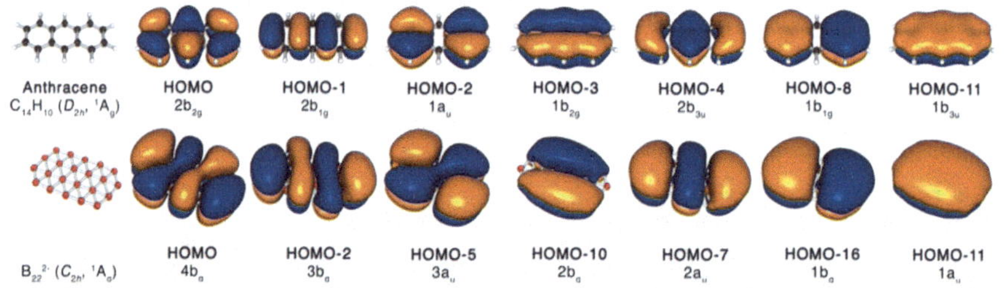

Fig. 4.87 Comparison of the π molecular orbitals in the flattened B_{22}^{2-} cluster with those in anthracene. This figure was reproduced with kind permission of the Journal of the American Chemical Society [120]

VDE values 3.51 and 3.51 eV, calculated for the quasi-planar C_2 structure of B_{22}^- and obtained at the TD-B3LYP and TD-PBE0 level of theory, agree very well with 3.48 eV, the measured value at at the X band.

Chemical Bonding of B_{22}^-: The structure of the B_{22}^- isomer II with the C_2 symmetry in Fig. 4.85 is not planar and therefore the separation of the electron density into σ and π can be carried out only upon artificially flattening the structure to prevent mixing of σ and π orbitals. That can be occurred when the out-of-plane distortion of the isomer II structure is relatively small and a perfectly planar system is used for the chemical bonding analysis in the C_2 B_{22}^- species. The analysis can be done on the basis of the canonical molecular orbitals (CMOs). Consequently, the closed-shell planar structure of the B_{22}^{2-} species with C_{2h} symmetry enables the analysis. The authors found that the molecular orbitals of the B_{22}^{2-} cluster are corresponding to those of the hydrocarbon molecule anthracene ($C_{14}H_{10}$), as presented in Fig. 4.87. Therefore, the quasi-planar B_{22}^- cluster can be considered as an all-boron counterpart to the hydrocarbon molecule cation of anthracene ($C_{14}H_{10}$).

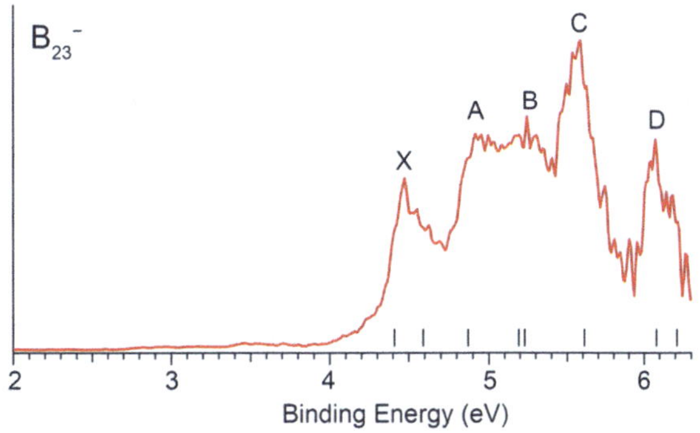

Fig. 4.88 Photoelectron spectrum of B_{23}^- at 193 nm. Vertical bars represent calculated VDEs at TD-PBE0/6-311+G(2df)//PBE0/6-311+G*. This figure was reproduced with kind permission of the Journal of the American Chemical Society [120]

Photoelectron spectrum of B_{23}^-: In comparison to B_{22}^-, B_{23}^- has a relatively simpler spectrum (Fig. 4.88) and a higher binding energy. The first photodetachment band, X, is well resolved and seems to contain fine features due to either possible vibrational structures or overlapping detachment channels. The X band yields a VDE of 4.46 eV and an ADE of 4.32 eV. The next two features, A (VDE: 4.95 eV) and B (5.22 eV), are broad and not well resolved, indicating a possible overlap of additional detachment bands. At the high binding energy side of the spectrum, a strong and well defined resolved band C with a VDE of 5.56 eV were observed, followed by a well-resolved band D with a VDE of 6.06 eV. The calculated TD-PBE0 VDE value 4.41 eV for the global minimum of B_{23}^- (Fig. 4.89, structure V), is in excellent agreement with the observed value 4.46 eV.

The Global Minimum of B_{23}^-: For the search of the global minima of the B_{23}^- clusters, Sergeeva et al. [120] used the same methods as in the case of the B_{17}^-, B_{18}^-, and B_{22}^- clusters. They are the coalescence kick (CK) and Cartesian walking (CW) methods. The The global minimum of B_{23}^- is found to be perfectly planar structure (see Fig. 4.89). The B_{23}^- cluster exhibits a heart-like form with a C_{2v} symmetry and is composed of 15 peripheral atoms and 8 interior atoms as well as of a pentagonal hole. The calculations confirmed that there is no any close-lying isomer, which is in competition with the B_{23}^- cluster. This fact implies that the "nanoheart" B_{23}^- cluster is highly stable and its delocalized π-bonds are also similar to those of the hydrocarbon molecule phenanthrene. Thus, the B_{23}^- is the largest boron cluster anions characterized heretofore. The high stability of the nanoheart B_{23}^- cluster can be related to the double aromaticity in its perfectly planar structure. The analogy between the aromatic hydrocarbons and the aromatic B_{22}^- and B_{23}^- boron clusters by expanding the family of all-boron motivates the researchers to explore new structures and properties. More details about the chemical bonding of B_{23}^- are given in [120].

Photoelectron Spectra of B_{24}^-: Popov et al. [121] continued their investigations and studied experimentally and theoretically the neutral and anionic boron clusters B_{24}^- using a joint photoelectron spectroscopy (PES) and computational methods. Experimentally, the anions were generated by a laser vaporization technique, then been carried by He and Ar gas undergoing a supersonic expansion. The produced

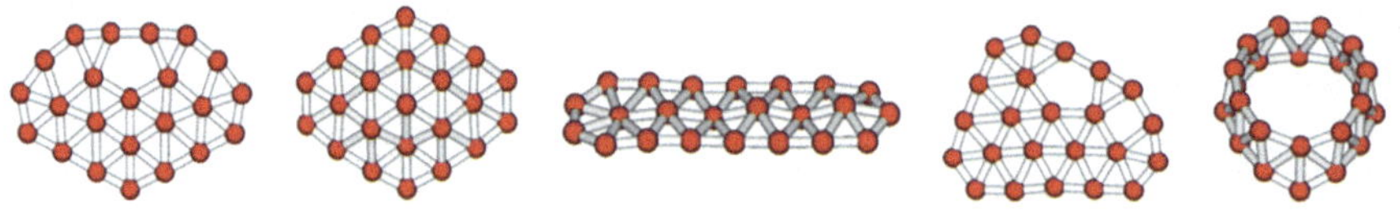

Fig. 4.89 Global minimum and other higher energy isomers of B_{23}^-. Relative energies are given in kcal/mol at the PBE1PBE/6-311+G* and B3LYP/6-311+G* (in parentheses) levels of theory. This figure was reproduced with kind permission of the Journal of the American Chemical Society [120]

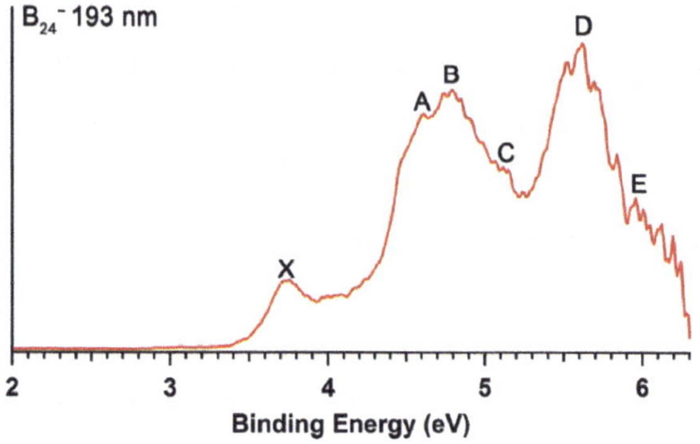

Fig. 4.90 Photoelectron spectrum of $B^-{}_{24}$ at 193 nm (6.424 eV). This figure was reproduced with kind permission of the Journal of Chemical Physics of the American Institute of Physics [121]

cluster beam was analyzed and mass-selected by a time-of-flight mass spectrometer. The selected B_{24}^- cluster was intercepted by a detachment laser beam. Two laser detachment energies were used in order to study the B_{24}^- cluster. The first energy is at 193 nm (6.424 eV) created from an ArF excimer laser. The second energy is at 266 nm (4.661 eV) created from a Nd:YAG laser. Because no fine features for the 266 nm spectrum were resolved, only the PES of B_{24}^- for the 193 nm spectrum is considered and presented in Fig. 4.90.

Global Minimum of B_{24}^-: In similar way, Popov et al. [121] used the Cartesian Walking (CW) and Coalescence Kick (CK) methods (see [18] and [19] in [121]) in order to search for the global minimum structure of the B_{24}^- cluster. For the energy and gradient calculations, the authors applied the hybrid density functional B3LYP and the 3-21G basis set for the CK search. Moreover, the same basis set 3-21G but the PBE0 functional were used for the initial CW search. Furthermore, the computations were carried out using two different density functionals and corresponding basis sets; the TPSSh/6-311+G(2df)//TPSSh/6-311+G(d) and the PBE0/6-311+G(2df)//PBE0/6-311+G(d), to determine single point calculations for all structures. Thus, additional calculations at a higher level of theory were undertaken; the ROCCSD(T)/6-311+G(d)//PBE0/6-311+G(d) as well as ROCCSD/6-311+G(d)// PBE0/6-311+G(d), in order to obtain most accurate single point ground state energies. For the purpose of achieving a proper $\langle S^2 \rangle$ value for the open-shell cluster B_{24}^-,

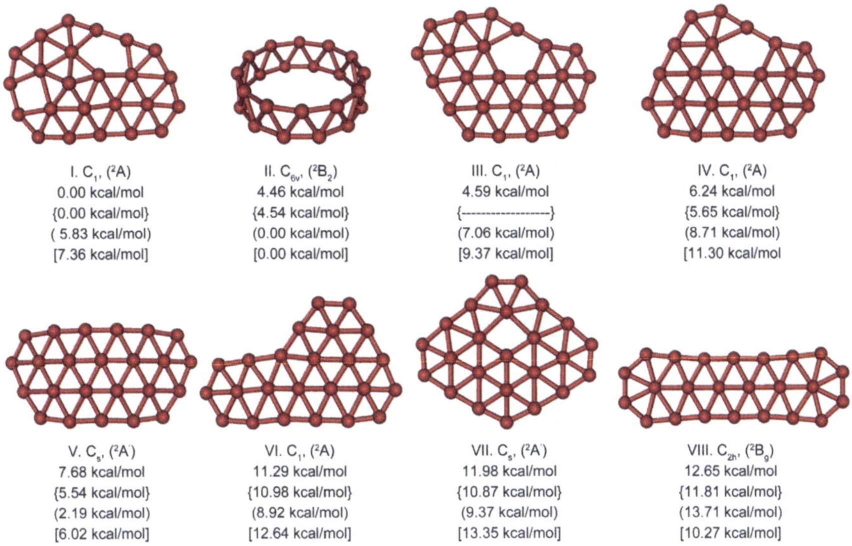

Fig. 4.91 The global minimum and low-lying isomers of B_{24}^-. Relative energies are shown from single-point calculations at ROCCSD/6-311 + G(d)//PBE0/6- 311 + G(d) and ROCCSD(T)/6-311 + G(d)//PBE0/6-311 + G(d) (in curly brackets), as well those from PBE0/6-311 + G(2df)//PBE0/6-311 + G(d) (in parentheses) and TPSSh/6-311 + G(2df)//TPSSh/6-311 + G(d) (in square brackets). This figure was reproduced with kind permission of the Journal of Chemical Physics of the American Institute of Physics [121]

a formalism for restricted open-shell system was used. The global minimum of B_{24}^-, obtained at the ROCCSD and ROCCSD(T) level of theory, was found to exhibit quasi-planar structure I without symmetry, as already shown in Fig. 4.91. The next low-lying isomer has a tubular form with a double-ring structure and C_{6v} symmetry, but it has never been observed experimentally. The central atom of the global minimum structure I (C_1, ^{2}A) is sticking out of the molecular plane by approximately 0.9 Åand hence the geometry is slightly puckered and composed of filled and empty pentagonal moieties. For comparison, the same filled pentagonal moiety is found in B_{21}^- (Fig. 4.83) with an out of plane distortion of about 0.5 Å, whereas in B_{19}^- (Fig. 4.70) the filled pentagonal unit is found to be perfectly planar.

Comparison VDE Values of B_{24}^-: By means of the PBE0 and TPSSh density functional methods, photoelectron spectra of the B_{24}^- isomer I in Fig. 4.91 were simulated and compared with the measured spectrum as presented in Fig. 4.92. The band structure of the experimental data as well as the patterns at both functionals are in excellent agreement. Because the anionic cluster B_{24}^- is an open-shell system, the electron detachment of the singly occupied HOMO (37a) leads to the singlet ground state ^{1}A of the neutral cluster. Therefore, this electron removal corresponds to the first VDE, which is about 3.79 eV at the PBE0 level of theory. This PBE0 value differs very slightly from the experimental value 3.75 eV. Both VDE values are in

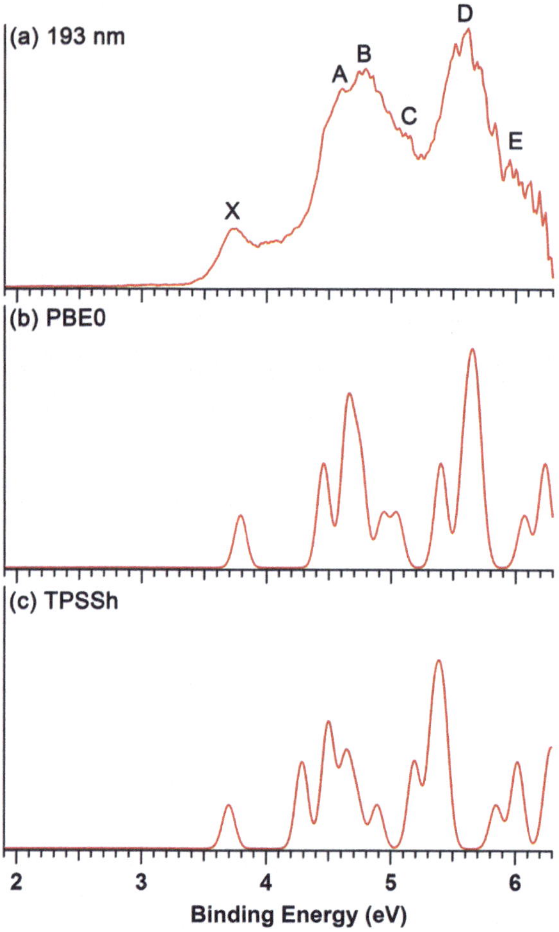

Fig. 4.92 Comparison of the experimental PES spectrum **a** with the simulated PES spectra for the global minimum isomer I of B_{24} at PBE0 (**b**) and TPSSh (**c**). This figure was reproduced with kind permission of the Journal of Chemical Physics of the American Institute of Physics [121]

excellent agreement. Removal of an valence electron from HOMO-1 (36a) instead from HOMO (37a) leads to two open-shells producing the first triplet state of the neutral cluster. The calculated VDE value 4.46 eV of this detachment is in a good agreement with the experimental VDE value 4.61 eV, obtained at the feature A. As mentioned above, the tubular isomer II has never been observed experimentally and hence consequently the experimental features do not agree with the calculated VDEs. In other words, there is no any feature in the experimental spectrum could correspond to the calculated VDEs. The observed first VDE values for the double-ring structure are much higher than the calculated 2.88 and 2.74 eV, obtained respectively at the PBE0 and TPSSh methods. Therefore, the observable features in the experimental spectrum totally disagree with the computed first VDE values. For this reason the presence of the double-ring structure can be ruled out from the cluster beam.

Delocalized σ- and π-Bonds of B_{24}^-: In order to analyze the σ- and π-bonding of the B_{24}^- system, the authors used the AdNDP method, which was developed by Zubarev and Boldyrev ([47–49] in [121]) and successfully used to analyze the chemical bondings of smaller 2D boron clusters mentioned in the previous sections. In order to obtain the local block eigenfunctions and the optimal convergence properties for an electron density description, the authors used the reduced density matrix of the first-order to execute the AdNDP analysis. Due to the fact that the current version of the AdNDP allows performing the analysis of closed-shell systems only, it was not possible to assess the chemical bonding in the open-shell anionic cluster B_{24}^-. Therefore, the AdNDP analysis was performed for both closed-shell neutral B_{24} and closed-shell dianion B_{24}^{2-} systems, Fig. 4.93a, b, but of course at the geometry of B_{24}^- anion. However, this figure shows 2c-2e B-B σ-bonds localized between the 15 peripheral boron atoms in B_{24}^-. In contrast, at the 9 inner atoms there are no any associated 2c-2e B-B σ-bonds were found, indicating the delocalization from 2c-2e to 3c-2e or higher multi-center-two-electron σ-bonds. Figure 4.93a also shows eleven 3c-2e and three 4c-2e σ-bonds with different values of the occupation number (NO). The first three plots at the second row of the Fig. 4.93a present 3c-2e σ-bonds localized inside the filled pentagon. The bonding between the filled pentagonal and the surrounding atoms are characterized by the four 3c-2e and two 4c-2e σ-bonds. The authors found that there are six 4c-2e and one 6c-2e delocalized π-bond, which are bonding the 9 interior boron atoms with the peripheral atoms of the neutral B_{24} cluster. Furthermore, Fig. 4.93b presents the delocalized 5c-2e π-bond of the dianionic cluster B_{24}^{2-}. It is worth to mentioned that the 4c-2e π-bonds in B_{21}^- (Sect. 4.4.2) and in B_{24}^- at the vertices of the filled pentagon in both anions look very similar. In addition, further multi-center bonds in both B_{24}^{2-} and B_{21}^- clusters are very similar. Their unique 6c-2e π-bonds are delocalized over the six inner B atoms comprising the B-centered pentagon, as already shown in Fig. 4.84.

Charge Effect in B_n for (n = 20, 22, and 24): Pham et al. [122] studied theoretically a series of boron cluster isomers B_n which possess two-dimensional (2D) quasi-planar, and three-dimensional (3D) dovetailed double-ring structures, in a range of $n = 20$–24 atoms. The effective net charge of these boron isomers are varying between -2 (dianion) and +2 (dication) charged clusters. They confirmed that by adding extra electrons to the 2D and 3D isomers, the stability of the 2D isomers is enhanced over that of the 3D isomers, independent of the existing electrons. The authors applied beside the stochastic search method the popular hybrid functionals B3LYP and TPSSh of the density functional theory (DFT), starting the initial computations with the 6-31G(d) basis set then ending the final structural and energetic optimizations using the extended basis set 6-311+G(d). The comprehensive benchmark calculations indicate that the B3LYP functionals seem to undervalue the estimated energy differences between the tubular and planar clusters. The TPSSh functional was also calibrated and shown to give better relative energies for boron clusters than other available functionals. The global minimum of charged clusters, obtained from an extensive structural search using the genetic algorithm, are presented in Fig. 4.94. The DFT calculations at the TPSSh/6-31+G(d) level show that

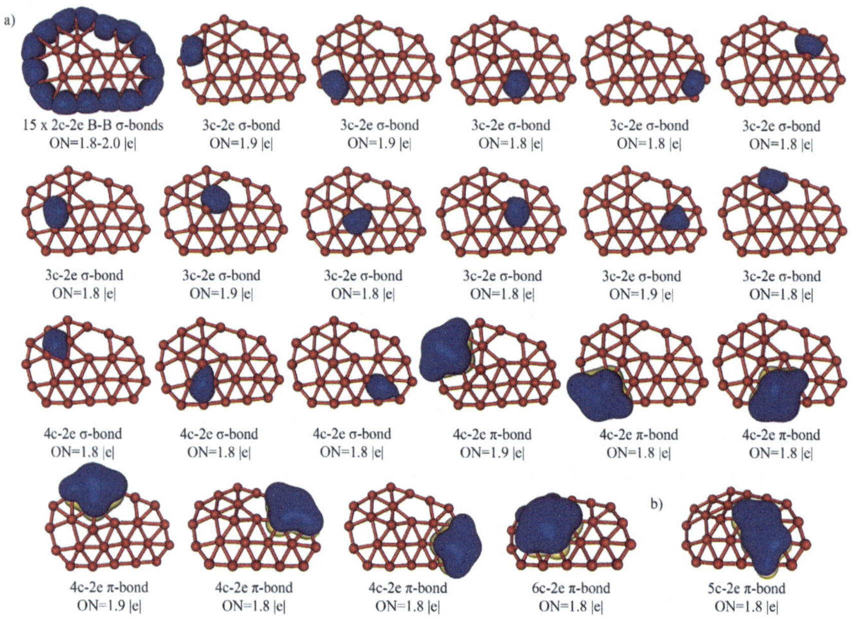

Fig. 4.93 AdNDP analysis of closed-shell neutral B_{24} (**a**), dianion B_{24}^{2-} (**b**), at the geometry of B_{24}^{-}. ON stands for occupation number and is equal to 2.0 |e| in an ideal case. This figure was reproduced with kind permission of the Journal of Chemical Physics of the American Institute of Physics [121]

the pairs of low-lying 2D QP and 3D SDR isomers for B_{20} (Fig. 4.94a) are similar to those of Kiran et al. [115]. Similar results on 2D-3D relative stabilities are also obtained with calculated relative energies presented in Table 4.3. The global minimum of either the neutral B_{22} or the positively charged B_{22}^{+} boron clusters have a 3D SDR structure (Fig. 4.94b). Similarly to B_{22} system, the 2D QP isomer in the anionic B_{22}^{-} clusters is nearly iso-energetic with the 3D SDR counterpart, e.g., energy difference being equal to about 1.0 kcal/mol (see Table 4.3). As one more electron is added to form the dianionic state, the elongated isomer B22-QP-1 becomes much more stable than 3D SDR one, the former being 18.0 kcal/mol lower in energy than the latter. The second 2D isomer of the dianion B22-QP-2 remains even iso-energetic with the 3D SDR structure. In order to observe geometrical changes of 2D isomers as being charged, the peripheral bonds and side-projected geometry of 2D QP of the B_{24} clusters are considered as for an example. Geometrical and structural parameters of B_{24} are depicted in Fig. 4.95. The planarization with conservative boundary obviously increases the atomic repulsion significantly. Such a pattern is expectedly stabilizing the structure by maximizing the exchange energy obtained from the π-electron delocalization within molecular planes. In a sense, the planarity is favored as more electrons are added. Finally, the computations show the evidence that the isomers of the respectively negative singly and doubly charged clusters $B_{20}^{-/2-}$, $B_{22}^{-/2-}$, and $B_{24}^{-/2-}$ possess planar structures, while the neutral B_{20} has tubular form with dove-

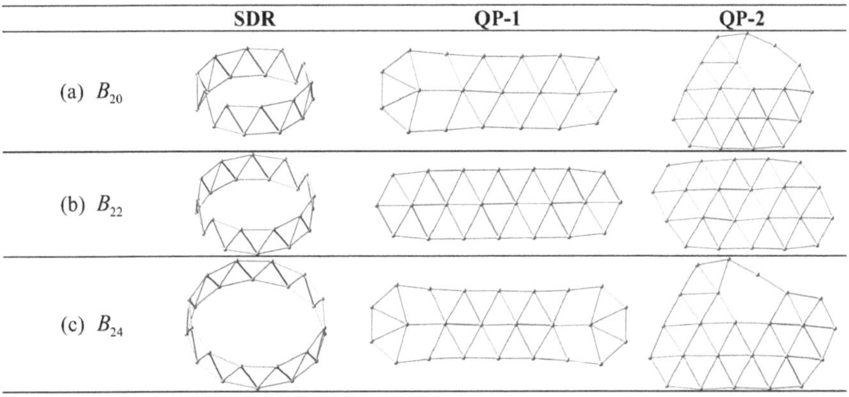

Fig. 4.94 Geometries of the 2D quasi-planar (QP) and 3D staggered double-ring (SDR) isomers of boron clusters. This figure was reproduced from the Chemical Physics Letters with kind permission of the Elsevier Ltd [122]

Table 4.3 The relative energy of 3D-2D (ΔE in kcal/mol) (QP quasi-planar relative to SDR staggered double-ring) with respect to implemented charges. For example, a 2D QP structure is more stable than the corresponding 3D SDR isomer when indicated by a negative relative energy [122]

Charge	B20-QP-1	B20-QP-2	B22-QP-1	B22-QP-2	B24-QP-1	B24-QP-2
2	22.6	30.4	36.9	32.4	41.6	38.5
1	18.5	24.6	30.6	20.4	36.2	37.8
0	14.1	9.9	14.9	7.2	26.7	32.7
−1	−3.6	−5.0	−0.9	−0.9	8.5	10.2
−2	−20.9	−18.4	−17.8	−17.8	−11.0	−18.7

tailed double-ring structure. Herewith, one can conclude that the excess of electrons modifies the structural configuration of boron clusters and pushes the structures to extend the planarity.

All-Boron Anions of B_n^- for ($3 \leq n \leq 24$): Sergeeva et al. [123] reported once again the joint studies of ab initio quantum chemical approaches and photoelectron spectroscopy. Both theoretical and experimental methods can enable the identification of the global minima of the negatively charged boron clusters being with their planar forms the most stable structures, as presented in Fig. 4.96. This figure shows clearly that the negatively charged boron clusters up to at least 24 atoms continue keeping thier planar or quasi-planar structures. Sergeeva et al. [123] discussed also the observed B_{20} double-ring found by Kiran et al. [115]. Sergeeva found that there is an analogy between the chemical bonding of the B_{20} cluster and the carbon ring cluster C_{10}, as shown in Fig. 4.97. The chemical bonds in the C_{10} ring cluster are composed of ten 2c-2e σ-bonds and 10 π-bonds. The σ-bonds are localized between the carbon atoms, while the π-bonds are delocalized over the ring structure, as shown in Fig. 4.97b, c. These σ-, and π-bonds produce double aromaticity. In comparison

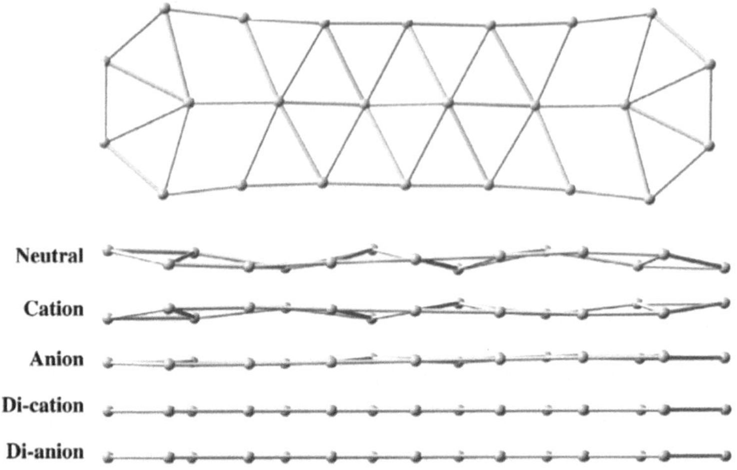

Fig. 4.95 The structure of 2D Qp B_{24} boron cluster presented from the top and side of neutral and charged isomers. This figure was reproduced from the Chemical Physics Letters with kind permission of the Elsevier Ltd [122]

to the C_{10} cluster, each B-B unit in the double ring of the B_{20} cluster provides four valence electrons producing 20 2c-2e σ-bonds localized between the atoms in each ring, while the remaining two valence electrons are delocalized. However, it is obviously to notice that there is a strong analogy between the bonds in both double-ring boron and single-ring carbon systems. In both systems there are localized X-X σ-bonds for (X = B or C) as shown in Fig. 4.97c and delocalized σ-Bonds can be seen in Fig. 4.97d. The delocalized π-bonds of the both C_{10} and B_{20} clusters as presented in Fig. 4.97e and are of course pretty striking.

Low-lying Isomers of B_{25}^-: Piazza et al. [124] forwarded their investigations on boron clusters studying experimentally and theoretically the electronic and structural properties as well as the chemical bonds of the anionic isomers of the B_{25}^- cluster, similar to the previous study on B_{24}^-, mentioned in the last section. Using the density functionals PBE0 and the 6-311+G(d) basis set, the authors found by the search for the global minimum around 51 isomers whose energies are laying within 35 kcal/mol only. At the higher level of theory CCSD(T)/6-311+G(d), they started the structural optimization from initial geometries obtained at PBE0/6-311+G(d) level and selected 20 low-lying isomers. An abbreviated assembly of these B_{25}^- anions is shown in Fig. 4.98 stacked after their energies. This figure shows that there are three low-lying energetically degenerate B_{25}^- isomers within 1 kcal/mol only. The energetically lowest two isomers of B_{25}^- are quasi-planar structures with a pentagonal hole, while the third lowest isomer has an elongated planar structure.

Photoelectron Spectrum of B_{25}^-: The PES of B_{25}^-, obtained at the 193 nm, is presented in Fig. 4.99. The B_{25}^- spectrum exhibits three broad spectral bands in the low binding energy region, but above 5.2 eV in the high binding energy area there are some

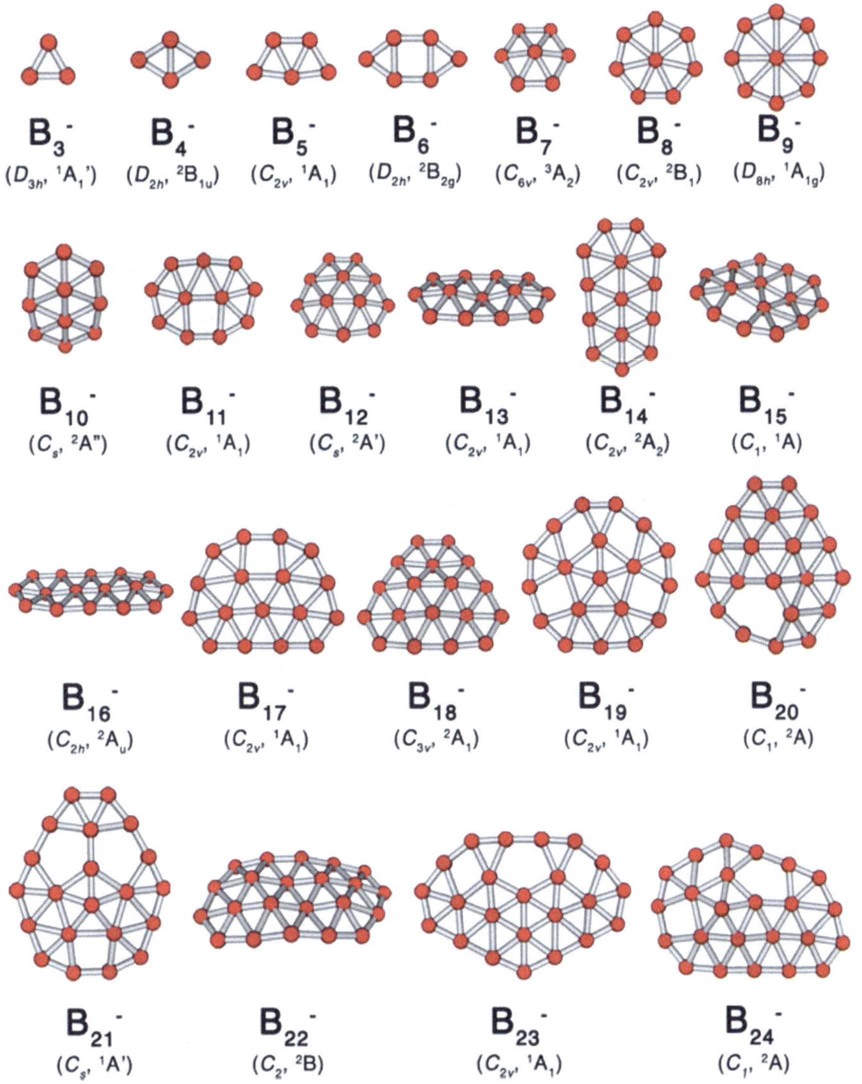

Fig. 4.96 The structures of the major contributors to the photoelectron spectra of B_n^- ($n = 3-24$) with the corresponding symmetry and ground state. This figure was reproduced with kind permission of the Accounts of Chemical Research of the American Chemical Society [123]

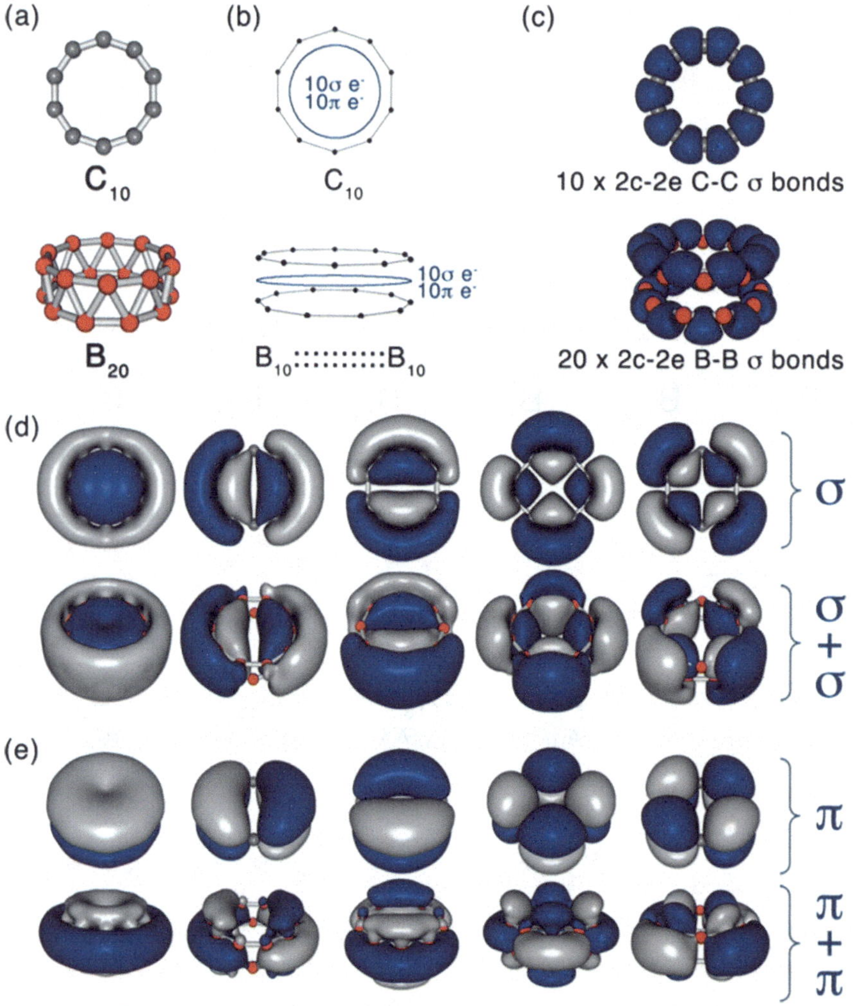

Fig. 4.97 Elucidation of the analogy between the C_{10} ring and the B_{20} double ring. Comparison of **a** structures, **b** schematic representations of bonding, **c** localized Lewis bonds at the periphery, **d** delocalized σ bonds, and **e** delocalized π bonds. The top (bottom) rows correspond to C_{10} (B_{20}). This figure was reproduced with kind permission of the Accounts of Chemical Research of the American Chemical Society [123]

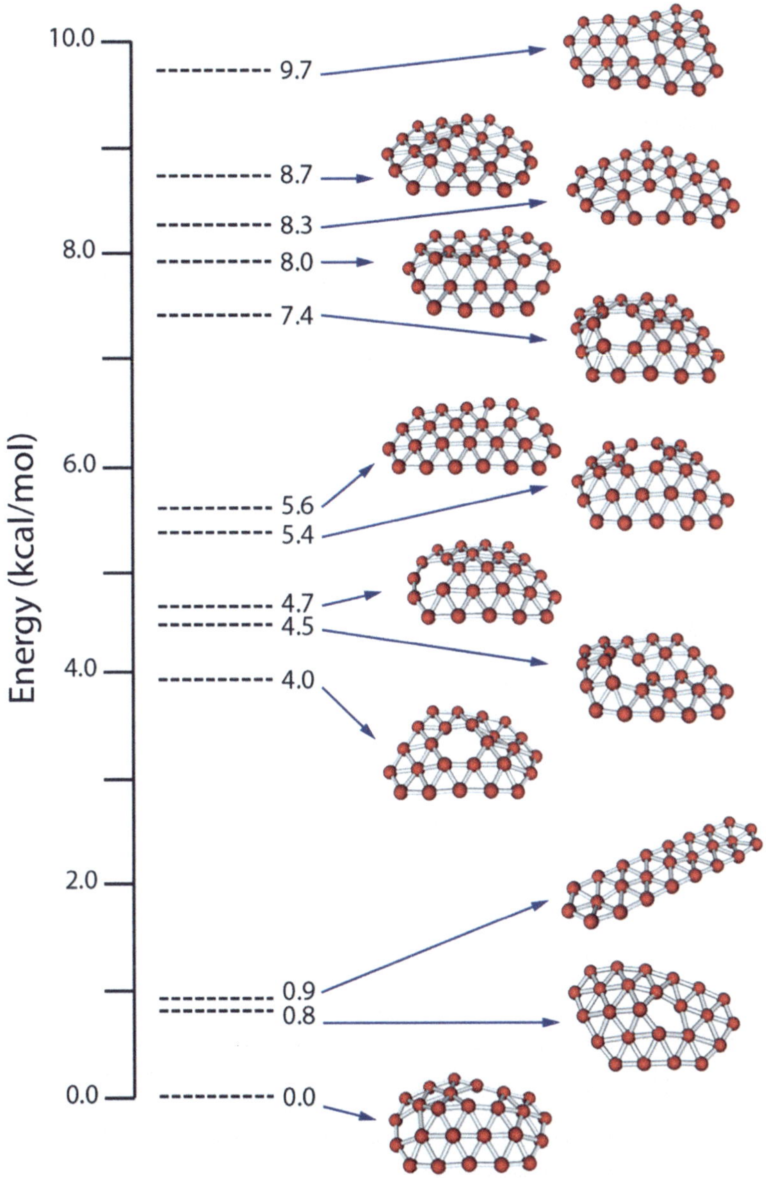

Fig. 4.98 Relative energies in kcal/mol of the lowest lying isomers of B_{25}^- obtained from global minimum searches based on the CCSD(T)/6-311+G(d)//PBE0/6-311+G(d) method. Sticks drawn between atoms represent inter-atomic distances <2.0Å; they do not necessarily represent single B-B σ bonds here and elsewhere. This figure was reproduced with kind permission of the Journal of Chemical Physics of the American Institute of Physics [124]

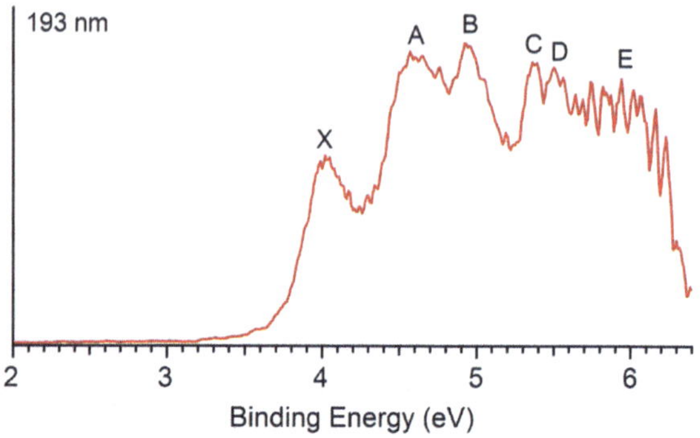

Fig. 4.99 The PES photoelectron spectrum of B_{25}^{-}, observed at 193 nm (6.424 eV). This figure was reproduced with kind permission of the Journal of Chemical Physics of the American Institute of Physics [124]

spectral bands with congested features. At the low binding energy region and for the vertical electron detachment transition to the neutral ground state, the VDE value of 4.02 ± 0.06 eV was obtained at the X band. From the leading edge of the X band the ADE value of 3.8 ± 0.1 eV was estimated for the adiabatic detachment energy of the ground state transition.

Global Minimum of B_{25}^{-}: Piazza et al. [124] used three methods to search for the global minimum and low-lying isomers of B_{25}^{-}. These methods are the Coalescence Kick (CK), the Cartesian walking (CW) and the basin-hopping (BH). The CK method briefly creates by randomly generated coordinates a population of trial structures, then eliminates the remaining fragmented cluster from the population, when the procedure of gradually pushing atoms towards the molecular center of mass is done. As soon as the CK procedure is over, the obtained trial structures will be optimized looking for the utmost minima, using the density functional theory formalism for energy evaluations. In order to improve the structural optimization of the low-lying B_{25}^{-} clusters, with the ambition for the search of global minimum, the authors increased the accuracy by using two PBE057 and TPSSh density functionals and exerting the more expansive 6-311+G(d) basis set. Using the highest level of theory, the CCSD(T)/6- 311+G(d), the authors achieved three nearly degenerate structures of the B_{25}^{-} cluster, separated only by just 0.9 kcal/mol and presented in Fig. 4.98. As shown in Fig. 4.100, the three nearly degenerate low-lying isomers of B_{25}^{-}, labeled as I, II, and III, are assigned by the calculated relative energies computed at four levels of theory. The ranking of the three B_{25}^{-} isomers depends of course on the employed theoretical method. However, the isomer III seems by the two DFT methods,[8] and by the coupled cluster method[9] to be favored over the other isomers I and II. Nevertheless, at the highest level of theory[10] remains isomer I the most stable cluster.

Comparison of Simulated and Observed PES of B_{25}^{-}: Piazza et al. [124] compared the simulated spectra of the isomers I, II, and III with the experimental spectrum, as

[8] PBE0/6-311+G(2df)//PBE0/6- 311+G(d), TPSSh/6-311+G(2df)//TPSSh/6-311+G(d).

[9] CCSD/6-311+G(d)//PBE0/6-311+G(d).

[10] The coupled cluster method CCSD(T)/6-311+G(d)//PBE0/6-311+G(d).

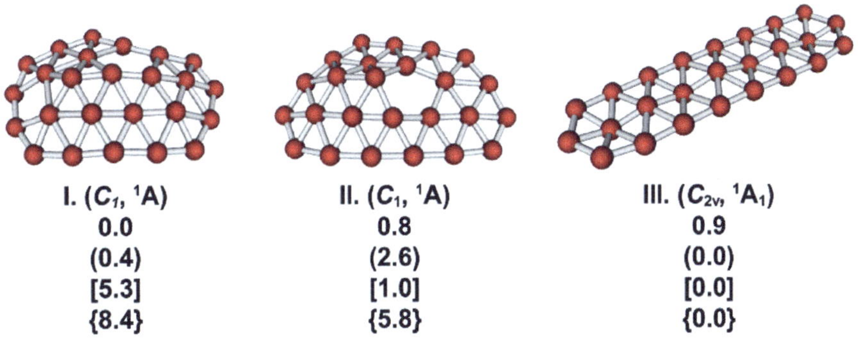

Fig. 4.100 The ground electronic states and relative energies of the three lowest lying isomers of B_{25}^-. Energies are given at CCSD(T)/6-311+G(d)//PBE0/6-311+G(d), CCSD/6-311+G(d)//PBE0/6-311+G(d) (in parenthesis), PBE0/6-311+G(2df)//PBE0/6-311+G(d) (in square brackets), and TPSSh/6-311+G(2df)//TPSSh/6-311+G(d) (in curly brackets) in kcal/mol. This figure was reproduced with kind permission of the Journal of Chemical Physics of the American Institute of Physics [124]

presented in Fig. 4.101. They fitted a Gaussian function to each obtained VDE considering 0.1 eV FWHM (full width at half maximum). They found that the calculated and the observed VDEs of isomer II are in excellent agreement. Thus, the spectral pattern of the experimental PES obtained for isomer II imparts respectable reliability that this isomer II of B_{25}^- belongs truely to the aspired global minimum. However, the isomers I and III cannot be ruled out due to their considerable contributions and to their non-vanishing population in the beam of clusters. The measured VDE 4.02 eV of the isomer II, observed at the X band, is in excellent agreement with the calculated value 3.97 eV, determined at the TD-PBE0 level of theory. The latter value is the first vertical detachment for the B_{25}^- cluster, which was obtained by removing one electron from the doubly occupied HOMO of isomer II. Furthermore, the electron detachments from the HOMO-1 and HOMO-2 of isomer II are labeled as the second and third VDEs and being calculated by 4.51 and 4.70 eV, respectively. The latter VDE values are laying quite close together in a good agreement with the 4.6 eV value, obtained at the broadband A.

Delocalized σ- and π-Bonds of B_{25}^-: As already given in previous sections, the σ- and π-bonding of the B_{25}^- system can be analyzed using the AdNDP method. Some of the inner atoms in both isomers I and II are building a pentagonal-pyramid-like unit, which cambers from the molecular plane causing out-of-plane distortions. In analogous to the previous bonding patterns, determined for smaller sized of boron clusters, the peripheral 2c-2e B-B σ-bonds of the isomer II are definitely the classical chemical σ-bonds with ON = 1.8–1.9 |e|, as shown in Fig. 4.102. The AdNDP chemical bonding analysis of the isomer II shows that the elements in the σ-, and π-bonding are quite similar and deviate from each other very slightly. All σ-bonds of the ten inner atoms seem to be delocalized in twelve 3c-2e σ-bonds with ON = 1.7–1.9 |e|. Three 3c-2e σ-bonds of the latter 12 bonds are localized between the

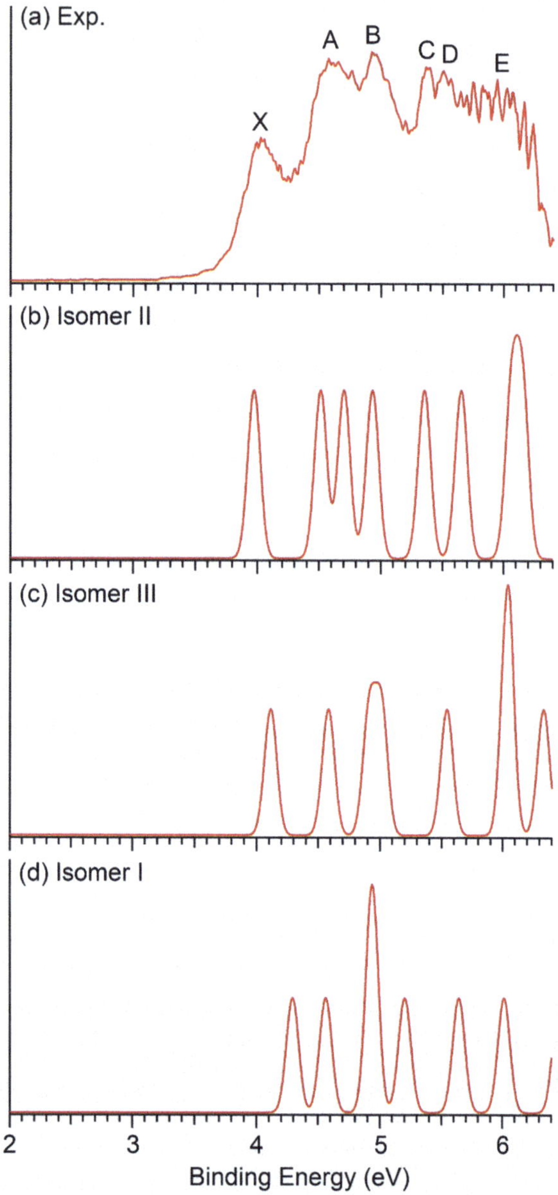

Fig. 4.101 Comparison between simulated PES for isomers I, II, III with the experimental spectrum. The simulated spectra were created by fitting Gaussian functions with a 0.1 eV full width at half height to TD-PBE0 VDE values. This figure was reproduced with kind permission of the Journal of Chemical Physics of the American Institute of Physics [124]

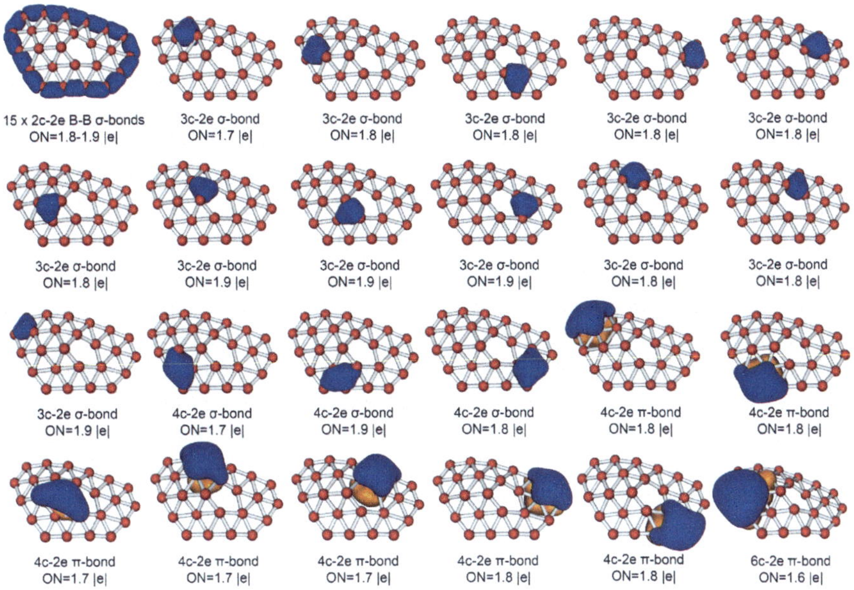

Fig. 4.102 AdNDP analysis for isomer II of B_{25}^-. This figure was reproduced with kind permission of the Journal of Chemical Physics of the American Institute of Physics [124]

atoms of the pentagonal-pyramid-like unit inside the isomer II. In addition, there are three 4c-2e σ-bonds with NO = 1.7, 1.8, and 1.9 |e| are responsible for the bonding between the inner and the peripheral atoms. Furthermore, the AdNDP analysis of the isomer II also shows eight delocalized π-bonds. They are separated in seven 4c-2e π-bonds with ON = 1.7–1.8 |e| and just one 6c-2e π-bond with ON = 1.6 |e|. One can notice that all ten inner atoms are participating in the formation of eight delocalized π-bonds. The shifting of the 6c-2e π-bond to the edge of the cluster can be related to the longer B-B bond in the pentagonal-pyramid-like unit causing an electron density rearrangement, as can be seen in Fig. 4.102.

Chemical Bonding of B_{25}^-: Piazza et al. [124] investigated the chemical bonding pictures of the low-lying isomers I, II, and III of the B_{25}^- cluster. Beside the planar isomer III, as already mentioned above, we have seen that the structures of the other both two isomers I, and II are exhibiting out-of-plane distortion. Therefore, the σ and π bonding can be assigned, but only approximately. Due to the presence of both delocalized and localized bonding features, the AdNDP method is able to decipher these chemical bonding in the 2D boron clusters. In the case of the isomer III in Fig. 4.100, which has quasi-planar structure with C_{2v} symmetry, the π molecular orbitals are presented by flattening the quasi-planar cluster to a planar one with a D_{2h} symmetry. Then the MOs were compared with those of the hydrocarbon molecule $C_{16}H_{10}$ (dibenzopentalene), as shown in Fig. 4.103. The execution of flattening the non perfectly planar C_{2v} into the planar D_{2h} structure, allows to facilitate the division of the electron density into σ-, and π-bonds and also to prevent the mixing of the σ and π orbitals in isomer III. However, due to the fact that the out-of-plane distortions of the inner atoms in isomer III are relatively neglectable small (~0.2 Å), the artificial flattening helps their visualization without changing the number of the CMOs or any of the nodal planes. It is worth to mention that the canonical molecular orbitals (CMOs) are the basis for the chemical bonding analysis of the ribbon-like isomer III. As can be seen in Fig. 4.103, the eight π canonical molecular orbitals, found in the elongated flattened structure of the isomer III of the B_{25}^- cluster, are reminiscent of those in the dibenzopentalene structure of ($C_{16}H_{10}$). This hydrocarbonal molecule is consisting of a bicyclic pentalene fragment connected by two benzene rings. Therefore, the analogy between the dibenzopentalene molecule and the all-boron of B_{25}^- cluster can be related to the striking similarities between the π-CMOs of the hydrocarbon and of the ribbon-like isomer III. Further AdNDP analysis (not shown) discloses that the bonding between the 18 peripheral boron atoms are depicted by typical 2c-2e σ-bonding with ON = 1.8–1.9 |e|. However, the bonds between the interior and the peripheral atoms of isomer III, arising from delocalized σ- and π-bonds, are provided by the remaining electron density.

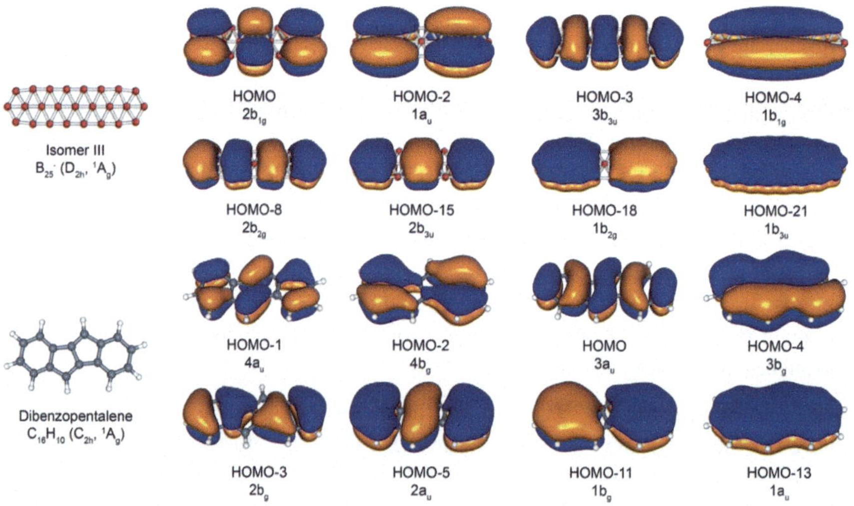

Fig. 4.103 Comparison of the π molecular orbitals of a flattened isomer III (D_{2h}) with those of dibenzopentalene $C_{16}H_{10}$. This figure was reproduced with kind permission of the Journal of Chemical Physics of the American Institute of Physics [124]

4.4.2 Boron Clusters B_n ($26 \leq n \leq 36$)

Electronic Structures of Boron Clusters $B_{26-29}^{0/-}$: Tai and Nguyen [125] performed density functional theory methods (DFT) with the hybrid functional B3LYP and the basis set 6-311+G(d) in combination with a stochastic search algorithm and the high accuracy quantum chemical approach G3B3, to study the geometrical and electronic structures of the neutral and anionic boron clusters B_n for ($n = 26$–29). It is obviously clear that the structural and electronic properties of these boron cluster sizes can be determined as soon as the geometrical characteristics are already known. However, the study of such relatively large boron cluster sizes gives insight into a general growth mechanism of boron clusters but also fulfills the missing gap of B_n clusters for ($n =$ 26–29). Thus, Tai and Nguyen [125] investigated all electronic structure calculations of the neutral and anionic boron clusters B_{26-29}. The authors used their own stochastic search algorithm for the initial searches for all possible lower-lying isomers of B_n for ($n = 26$–29) clusters. Using the kick method, the randomly generated possible structures of the clusters were considered by the optimization procedure applying the density functional theory within the functional hybrid PBE as well as the small basis set 3-21G. Further optimization were carried out at the PBE/6-311+G(d) level to investigate the local minima structures with relative low-lying energies located within the range of 0.0–5.0 eV. The low-lying isomers, obtained at the PBE/6-311+G(d) level, are recomputed at a higher level of theory, the coupled-cluster theory CCSD(T) and the basis set 6-311G(d). Finally, beside using the same functional and basis set PBE/6-311+G(d), the time-dependent density functional theory (DT-DFT) was also applied to simulate the photoelectron spectra of B_n^-.

Neutral and Anionic Boron clusters $B_{26}^{0/-}$: Tai and Nguyen [125] reported the final structures of the low-lying isomers B_{26} and B_{26}^- and their relative energies and presented in Fig. 4.104. This figure shows that the neutral cluster B_{26} with the double-ring tubular form 26n.1 is the most stable isomer. This structure is composed of two connected 13-membered rings forming an antiprism bonding motif similar to the structures of the B_{20} and B_{24} clusters. The next stable isomer of the neutral cluster B_{26} 26n.2 (C_1, ^1A) is by only 2.1 kcal/mol above the quasi-planar isomer 26n.1 (Fig. 4.104a), obtained at CSD(T)/6-311G(d)//PBE/6-311+G(d). By the anionic clusters B_{26}^-, the stability of the anions is reversed. The most stable B_{26}^- isomer exhibits a quasi-planar structure 26a.1 (C_1, ^2A), while the double ring structure 26a.2 (C_i, ^2A_u) is by 3.8 kcal/mol less table than the favored 26a.1 (Fig. 4.104b). It is obviously clear that as soon as the neural B_{26} cluster is negatively charged, the structure favors the 2D over the 3D configuration.

Vertical and Adiabatic Detachment of B_{26}^-: Tai and Nguyen [125] also determined the adiabatic and vertical detachment energies, (ADE) and (VDE), respectively, and subsequently simulated the photoelectron spectra (PE) of anionic clusters using the TD-DFT methods. They depicted the graph of the simulated PE spectra in Fig. 4.105. This figure shows that the first VDE of 26a.1 (C_1, ^2A) is about 3.71 eV, obtained by detaching an electron from its singly occupied molecular orbital (SOMO). It

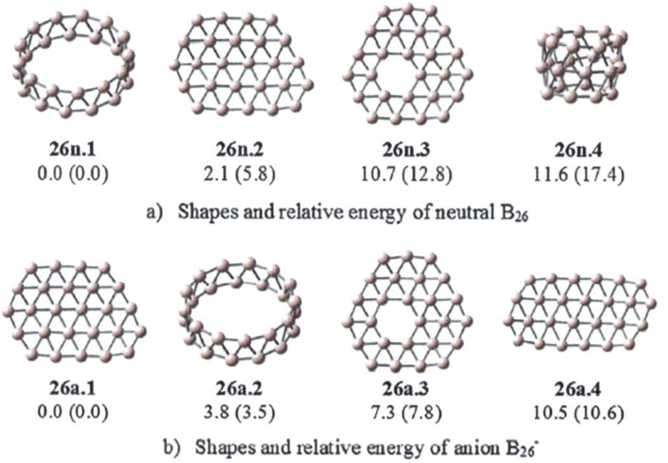

Fig. 4.104 Shapes and relative energies (kcal/mol) of the low-lying isomers of the **a** neutral B_{26} and **b** anionic B_{26}^- clusters, obtained at CCSD(T)/6-311G(d)//PBE/6-311+G(d) and PBE/6-311+G(d) (in parentheses). This figure was reproduced from the Physical Chemistry Chemical Physics with kind permission of the Royal Society of Chemistry [125]

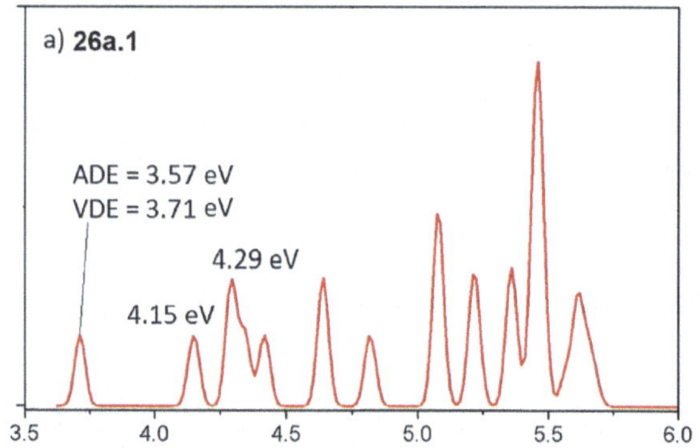

Fig. 4.105 Simulated photoelectron Spectrum of B_{26}^-, obtained at the TD-PBE/6-311+G(d) level. This figure was reproduced from the Physical Chemistry Chemical Physics with kind permission of the Royal Society of Chemistry [125]

is relative larger than the ADE value 3.57 eV. The difference between the both latter VDE and ADE values account only 0.14 eV assigning a small energy gap and interpreting a minimal relaxation energy for the transition from the anionic to the neutral structure. The first triplet state occurs by the next electron detachment from the HOMO-1 of the 26a.1 isomer and accounts around 4.15 eV. The energies observed in the range of 4.29–5.68 eV are corresponding to several singlet and triplet transitions of excited states.

Why Simulated Spectra are Important? The ascertained VDE and ADE values of anionic boron clusters are very important indications for the structural configurations and features of the anionic and neutral clusters. Analytical experimental methods alone can not provide directly the structural identification of the atomic boron

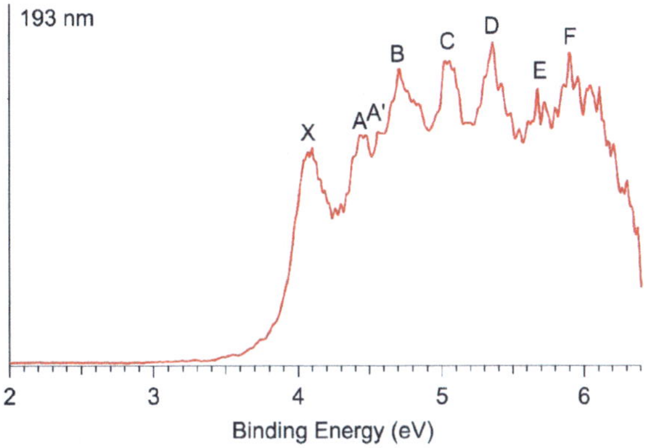

Fig. 4.106 Experimental photoelectron Spectrum of B_{27}^-, observed at 193 nm (6.424 eV). This figure was reproduced with kind permission of the Journal of Chemical Physics of the American Institute of Physics [126]

clusters. Hence, they are combined with theoretical methods in order to determine indirectly the quested formations, as was the case in all anionic clusters mentioned in previous sections. The simulated spectra of theoretically predicted low-lying energy minima located at the potential energy surface in comparison, either with the infrared spectra (IR), or with photoelectron spectra (PE), are nothing else as fingermarks for the cationic and neutral clusters, or for the anionic clusters, respectively [125].

Photoelectron Spectrum of B_{27}^-: Li et al. [126] carried out photoelectron spectroscopy in combination with ab initio calculations to study the geometrical and electronic structures of the anionic cluster B_{27}^-. Experimentally, they used the same apparatus which was applied for the B_{20}^- cluster, as described in Sect. 4.4.1. The photoelectron spectrum of the mass-selected B_{27}^- anion is obtained at 193 nm and presented in Fig. 4.106. This figure shows different detachment transitions from the anion ground state to the neutral ground (X) state and excited states (A, B, ..). The photodetachment spectrum has blocked well-defined features. The vertical detachment energy VDE was measured at 4.09 eV in corresponding to the ground state transition (X). The adiabatic detachment energy ADE is ~3.9(1) eV, which is slightly smaller than the VDE value. This small energy gap of ~0.18 eV can be interpreted by the structural transition from the anionic to the neutral isomer that needs minimal relaxation energy. The ADE value was estimated from the onset of band X, obtained by drawing a straight line along the leading edge. The next following band A is located after band X by an small energy gap of around ~0.35 eV and the observed value is about 4.4 eV. After the feature A follows the fairly broad band B, which is measured at 4.71 eV. This broad band B can be connected to multiple detachment channels which could be the reason for its appearance. The next observed bands C and D measured respectively at 5.05 and 5.35 eV, are exhibiting intense characteristics. However, two additional bands are tentatively identified at the 5.68 and 5.89 eV characterized by E and F indicating the high binding energy. As shown in Fig. 4.106 there is a VDE feature at 4.56 eV, assigned by A′, which is laying between the bands A and B. This A′ could be related to another low-lying isomer of B_{27}^-.

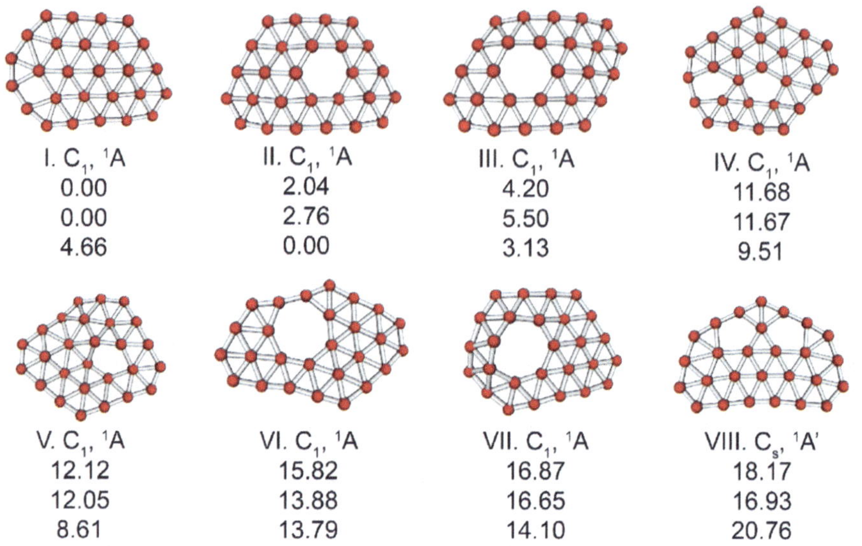

Fig. 4.107 Relative energies in kcal/mol of the low-lying isomers of B_{27}^- based on the PBE0/6-311+g(2df) (top), TPSSh/6-311+g(2df) (middle), and CCSD(T)/6-311-g(d) (bottom) methods. This figure was reproduced with kind permission of the Journal of Chemical Physics of the American Institute of Physics [126]

Global Minimum of B_{27}^-: Li et al. [126] used both the Cartesian walking (CW) and the basin-hopping (BH) algorithms to perform the global minimum searches for B_{27}^- isomers. The initial structures generated by the CW method were optimized using the exchange correlation functional (PBE0) and the 3-21G basis set. The authors used also the density functional TPSSh[11] and the 3-21G basis set to optimize the initial structures generated by the BH method. The obtained structures of the B_{27}^- isomers within 40 kcal/mol were reoptimized at the both hybrids PBE0 and TPSSh using larger 6-311+g* basis set. Afterwards, further computations were carried out using the more expensive basis set 6-311+g(2df) to determine the final single-point calculations. Figure 4.107 shows eight structures of low-lying B_{27}^- isomers. It is interesting to assert that all of these isomers possess 2D structures, but with empty polygonal moieties. As can be seen in Fig. 4.107, the isomers I and II have tetragonal and hexagonal holes, respectively. The remaining isomers are assigned by pentagonal, hexagonal, and heptagonal holes. At the PBE0 and TPSSh level of theory, the isomer I seems to be the global minimum with a 2D structure, while the low-lying isomers II and III are respectively just 2.04 and 4.20 kcal/mol higher in energy. In contrast, the isomer II is the global minimum at the most accurate CCSD(T) method, while the isomer I is by 4.66 kcal/mol less stable.

[11]The hybrid exchange functional of Tao, Perdew, Staroverov, and Scuseria (TPSSh).

Simulated and Experimental Spectra of B_{27}^-: Li et al. [126] compared the simulated spectra of three low-lying isomers with the observed spectrum of anion B_{27}^-, as presented in Fig. 4.108. Since the low-lying isomers of the B_{27}^- anions being closed shell systems, it is obviously to expect that after an electron detachment the final systems will only be in doublet states. However, due to the theoretical calculations the first three low-lying isomers within 4.20 kcal/mol are energetically nearly degenerate and thus the predication of the global minimum by theory is not sufficient. Therefore, the experimental data is critical and should help to distinguish the real global minimum. Concerning isomer I, the first vertical electron detachment occurs from the singlet state of the doubly occupied HOMO (41a) of the anion B_{27}^- resulting into the doublet ground state ^2A of the neutral B_{27}. The observed experimental VDE value of 4.09 eV is in excellent agreement with the theoretical value 3.98 eV, calculated for the first VDE at the PBE0 level of theory. The detachment of the next electron from the molecular orbital HOMO-1 is computed at 4.46 eV and agrees excellently with the observed value 4.44 eV of the band A. Therefore, the most of the calculated upper VDEs from the isomer I are in excellent agreement with the experimental data. The comparison of the simulated spectrum of isomer I in Fig. 4.108b with the reproduced experimental spectrum (a) in Fig. 4.108 shows how good is the agreement between the theory and the experiment. Thus, one can assert that the analogy between the simulated spectra and the experimental data tends rather to the isomer I but not to the isomers II and III implying that the observed photoelectron spectrum is induced by the isomer I, despite the experimental presence of the isomers II and III.

Delocalized σ- and π-Bonds of B_{27}^-: Concerning σ- and π-bonding of the B_{27}^- system, Li et al. [126] carried out the chemical analysis of the first three low-lying isomers I, II, and III, applying the AdNDP method, proposed by Zubarev and Bodyrev [103]. The localized and delocalized σ-, and π-bonds of these three isomers I, II, and III are presented in Fig. 4.109 at (a), (b) and (c), respectively. The figure at (a) shows the localized classical 2c-2e σ-bonds of the first isomer I are localized on the peripheral atoms, in analogous to the σ bondings found by all smaller 2D clusters. The corresponding occupation numbers of the isomers I are laying in the range between 1.70 |e| and 1.95 |e| indicating that the larger ON value is approaching the ideal case of 2.00 |e|. Because the peripheral B-B Atoms in isomer I possess slightly different bond lengths, the occupation numbers behave appropriately and are varying accordingly. In contrast to the localized 2c-2e σ-bonds of the peripheral atoms, the σ-bonds of the inner atoms are delocalized in eight 3c-2e, six 4c-2e, and two 5c-2e σ-bonds, assigning the density distribution around the neighboring atoms of the cluster plane. The nine delocalized π-bondings in isomer I are distributed in six 4c-2e, one 5c-2e, and two 6-2e π-bonds. The pattern of these multi-central π-bonds are similar to the π-bondings of the anion B_{25}^- [124], which has eight delocalized π-bonds, and just one less 6c-2e π-bonds, naturally due to two fewer boron atoms. Similar σ and π electron density analysis for the other two isomers II and III can be seen in Fig. 4.109 at (b) and (c), respectively. However, the ratio of valence electrons involved in delocalized σ-bonds and π-bonds of 2D boron clusters with triangular lattices can be estimated by approximately 2:1.

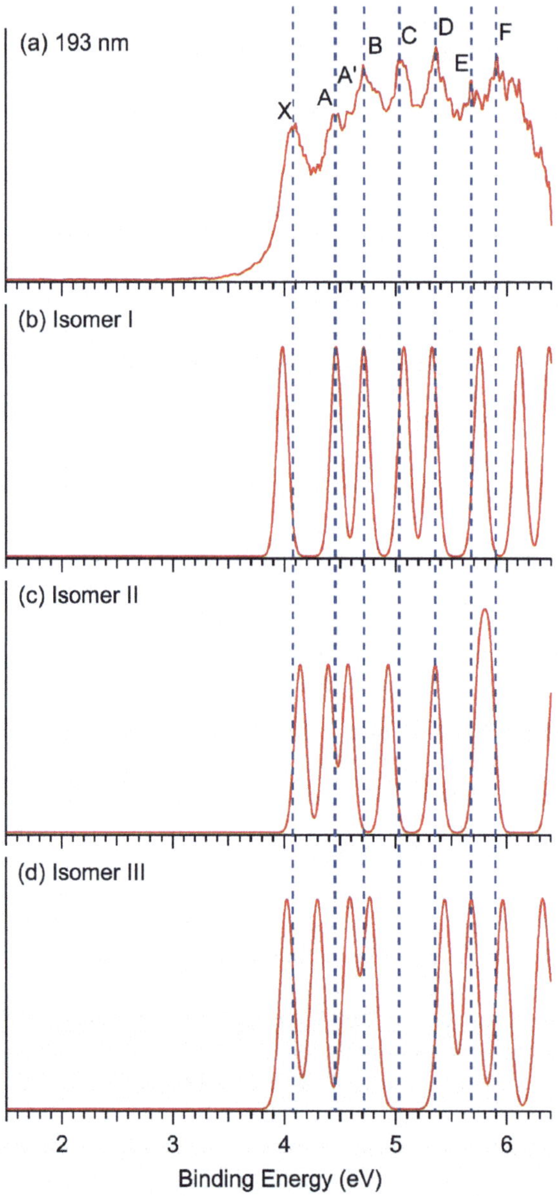

Fig. 4.108 Comparison between the simulated spectra for isomers I (**b**), II (**c**), and III (**d**), (Fig. 4.107), with the experimental spectrum (**a**). The simulated spectra were created by fitting Gaussian functions with a 0.26 eV full width at half height to the TD-PBE0 VDE values. This figure was reproduced with kind permission of the Journal of Chemical Physics of the American Institute of Physics [126]

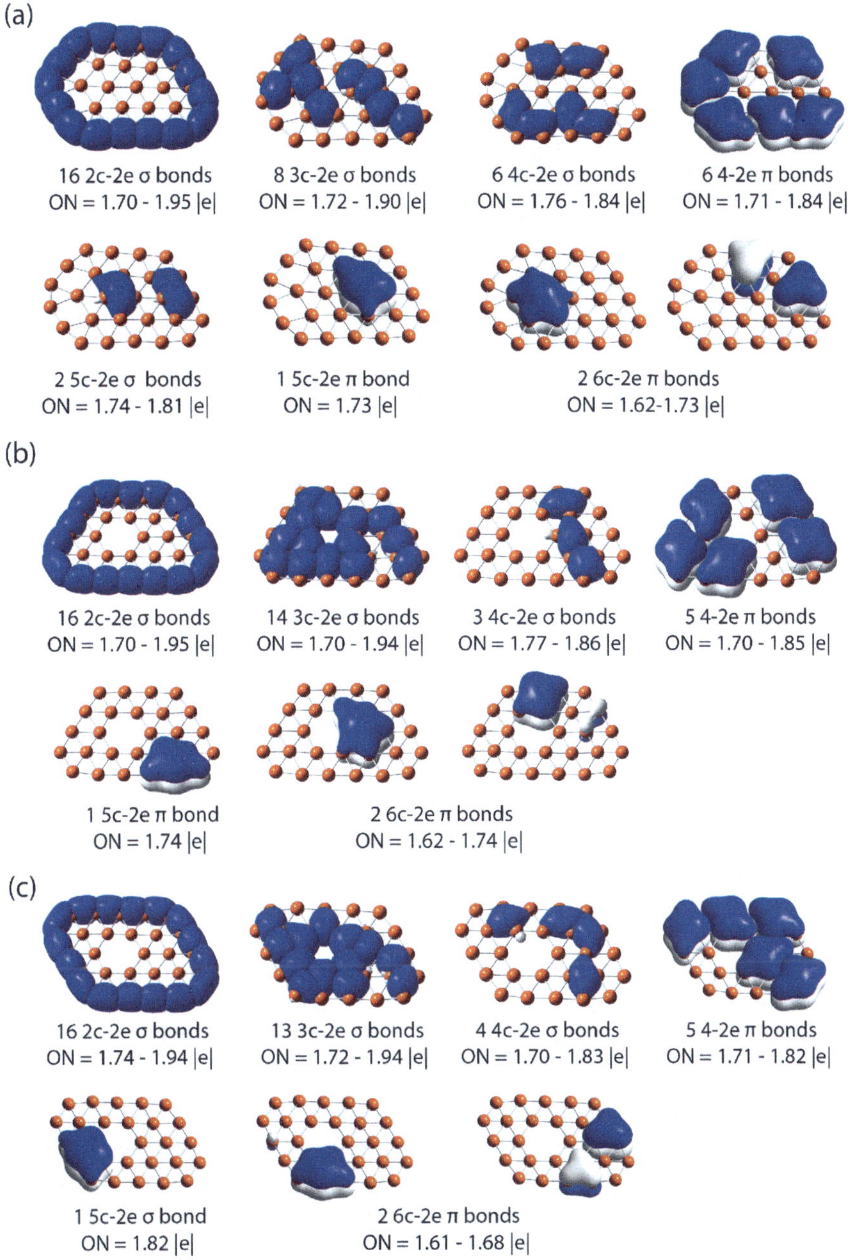

Fig. 4.109 AdNDP analysis of B_{27}^- for the isomers I (**a**), II (**b**), and III (**c**). ON represents occupation number. This figure was reproduced with kind permission of the Journal of Chemical Physics of the American Institute of Physics [126]

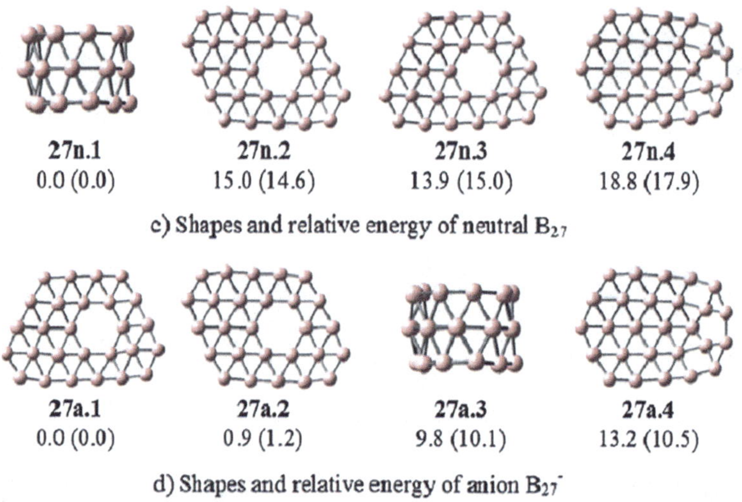

Fig. 4.110 Shapes and relative energies (kcal/mol) of the low-lying isomers of **c** neutral B_{27} and **d** anionic B_{27}^- clusters, obtained at CCSD(T)/6-311G(d)//PBE/6-311+G(d) and PBE/6-311+G(d) (in parentheses). This figure was reproduced from the Physical Chemistry Chemical Physics with kind permission of the Royal Society of Chemistry [125]

Neutral and Anionic Clusters of B_{27} and B_{27}^-: Tai and Nguyen [125] determined also the structures of the neutral and anionic boron clusters B_{27} and B_{27}^-, using the same theoretical methods applied by the B_{26} and B_{26}^- clusters, mentioned above. The obtained structures and the calculated relative energies for neutrals and anions are presented in Fig. 4.110. This figure exhibits that the tubular structure of the three together connected nine-membered rings 27n.1 (C_1, ^2A) is the global minimum of the neutral cluster B_{27}. The next two 2D isomers, 27n.2 (C_1, ^2A) and 27n.3 (C_1, ^2A) are respectively about 15.0 and 13.9 kcal/mol less stable than the 27n.1 isomer. However, the structures of the latter both planar isomers were found in the anionic state as the most stable anions 27a.1 (C_1, ^1A) and 27a.2 (C_1, ^1A) of B_{27}^-. Both anions of B_{27}^-, each contains one hexagonal hole, have an energy difference of only 0.9 kcal/mol, and thus they are degenerate in energy.

Simulated Photoelectron Spectra of B_{27}^-: Besides the structures, the photoelectron spectra simulation of the B_{27}^- isomers was also investigated by Tai and Nguyen [125] using the TD-DFT methods. They also determined the corresponding vertical and adiabatic values, VDE and ADE. As can be seen in Fig. 4.111, there are many close and intense peaks in the simulated photoelectron spectra of the nearly degenerate 27a.1 (C_1, ^1A) and 27a.2 (C_1, ^1A) neutral clusters. The isomers 27a.1 and 27a.2 are different in form but they are exhibiting a tiny energy gap. The calculated first VDE and the ADE of the 27a.2 isomer 3.88 and 3.78 eV, respectively, are slightly smaller than the corresponding VDE and ADE values 3.98 and 3.85 eV of the 27a.1 isomer. The computed first VDEs of both isomers 27a.1 and 27a.2 are obtained by detaching

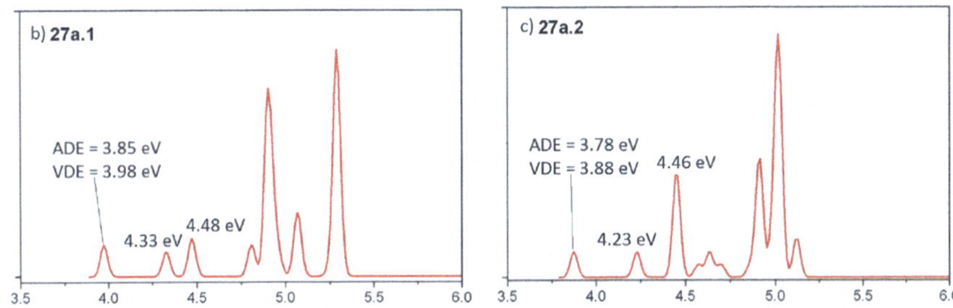

Fig. 4.111 Simulated photoelectron spectra of the B_{27}^- isomers 27a-1 and 27a-2, obtained at the TD-PBE/6-311+G(d) level. This figure was reproduced from the Physical Chemistry Chemical Physics with kind permission of the Royal Society of Chemistry [125]

one valence electron from their doubly occupied HOMOs. The second VDEs of both 27a.1 and 27a.2 isomers are respectively 4.33 and 4.23 eV, obtained by removing one electron fron the open shell HOMO-1. Similarly, the third VDEs of the 27a.1 and 27a.2 isomers are computed respectively by 4.48 and 4.46 eV. They are obtained by the removal of one valence electron from their closed shell HOMO-2 and seem to be very closely. However, the remaining peaks, located in the range of 4.58-5.30 eV, look very different.

Neutral and Anionic Clusters of B_{28} and B_{28}^-: Tai and Nguyen [125] determined also the structures of the neutral and anionic boron clusters B_{28} and B_{28}^-, using the same theoretical methods applied by the B_{26}, B_{26}^-, B_{27}, and B_{27}^- clusters, mentioned above. The structures of the low-lying isomers and the calculated relative energies for neutrals and anions are depicted in Fig. 4.112. This figure shows that the 2D quasi-planar structures of the neutral and anionic clusters, 28n-1 (C_s, $^1A'$) and 28a-1 (C_2, ^2A) are considered to be the global minima. In contrast, the tubular double-ring structure 28n.2, which is composed of two 14-membered connected rings with antiprism bonding motif, is by 4.9 kcal/mol less stable than the 28n.1 isomer and thus no more favorable. The Fig. 4.112 also shows that the next low-lying neutral clusters 28n.3 (C_2, ^1A) and 28n.4 (C_s, $^1A'$) are by 7.9 and 10.9 kcal/mol less table than the 28n.1 cluster and are exhibiting quasi-planar structures. The authors found that even when the neutral cluster 28n.1 of B_{28} adopts one excess electron, the B_{28}^- anion still prefers to keep the quasi-planar structure. Three of the low-lying isomers of the B_{28}^- anions possess 2D shapes, while the fourth one exhibits a tubular structure composed of a double rings. However, only 2.6 kcal/mol is the energy gap between the first two isomers 28a.1 (C_2, ^2A) and 28a.2 (C_1, ^2A) and therefore they are considered as nearly degenerate clusters. Also, it is interesting to note that the structure of the anion 28a.3 (C_s, $^2A'$) is in fact the same of the neutral isomer 28n.1. The isomer 28a.3 is by 6.2 and 2.9 kcal/mol higher in energy than the isomer 28a.1, respectively, obtained at the CCSD(T)/6-311G(d) and PBE/6-311+G(d) levels of theory. Therefore, one can expect that the experimental observations should include at least one of the three low-lying isomers.

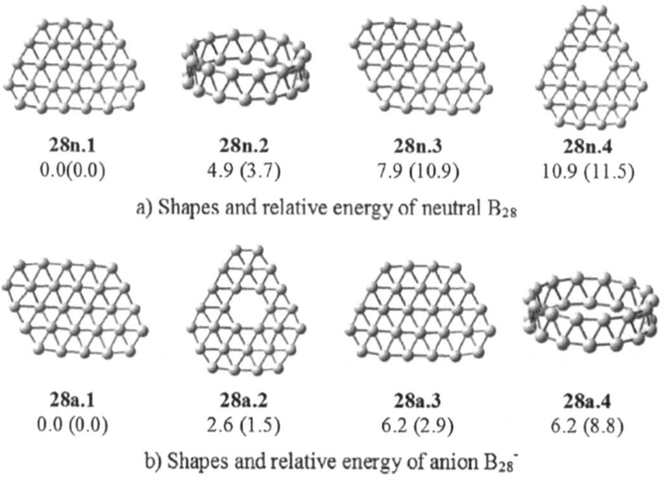

Fig. 4.112 Shapes and relative energies (kcal/mol) of the low-lying isomers of the **a** neutral B_{28} and **b** anionic B_{28}^- clusters, obtained at CCSD(T)/6-311G(d)//PBE/6-311+G(d) and PBE/6-311+G(d) (in parentheses). This figure was reproduced from the Physical Chemistry Chemical Physics with kind permission of the Royal Society of Chemistry [125]

Simulated Photoelectron Spectra of B_{28}^-: Tai and Nguyen [125] simulated the PE spectra of the low-lying isomers of B_{28}^- using the TD-DFT methods. Similar to the B_{26}^- and B_{27}^-, they determined also the related vertical and adiabatic values, VDE and ADE, and expected that both isomers 28a.1 (C_2, ^2A) and and 28a.2 (C_1, ^2A) will be presented in the experimental spectra. The calculations of the relative energies of both B_{28}^- anions 28a.1 and 28a.2, determined at the CCSD(T)/6-311G(d) and PBE/6-311+G(d) levels of theory are respectively only 2.6 and 1.5 kcal/mol and therefore they are almost degenerate, as can be seen in Fig. 4.112b. By an electron detachment from the singly occupied molecular orbital (SOMO) of the 28a.1 isomer, one obtains 3.68 eV for the first VDE which is by 0.09 eV slightly larger than the first VDE value of 3.59 eV determined for the isomer 28a.2. In the case of detaching one electron not from the SOMO but directly from the next laying orbital HOMO-1, then one receives the first triplet state with the corresponding second VDEs of 3.93 and 4.12 eV for the isomers 28a.1 and 28a.2, respectively. As already shown in Fig. 4.113, the next following five VDE values of the 28a.1 and 28a.2 isomers are closed together, and therefore the experimental spectrum is expected to include some of them.

Neutral Boron Clusters of B_{28}: Zhao et al. [127] studied also the structures of the neutral boron clusters B_{28} using an unbiased search on the PES of these clusters carrying out with simulated annealing (SA) procedure incorporated with ab initio molecular dynamics. The final structure from SA-AIMD search was further refined by the basin hopping (BH) algorithm also combined with ab initio calculations. These methods in both SA and BH searches were applied using the density functional theory (DFT) with the plane-wave basis (400 eV cutoff energy) and ultrasoft

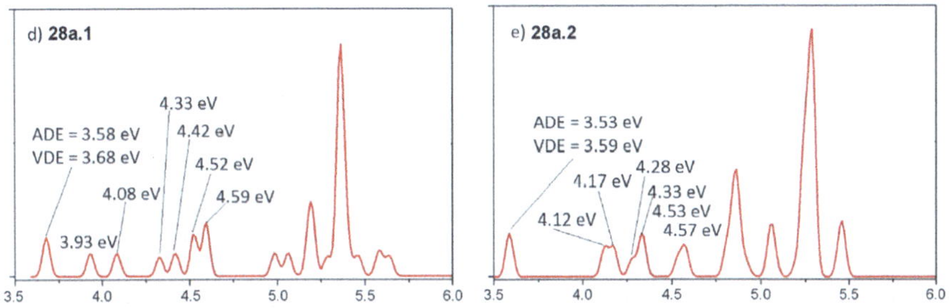

Fig. 4.113 Simulated photoelectron spectra of the B_{28}^- isomers 28a-1 and 28a-2, obtained at the TD-PBE/6-311+G(d) level. This figure was reproduced from the Physical Chemistry Chemical Physics with kind permission of the Royal Society of Chemistry [125]

pseudopotentials. The global searches yield four low-lying isomers, one with 2D isomer in form of quasi-planar structure, and three 3D isomers in form of a cage, tubular, and bowl structures, as presented in Fig. 4.114. The quasi-planar, double-ring and bowl isomers were 0.034, 0.173 and 1.766 eV above the global minimum of the cage, obtained at the PBE0/6-311+G(d) level, while they were about 0.029, 0.179, and 1.854 eV above the global minimum of the cage, obtained at the CCSD(T)/6-311+G(d) level, respectively. In addition they simulated the photoelectron spectra for these four isomers at the PBE0/6-311+G(d) level, as can be seen in Fig. 4.114. They also investigated the anions B_{28}^- and found that the global minimum for the anion is still the cage structure followed by the double-ring structure by 0.112 eV above the global minimum, then by the quasi-planar structure by 0.182 eV, and then by the bowl structure by 1.461 eV, all obtained at the PBE0/6-311+G(d) level of theory. The HOMO-LUMO gap of these cage, quasi-planar, double-ring, and bowl isomers are 2.104, 2.489, 1.799, and 1.701 eV, respectively. However, the largest gap is around 2.5 eV which belongs to the quasi-planar structure denoting its high stability.

The Aromaticity of B_{28}: Zhao et al. [127] investigated further the nucleus independent chemical shift (NICS) as a quantitative measure of aromaticity. They computed with the PBE0/6-311+G(d) method the NICS values at the centers for various cluster isomers. The B_{28} cage from the global search possesses an appreciable negative NICS value of about 38 ppm, suggesting that it is strongly aromatic. A rough correlation between the NICS values and relative energies is observed for the B_{28} isomers. The meta-stable double-ring isomer that is close to the ground state cage also has a large negative NICS value of 38.8 ppm, whereas the energetically favorable quasi-planar structure has smaller NICS value of −24.2 ppm, while the energetically unfavorable bowl-like structure is even anti-aromatic with a positive NICS value of +5.82 ppm.

The Components of Cage B_{28}: Zhao et al. [127] also found that the B_{28} cage can be built or sandwiched from two B_{12} clusters linked by two B atoms and one B_2 dimer, evenly enclosing three polygonal holes, as shown in Fig. 4.115. The stability of the B_{28} cage, can be related to the quasi-planar B_{12} which is a unique doubly aromatic system of σ-, and π-bonds.

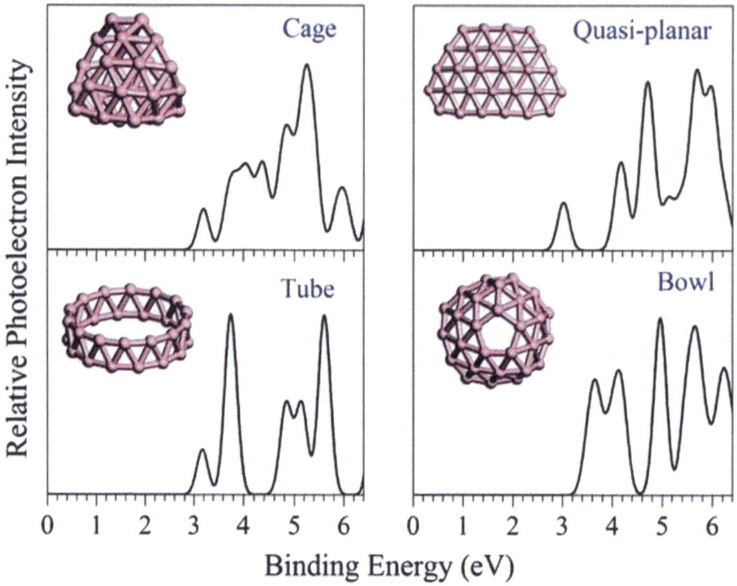

Fig. 4.114 The low-lying isomers of B_{28} of cage, quasi-planar, double-ring, and bowl structures with the corresponding simulated photoelectron spectra, obtained at the PBE0/6-311+G(d) level of theory. A uniform Gaussian broadening of 0.12 eV was adopted. This figure was reproduced from the Nanoscale with kind permission of the Royal Society of Chemistry [127]

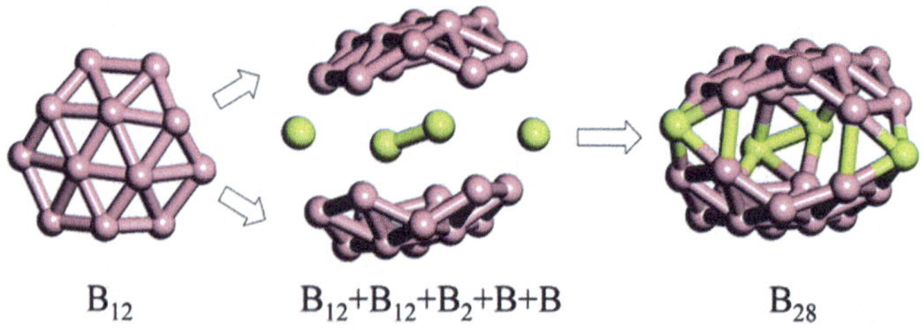

Fig. 4.115 Routine for constructing the B_{28} cage from two quasi-planar B_{12} units and three linkers ($B_2 + B + B$) highlighted in yellow. This figure was reproduced from the Nanoscale with kind permission of the Royal Society of Chemistry [127]

Neutral and Anionic Clusters of B_{29} and B_{29}^-: Tai and Nguyen [125] extended their investigations to study the structures of the neutral and anionic boron clusters B_{29} and B_{29}^-, using the same theoretical methods applied by the $B_{26}^{0/-}$ to $B_{28}^{0/-}$ clusters, mentioned above. The structures of the low-lying isomers and the calculated relative energies for neutrals and anions are presented in Fig. 4.116c, d. This figure shows that the quasi-planar structure of the isomer 29n.1 (C_2, ^2A) exhibits one hexagonal hole and is the global minimum of the neutral B_{29} clusters, obtained by a good agreement between CCSD(T) and DFT computed results. The next most stable isomer 29n.2

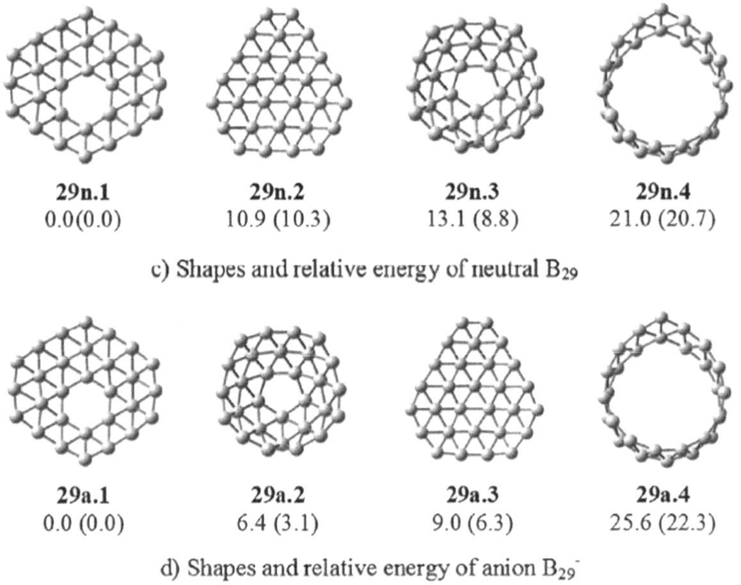

Fig. 4.116 Shapes and relative energies (kcal/mol) of the low-lying isomers of **c** neutral B_{29} and **d** anion B_{29}^- obtained at CCSD(T)/6-311G(d)//PBE/6-311+G(d) and PBE/6-311+G(d) (in parentheses). This figure was reproduced from the Physical Chemistry Chemical Physics with kind permission of the Royal Society of Chemistry [125]

(C_1, ^2A) exhibits also a quasi-planar structure and is likely composed by triangular B_3 units and with around 10.9 kcal/mol higher in energy as the isomer 29n.1. The third low-lying isomer 29n.3 (C_s, $^2A'$) has a bowl structure but with a central hole. It is similar to the bowl structure of the B_{30} IX isomer with a one less peripheral boron atom, as given in Fig. 4.120. The isomer 29n.3 is by 13.3 kcal/mol higher in energy than the isomer 29n.1, while the tubular isomer 29n.4 (C_s, $^2A'$) is by 21.0 kcal/mol is much higher in energy than 29n.1. Thus, the structure of the most stable neutral boron cluster B_{29} is the global minimum 29n.1, which can be considered as the smallest unit assigned by a hexagonal hole defect. The structure of the most stable anionic boron cluster B_{29}^- is the global minimum 29a.1, which has the same shape of the neutral isomer 29n.1. The next anionic isomer 29a.2 (C_s, $^1A'$) has the same form of the neutral 29n.3 and with 6.4 kcal/mol less stable than the 29a.1 isomer. The other low-lying anions 29a.2 (C_s, $^1A'$) and 29a.3 (C_s, $^1A'$) are by 9.0 and 25.6 kcal/mol less stable than the 29a.1 isomer, respectively.

Simulated Photoelectron Spectra of B_{29}^-: Tai and Nguyen [125] simulated the photoelectron spectra of the anionic isomer 29a.1 of the B_{29}^- cluster, which is presented in Fig. 4.116d, using TD-DFT methods. Similar to the B_{26-28}^-, the authors determined also the related vertical and adiabatic values, VDE and ADE. As shown in Fig. 4.117, the presented simulated photoelectron spectra of the anionic isomer 29a.1 (C_1, ^1A) are obtained at the TD-PBE/6-311+G(d) level of theory. The first VDE of 3.62 eV,

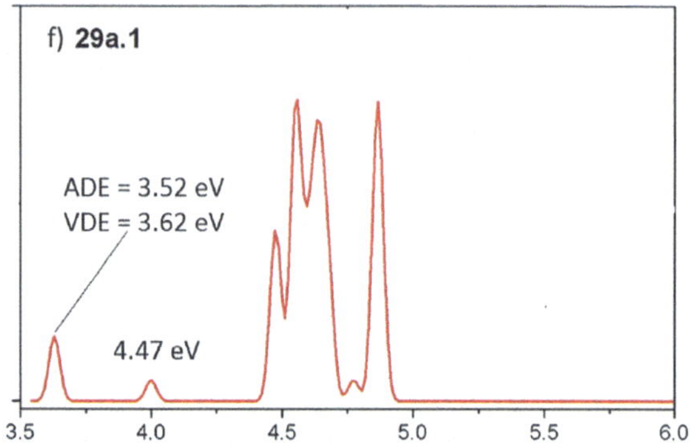

Fig. 4.117 Simulated photoelectron spectra of the B_{29}^- isomer 29a.1, obtained at the TD-PBE/6-311+G(d) level. This figure was reproduced from the Physical Chemistry Chemical Physics with kind permission of the Royal Society of Chemistry [125]

obtained by removing one valence electron from its doubly occupied HOMO, is as expected slightly larger than the ADE value 3.52 eV. This can be related to the minimal change in the geometry, namely from the anionic to the neutral structures. The second VDE of the 29a.1 isomer is obtained at 4.00 eV arose by an electron detachment from the corresponding molecular orbital HOMO-2. However, further peaks were found between 4.47 and 4.87 eV exhibiting high intensity.

Anionic Boron Clusters B_{30}^-: Wei et al. [128] studied the geometrical and electronic structures of the anionic B_{30}^- boron clusters using a joint study of photoelectron spectroscopy and of ab initio quantum chemical and density functional theory. Experimentally, the apparatus of the photoelectron spectroscopy, the production of the anionic boron clusters, and the characterization and identification of the mass-selected B_{30}^- isomers is already described in section "The Photoelectron Spectra of B_{20}^-", Sect. 4.4.1.

The Quasi-planar Chiral Boron Clusters of B_{30}^-: Wei et al. [128] found that the global minimum of the mass-selected isomer B_{30}^- exhibits a quasi-planar structure with a hexagonal hole located nearly the center of the anionic cluster, as can be seen in Fig. 4.118. This isomer is not only assigned by the hexagonal hole but also by its unique chiral Character.[12] The most stable structure of B_{30}^- seems to possess a pair of 2D enantiomers[13] characterized by symmetrical shapes possessing hexagonal holes in different axial symmetrical positions. This unique chiral form has never been observed in boron compounds except of the two boranes [B_4H_5] and [B_4H_5]$^-$, which were considered theoretically. For larger boron cluster sizes, it might exist such kind of chiral structures and of course it might experimentally be reproducible using size-selected cluster beams and definite surfaces.

[12]Chiral molecule or cluster that they exist in the form of a pair of stereo-isomers that are mirror images of each other, having identical physical and chemical properties unless they are interacting with other chiral species.

[13]An enantiomer is one of the two molecules that are mirror images of each other.

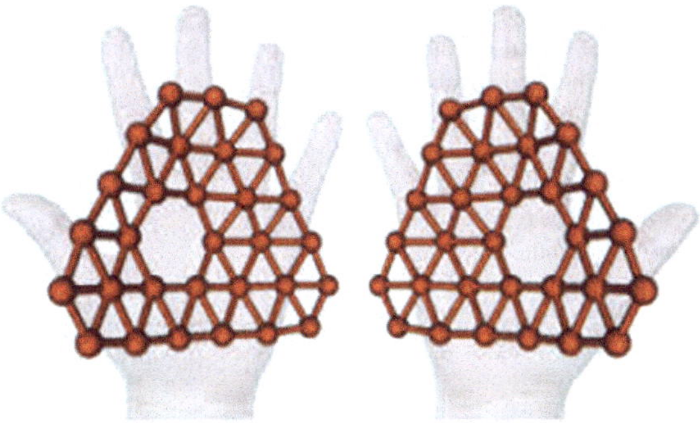

Fig. 4.118 The quasi-planar chiral B_{30}^- anionic cluster. This figure was reproduced from Angewandte Chemie with kind permission of Wiley-VCH Verlag [128]

Photoelectron Spectrum of B_{30}^-: Li et al. [128] determined the photoelectron spectrum of B_{30}^-. The observed spectral bands of the B_{30}^- anion, obtained at 193 nm (6.424 eV), are labeled with the letters (X, A to F) and displayed in Fig. 4.119a. The PES spectrum indicates the first four well-defined bands (X, A, B, C) in the area of the low binding energy region. In the high binding energy region there are few bands (D, E, F) assigned by less intensity. Due to their poorer signal-to-noise ratios, these latter bands are labeled only tentatively. The first measured VDE at the band X is located at 3.69(7) eV, followed by an energy gap of around 0.65 eV, then by three VDEs bands A, B, and C, which are respectively measured by 4.34(5), 4.61(5), and 5.03(8) eV. The authors calculated the VDEs of the low-lying isomers I and II (Fig. 4.120) using the DSCF-TDDFT method with the hybrid functionals PBE0 and TPSSh, in order to verify the global minimum structures of the anionic B_{30}^- cluster. The 46a SOMO[14] orbital of the doublet ground state anionic B_{30}^- cluster, from which occurs the first VDE vertical electron detachment, corresponds to the LUMO,[15] producing the singlet ground state neutral cluster B_{30} of the chiral isomers. The calculated VDE value 3.72 eV of this detachment channel, obtained at the PBE0 functional, is in excellent agreement with the experimental VDE of band X at 3.69(7) eV. Also the calculated ADE of 3.60 eV at the PBE0 functional is also in excellent agreement with the experimental ADE value of 3.59(7) eV. As shown in Fig. 4.119b, the simulated spectrum of the isomers I/II,[16] obtained at the PBE0 functional, is in a good agreement with the observed spectrum in Fig. 4.119a. In fact, the isomer III is energetically nearly degenerate with the isomers I/II, since it is only by 1.90 kcal/mol higher in energy and seems to be quite present in the beam of the boron clusters. However, the simulated spectrum of isomer III failed to agree with the experiment data, as given in Fig. 4.119c. The excellent agreement between the experimental and simulated spectrum of the chiral isomer I gives significant credibility to the theoretical and experimental results ensuring the chiral global minimum of B_{30}^- cluster.

[14]The Singly Occupied Molecular Orbital (SOMO).

[15]The Lowest Unoccupied Molecular Orbital (LUMO).

[16]The simulated spectra of the isomers I and II are identical.

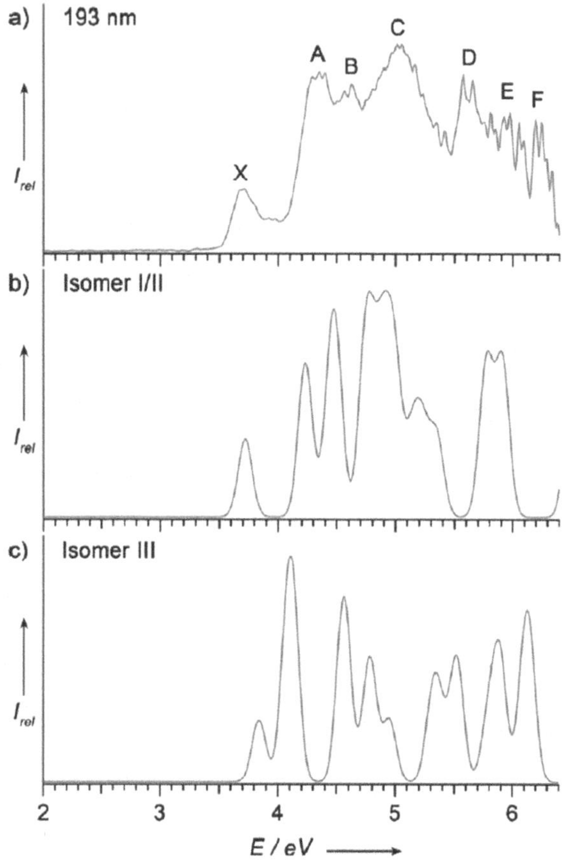

Fig. 4.119 **a** Photoelectron spectrum of Chiral B_{30}^{-} at 193 nm; Simulated spectrum of **b** isomer I/II, and **c** isomer III at TD-PEB0/6-311 + G(2df). This figure was reproduced from Angewandte Chemie with kind permission of Wiley-VCH Verlag [128]

The Isomers of B_{30}^{-}: Based on the Basin-Hopping algorithm [29] in [128]), Li et al. [128] developed the TG-Min code in order to perform the DFT-based global minimum searches of the B_{30}^{-} isomers of boron clusters. Executing the CP2K program, the authors generated and calculated more than 3600 configurations of the anionic isomers using the DFT formalism within the exchange-correlation functional PBE and the pseudopotentials of Goedecker-Teter-Hutter (GTH), associated with DZVP, which is known to be an accurate basis set with a double-ζ valence plus polarization. From these 3600 plausible possible configurations, generated by the TGMin code, the authors manually built different tubular forms and selected two quasi-planar isomers with hexagonal holes being as the global minimum of the B_{30}^{-} cluster. The resulting 11 low-lying isomers, computed at the PBE0 and LPNO-CCSD methods, are presented in Fig. 4.120. Six of these isomers exhibit quasi-planar structures assigned by hexagonal holes. Two isomer possess a bowl-like structures with a central pentagonal holes and two isomers have tubular structures composed of double-, and triple-rings. The last isomer has a 3D structure but without any visible symmetry. The most interesting

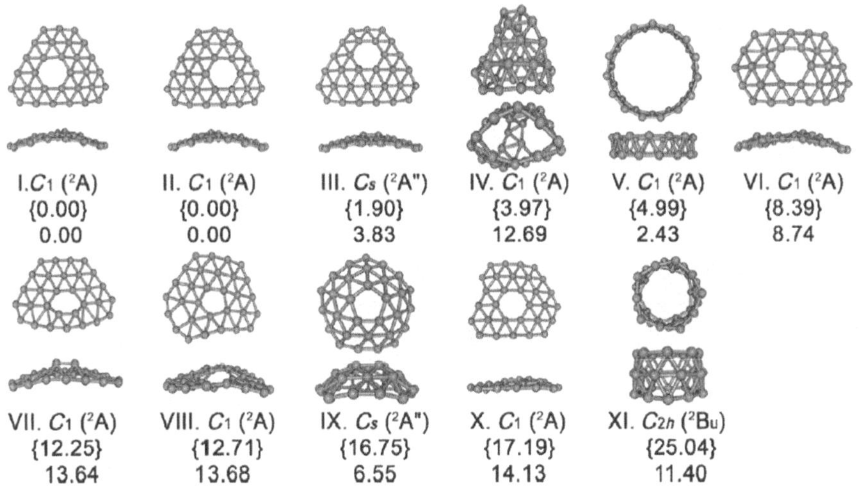

Fig. 4.120 The global minimum and low-lying isomers (I-XI) of B_{30}^-. The relative energies are given in kcal/mol at LPNO-CCSD (in curly brackets) and PBE0 levels of theory. The relative energies at PBE0 level are corrected by zero-point energies (ZPE). This figure was reproduced from Angewandte Chemie with kind permission of Wiley-VCH Verlag [128]

structures of these B_{30}^- clusters are those isomers I and II, which were found to be global minima of these anionic species, consistent with two degenerate quasi-planar chiral structures assigned by hexagonal holes and C_1 symmetry. The structures of both isomers I and II correspond to a pair of enentiomers exhibiting symmetrical positions of their hexagonal holes, as already shown in Fig. 4.118. Due to the fact that the electronic structures of both isomers are identical, therefore they must have the same photoelectron spectrum. However, the isomer III (C_s, $^2A''$) exhibits also a hexagonal hole and is only by 1.90 kcal/mol less stable than the global minima isomers I/II. The amorphous 3D isomer IV and the tubular double-ring structure V are respectively by 3.97 and 4.99 kcal/mol above the global minima of the isomers I and II.

Chemical Bonding of B_{30}^-: Li et al. [128] continued their investigation on the anionic B_{30}^- boron clusters and determined the chemical bonding of the chiral global minimum isomer I. They used the AdNDP method, proposed by by Zubarev and Boldyrev [103], in order to analyze the chemical bonding of the 2D B_{30}^{2-} and B_{30} clusters,[17] expressed by the σ-, and π-bonds and to give a picture of the electron closed-shell density distribution over the isomer, as already done by the previous smaller boron clusters. As can be seen in Fig. 4.121, the AdNDP analysis reveals three sets of σ-bonds. The first set contains seventeen 2c-2e σ-bonds, localized at the peripheral atoms of the quasi-planar isomer I. The second and the third set contain respectively thirteen 3c-2e σ-bonds and six 4c-2e σ-bonds delocalized at the inner atoms and

[17]The AdNDP method works only for closed-shell systems.

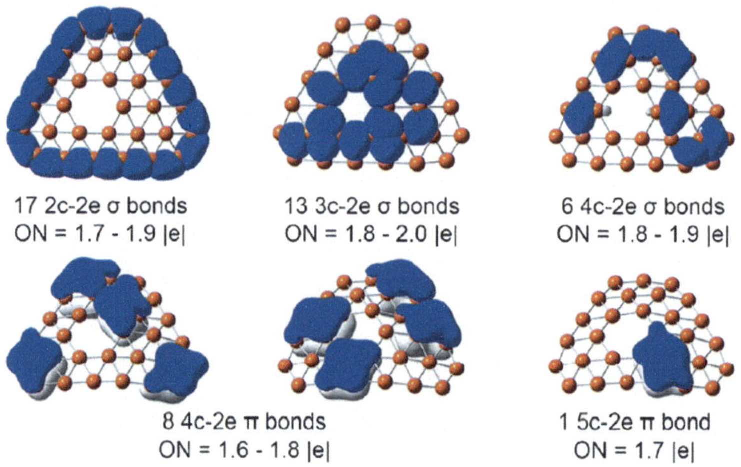

Fig. 4.121 AdNDP chemical bonding analysis of the closed-shell B_{30} (C_1, ^1A). ON = occupation number. This figure was reproduced from Angewandte Chemie with kind permission of Wiley-VCH Verlag [128]

also between the inner and peripheral atoms. The π electron density is distributed in delocalized eight 4c-2e π-bonds and one 5c-2e π-bonds. The occupation number ON of all σ-, and π-bonds lays between 1.6 and 1.9 |e|. The π-bonds deliver the aromaticity for the B_{30} cluster according to the (4N+2) rule of Hückel for the π electrons. However, the authors added an electron to the global minimum isomer I of the B_{30}^- cluster in order to carry out the AdNDP analysis for the planar B_{30}^{2-} cluster. They found that the chemical bondings of the B_{30}^{2-} and B_{30} are identical except the one of the multi-center 9c-2e π-bond in the dianionic B_{30}^{2-} cluster. Therefore, the delocalized π-bonds insert very important contribution for the stability of the planar boron clusters and the aromaticity reinforces their planarity. Nevertheless, despite the aromaticity of the planar B_{30}, this neutral cluster remains the high energy isomer in the potential energy surface.

The Transition Between Enantiomers of B_{30}^-: In order to understand the stability of two enantiomers of B_{30}^-, Li et al. [128] used the so-called dimer method in order to search for possible transition barriers between the chiral isomers I and II. As shown in Fig. 4.122, the lowest transition state is located between the two global minima of the isomers I and II indicating the inversion pathway. The calculated value of this inversion barrier is about 6.2 kcal/mol, determined at the CCSD/6-31g* level of theory. The authors carried out the vibrational frequency analysis of the transition-state geometry and found that there is only one complex eigenvalue at 71i cm^{-1}. The height of this energy barrier between the chiral two counterparts of the global minimum of B_{30}^- is in fact not staggering at all. Therefore, it is possible to overcome this energy barrier by upgraded temperatures, to enable the transition through the inversion of the out-of-plane structure. However, the finding of chiral boron cluster

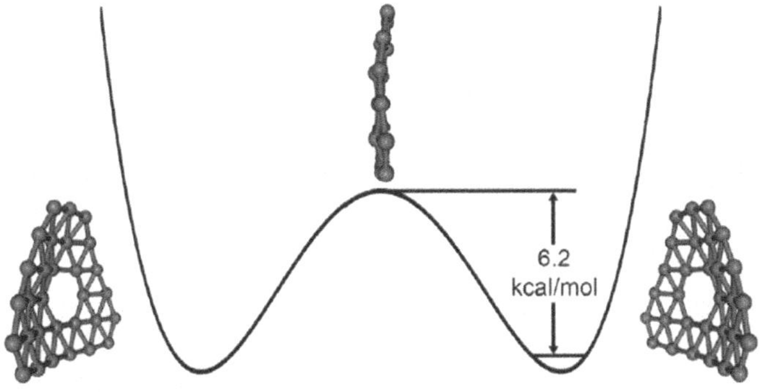

Fig. 4.122 A schematic drawing of the potential energy curve showing the transition state between the two enantiomers of B_{30}^- via a bending mode. This figure was reproduced from Angewandte Chemie with kind permission of Wiley-VCH Verlag [128]

B_{30}^- could open the door to search for larger boron clusters with chiral structures namely experimentally by deposition of size-selected clusters on suitable surfaces.

Neutral Boron Clusters B_{30}: Tai et al. [129] investigated 2D and 3D structures of B_{30} clusters using ab initio quantum chemical methods. The results show that the neutral B_{30} boron cluster favor the bowl over the both tubular structures of the double-, and triple-ring forms. B_{30} has a bowl rather than a double-ring or a triple-ring tubular structure. This bowl isomer exhibits disk-aromaticity similar to that found for B_{19}^- and B_{20}^{2-}. However, the bowl structure of B_{30} is the isomer IX of B_{30}^- of Fig. 4.120, determined by Li et al. [128]. The isomer bowl I is composed of an inner five-membered ring, surrounded by two larger ten- and fifteen-membered rings, as shown in Fig. 4.123. The HOMO-LUMO gap of I is equal to 2.2 eV and much larger as compared to the values of 1.3 eV for II and 0.2 eV for III. The PBE cohesive energy (E_c) of I amounts to 5.499 eV per atom and is close to the value of 5.640 eV per atom of the buckyball B_{80}. While the isomer II has a relatively high E_c value of 5.485 eV per atom, that of isomer III is lower E_c (III) = 5.461 eV per atom.

Disk Aromaticity of B_{30}: Tai *et al.* [129] determined the molecular orbitals of B_{30} and presented in Fig. 4.124. The authors show that the shapes selected molecular orbitals (MOs), reveal that I contains 20 valence π-electrons that fully occupy some MOs, as given in Fig. 4.124. Based on Hückel's rule, this species must be anti-aromatic (4N electrons, N = 5). However, according to a model of a particle in a box, the isomer I is a disk-aromatic system whose valence π-electrons fully occupy the lowest eigenstates in ascending order of $(1\sigma)^2(1\pi)^4(1\delta)^4(2\sigma)^2(2\pi)^4(1\phi)^4$. These predictions agree with NICS calculations. $NICS_{zz}$ values calculated at various positions show that I and II are highly aromatic species with highly negative $NICS_{zz}$ values ($NICS_{zz}(1)$ (I) = –57.8 to –71.1 and $NICS_{zz}(1)$ (II) = –4.1 to –29.8). In contrast, the double-ring tubular structure III is anti-aromatic with highly positive $NICS_{zz}$ values. The aromatic character of isomers I and II is presumably the main reason for their high stability, whereas the double-ring III becomes less stable due to its anti-aromatic character.

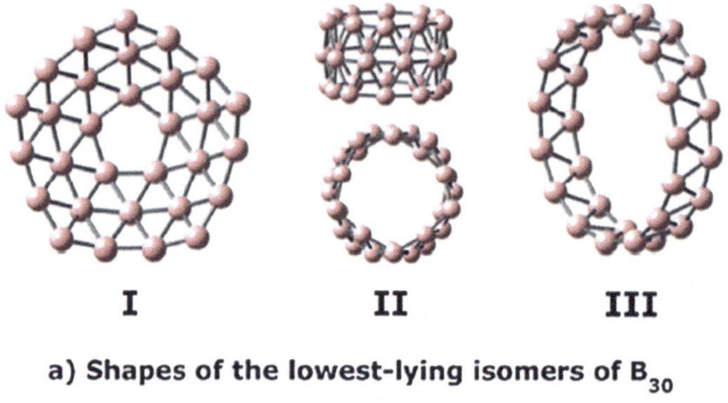

Fig. 4.123 The global minimum of the neutral B_{30} is the bowl structure. The low-lying isomers are the triple- and double-ring tubular structures. This figure was reproduced from the Chemical Communications with kind permission of the Royal Society of Chemistry [129]

Neutral Boron Clusters B_{32}: Boustani et al. [130] investigated thirteen isomers of 2D and 3D structures of B_{32} clusters using ab initio quantum chemical methods. They proposed reasonable initial geometries for the B_{32} clusters and optimized the structures of these isomers at the HF-SCF level using the small basis set STO-3G. Structural reoptimization was carried out using the extended basis set 3-21G and the direct-SCF procedure. However, due to the large number of electrons in B_{32} (160 electrons) the authors did not include correlation corrections to the HF-SCF/3-21G calculations. They found that the 3D spherical clusters are less stable than tubular and quasi-planar clusters. However, the most stable spherical B_{32} cluster has D_{4h} symmetry {isomer (5) in [130]} with a binding energy per atom of 3.10 eV. The most stable isomer of the quasi-planar B_{32} clusters has C_{2h} symmetry {isomer (7) in [130]} with a binding energy per atom of 3.17 eV. The most stable structures of the tubular isomers of B_{32} clusters are the double-rings {isomer (13) in [130]} and the quadruple-rings {isomer (12) in [130]} (Fig. 4.125), both with D_4 symmetry, with a binding energy per atom of 3.46 and 3.10 eV, respectively.

Anionic and Neutral Boron Clusters B_{35}^- and B_{35}: Li et al. [131] investigated the B_{35}^- and B_{35} clusters using a combination of photoelectron spectroscopy and ab initio quantum chemical and density functional methods, as already used in all previous studies on boron clusters, which were carried out by the groups of L.-S. Wang and A. I. Boldyrev. The authors [131] found that the both anionic B_{35}^- (C_s, $^1A'$) and neutral B_{35} (C_s, $^2A''$) boron clusters exhibit the same quasi-planar structures with a double-hexagonal vacancy (DHV), as presented in Fig. 4.126.

Photoelectron spectrum of B_{35}^-: Li et al. [131] used the photoelectron spectroscopy to determine the spectrum of the photoelectron detachment of the anion B_{35}^-, in the same experimental manner, which was already used for B_{16}^- and B_{20}^- and explained in Sects. 4.4 and 4.4.1, respectively. Four well-defined bands (X, A, B, C) were found in the photoelectron spectrum of the B_{35}^- anion, observed at the 193 nm, and presented

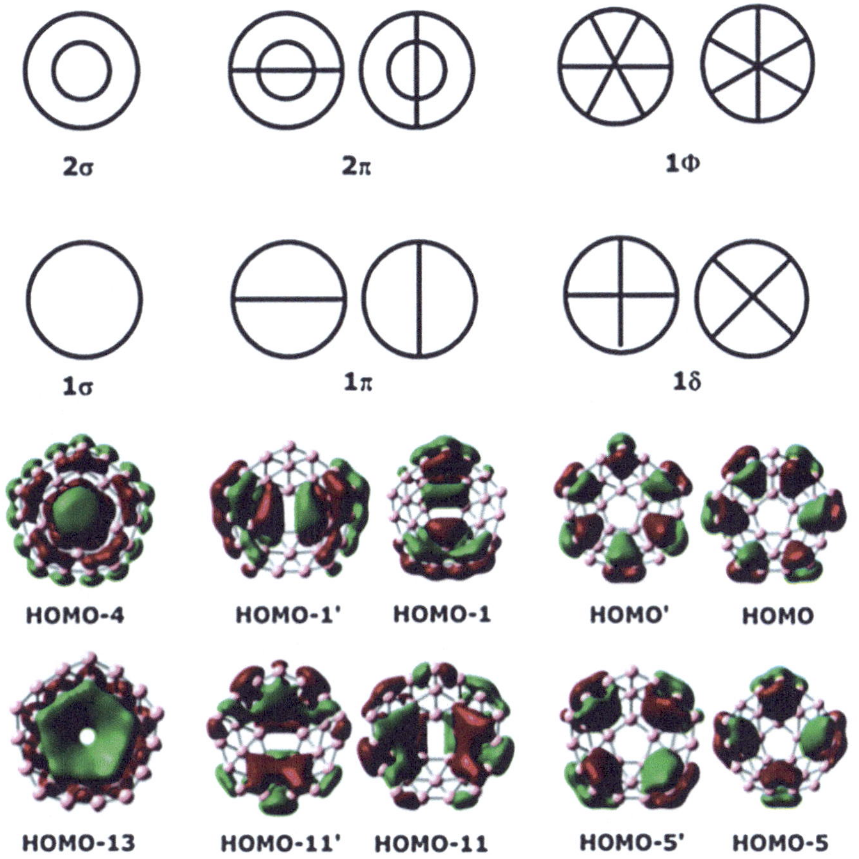

Fig. 4.124 The shapes of π-MOs of the bowl cluster B_{30}. The lowest-lying wavefunctions for a particle in a circular box. This figure was reproduced from the Chemical Communications with kind permission of the Royal Society of Chemistry [129]

Fig. 4.125 Tubular structure of B_{32}, composed of staggered four eight-membered rings. This figure was reproduced from the Chemical Physics Letters with kind permission of the Elsevier Ltd [130]

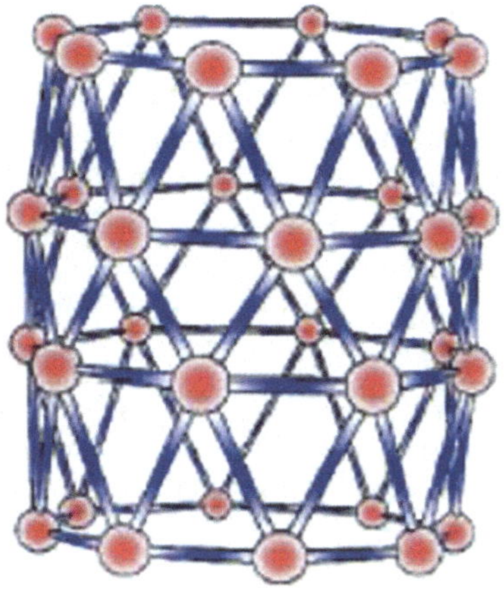

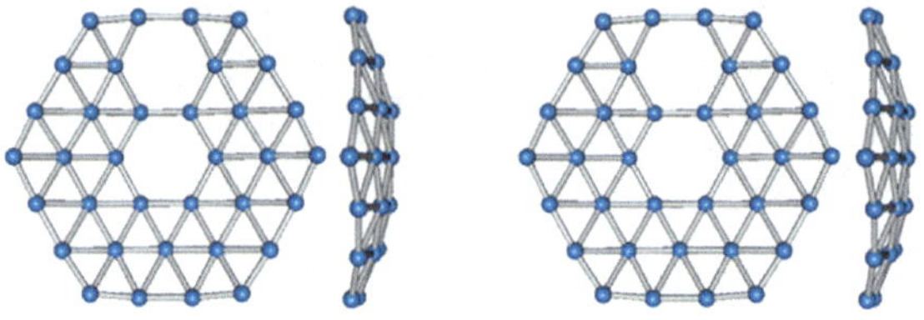

Fig. 4.126 Optimized global-minimum structures of B_{35}^- (C_s, $^1A'$) and B_{35} (C_s, $^2A''$), obtained at the PBE0/6-311+G* level. This figure was reproduced with kind permission of the Journal of the American Chemical Society [131]

in Fig. 4.127. The vertical detachment energy (VDE), observed at 4.06 ± 0.05 eV and lays at the low binding energy band X, is in fact the electron detachment transition from the ground state of the anion B_{35}^- to that of neutral B_{35}. The evaluated ADE value is 3.96 ± 0.05 eV, which represents the electron affinity of the neutral B_{35}, is obtained by drawing a straight line along the leading edge and then adding the instrumental resolution to the intersection with the binding energy axis. The VDE and ADE values of 4.06 and 3.96 eV respectively are close to each other indicating that there is a minor transition between the geometrical structures of B_{35}^- and B_{35} clusters pointing their similarities. After the band X follow three bands A, B, and C. They are higher in intensity than the band A, particularly the band B. However, the band A is separated from the band X by an energy gab of around ~0.9 eV and amounts ~5.0 eV. The prominent and broad band B, centered at 5.35 ± 0.05 eV, alludes to the assumption that it could contain several transitions of multiple electron detachments. At the high binding energy is the well-separated band C, observed at 5.93 ± 0.05 eV. Thus, Fig. 4.127 shows well-structured and relatively simple spectral pattern, which are providing for the B_{35}^- anion and the corresponding neutral B_{35} a representative electronic fingerprint. It is really exciting to know that the determined spectrum of the B_{35}^- anion is unbelievable similar to that one of the next size anion B_{36}^- anion[18] except the feature at a VDE of 3.3 eV, which is located in the low binding energy area. The similarity between the B_{35}^- and B_{36}^- anions is simply to understand namely by filling one of the two hexagonal holes in B_{35}^- with one boron atom to obtain the B_{36}^- anion only with one hexagonal hole. The authors used the basin-hopping program code named TGmin in order to search for the global minimum of B_{35}^-. They found that the related global minimum of the B_{35}^- anion and of the B_{35} neutral is a quasi-planar bowl structure with two hexagonal holes and symmetry C_s ($^1A'$) and C_s ($^2A''$), respectively, as presented in Fig. 4.126. It is to perceive that the quasi-planar structures are in fact triangular close-packed shapes, of course except these two neighboring hexagonal holes. Now arises the question, whether this minimum is the true global minimum for B_{35}^-. Therefore, the authors recalculated these VDEs using the time dependent density functional theory method (TDDFT) and the hybrid functional PBE0 (TDPBE0). The obtained results are then compared with experimental data. For the calculation of the diabetic detachment ADE, the authors used the total energy difference ΔSCF method, and for the computation of the VDEs for the anionic clusters, they used the combined ΔSCF-TDDFT approach. The calculated and experimental ground-state

[18]The anion B_{36}^- and related borophene of [132], will be discussed in Sect. 4.4.3.

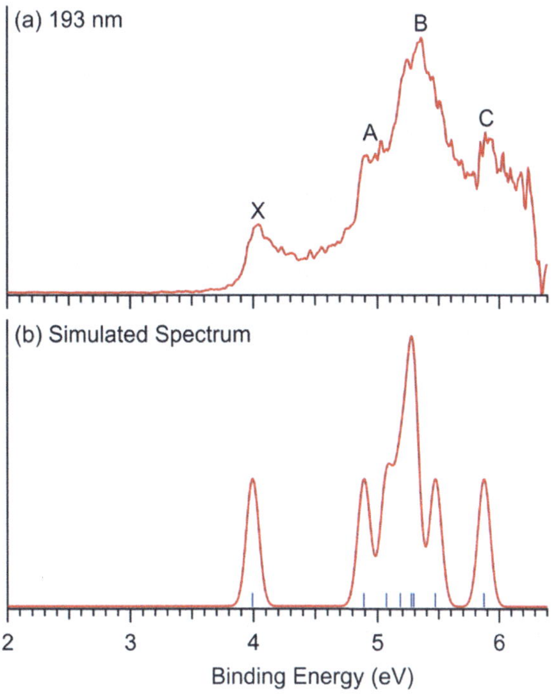

Fig. 4.127 Photoelectron Spectrum of B_{35}^-, observed at 193 nm (6.424 eV) (**a**), compared to the simulated spectrum at the PBE0 level (**b**). The simulation was done for the global-minimum C_s ($^1A'$) structure by fitting the calculated VDEs with unit-area Gaussian functions of 0.05 eV half-width. The vertical lines in **b** represent the calculated VDEs. This figure was reproduced with kind permission of the Journal of the American Chemical Society [131]

VDE values 3.99 and 4.06 eV respectively are in excellent agreement. The calculated and experimental ground-state ADE values 3.91 and 3.96 eV respectively also are in excellent agreement. Concerning the second detachment channel, the calculated and observed (at band A) VDE values 4.89 and ~5.0 eV are in good agreement.

Chemical Bonding of B_{35}^-: Li et al. [131] studied the chemical bonding of the closed-shell system of the anion B_{35}^- und used as before the adaptive natural density partitioning (AdNDP) method, in order to understand the stability of the 35-atom boron clusters. The AdNDP analysis shows that some of the molecular orbitals like HOMO-13, and HOMO-15, HOMO-22, HOMO-26, and HOMO-29 of the 11 σ and π canonical molecular orbitals (CMOs) are in combinations of bonding and partial bonding. As can be seen in Fig. 4.128a, the AdNDP analysis indicates 42 σ-bonds divided into three sets. The first one contains nineteen 2c-2e B-B σ-bonds localized between the eighteen peripheral atoms and one B-B σ-bond shared by the double hexagonal holes (DHV). However, it is rare to find one 2c-2e σ-bond localized in the interior of planar boron clusters. The second set is composed of nine 3c-2e σ-bonds, which are localized around the neighbouring hexagonal holes. The third set contains fourteen 4c-2e σ-bonds, which are distributed between the inner, middle and peripheral rings. Figure 4.128a shows the transformed CMOs into five five-center two-electron (5c-2e) π-bonds. These five π bonds are distributed between the outer B_{18} ring and the middle B_{11} ring. Figure 4.128a also shows two

sets of π systems, which can be transformed from six π CMOs. The first set contains just three 11c-2e π-bonds delocalized between the middle and inner rings, while the other set possesses three 35c-2e global π-bonds distributed over whole 35 atoms of the cluster. However, it is quite interesting to ascertain that each of these eleven π-bonds is obeying the (4n + 2) aromaticity rule of Hückel. Therefore, this triply π aromatic system of the B_{35}^- cluster is unique among all other aromatic π systems.
Comparison π-Bonding with $C_{22}H_{12}$: Li et al. [131] analyzed also the bonding in the hydrocarbon molecule $C_{22}H_{12}$ and compared it with those of the B_{35}^- cluster. The AdNDP analysis indicates that the π bonding of the polycyclic aromatic hydrocarbon, benzo(g,h,i)perylene $C_{22}H_{12}$ is nearly identical to that in the B_{35}^- cluster, as presented in Fig. 4.128b. Just as by the B_{35}^- cluster, the $C_{22}H_{12}$ possesses eleven π-bonds partitioned into three separate π systems. The first one contains five 5c-2e π-bonds localized between the peripheral C-C atoms and correspond to the five 5c-2e π-bonds found in the B_{35}^- cluster. The next π system is composed of three 6c-2e π-bonds distributed between the inner carbon ring, similar to those three 11c-2e π-bonds between the inner and middle rings in B_{35}^-. However, the π bonding in the B_{35}^- cluster, beside the π bonding in many previously investigated planar or quasi-planar boron clusters, provides an additional example for analogy to the π bonding in aromatic hydrocarbons.

The Building Block Clusters B_{35}: In analogous to the B_{36} cluster, which will be discussed by Piazza et al. [132] (Sect. 4.4.3), Li et al. [131] found that the planar B_{35} cluster, which is assigned by double hexagonal holes or vacancies (DHV), can also be considered as a building block cluster to construct boron sheets. In contrast to the C_{6v} B_{36}, which possesses a central hexagonal hole and exhibits a slight out-of-plane distortion,[19] the double hexagonal holes in B_{35}^- or B_{35} provoke less distortion and allow the cluster to flatten. However, the hexagonal vacancies seem to play a very important role in the stabilization of borophene. Therefore, the B_{35} and B_{36} clusters, respectively assigned by double and single hexagonal holes, can be considered as motifs or building blocks for boron sheets. The variations and arrangements of the hexagonal holes in monolayer boron sheets, facilitate the search for novel one-atom-thick boron sheets with different motives and electronic properties. Figure 4.129 shows for the B_{35} cluster schematically two of such arrangements, exhibiting borophenes with mixed hexagonal holes.

The Precursor of Borophene[20] Piazza et al. [132] continued their theoretical and experimental investigations and studied the anionic and neutral boron clusters of B_{36} and B_{36}^-. They used photoelectron spectroscopy combined with ab initio quantum chemical calculations, as already used in B_{35} clusters. Theoretically, they applied the Cartesian Walking (CW) method and TGmin and the constrained Basin-Hopping

[19]This out-of-plane distortion can be related to the peripheral effect, which affects the peripheral atoms and respectively the B-B bonds to be stronger and slightly shorter than those of the inner B-B bonds.

[20]Borophene is a name for boron nanosheets, proposed by L.S. Wang and co-workers et al. [132], considering B_{35} and B_{36} clusters as building block for boron sheets. B_{36}:

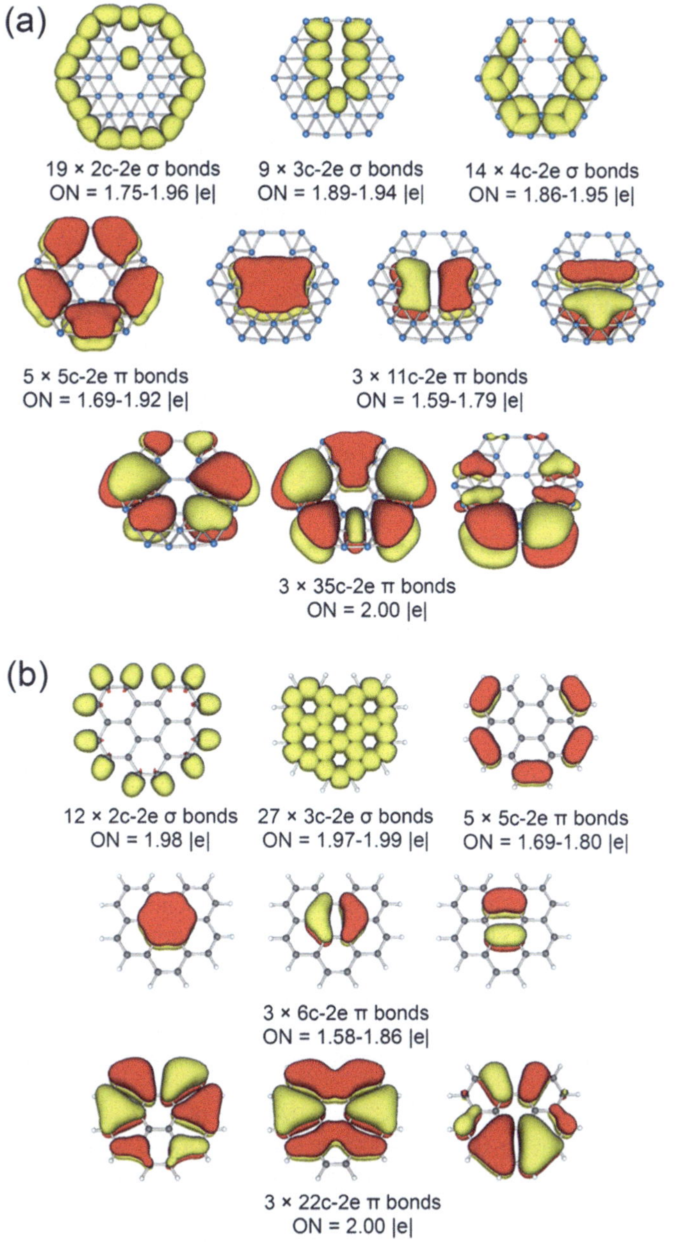

Fig. 4.128 AdNDP analysis for B_{35}^{-} C_s ($^1A'$) (**a**), and $C_{22}H_{12}$ (C_2, $^1A_1$) (**b**). Occupation numbers (ONs) are indicated. This figure was reproduced with kind permission of the Journal of the American Chemical Society [131]

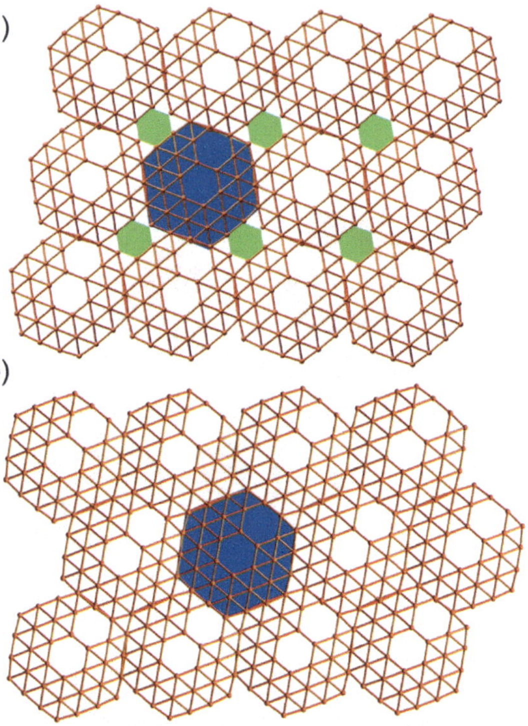

Fig. 4.129 Schematic drawings of borophenes constructed from two different arrangements of the planar hexagonal B_{35} motif. The blue shaded areas represent a single B_{35} unit and the green shaded areas in **a** indicate monohexagonal vacancies as a result of the arrangement of the B_{35} units. This figure was reproduced with kind permission of the Journal of the American Chemical Society [131]

code based on the DFT energies, to search for the low-lying isomers of B_{36}^- clusters. Among the 3000 generated configurations, they found for the B_{36}^- cluster a quasi-planar C_{2v} structure with a hexagonal hole. This structure seems to be the global minimum for the B_{36}^- cluster. The structure of the B_{36} neutral cluster exhibits also similar form with pseudo-C_{6v} and C_{6v} symmetry. The most low-lying isomers of B_{36} and B_{36}^- are presented in Fig. 4.130. The two anionic B_{36}^- isomers, with triple-ring and double hole structures, are by 10.5 and 14.7 kcal/mol higher in energy, obtained at the PBE0 functional. However, the related neutral clusters are by 20.03 and 32.30 kcal/mol much less stable.

The photoelectron spectrum of B_{36}^-: As shown in Fig. 4.131, Piazza et al. [132] determined experimentally the photoelectron spectrum of B_{36}^- cluster and compared with the theoretically simulated photoelectron spectra of low-lying B_{36}^- isomers. The adiabatic detachment energy is 3.12 eV, observed from the sharp onset of the first detachment band X in the 266 nm spectrum (not shown), represents the electron affinity (EA) of the neutral B_{36}. The measured electron affinity EA, revealed by the separation of the X-A bands, corresponds to the energy gap between both bands. The low value of EA indicates that there is a significant geometrical transition from the anion triplet ground state to the neutral singlet ground state. This change will be explained in the next paragraph in connection with the Jahn–Teller-Effect. However,

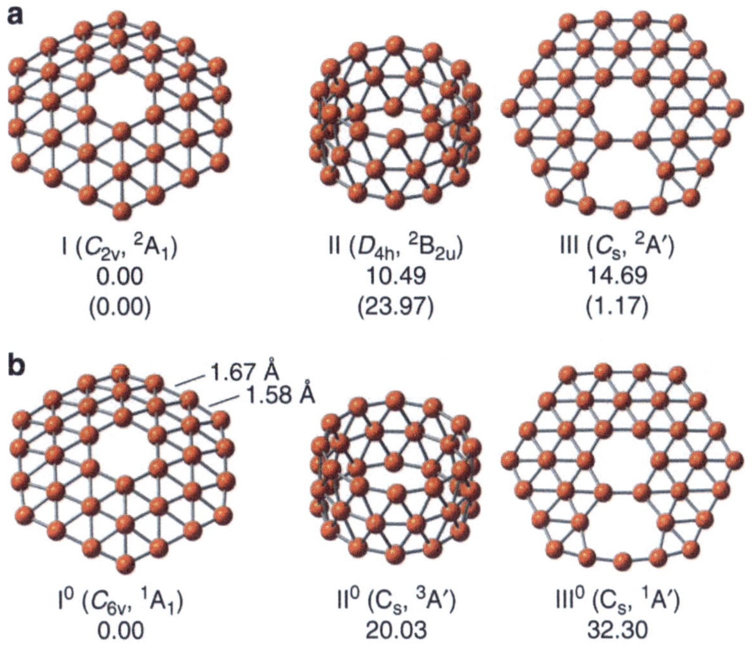

Fig. 4.130 The global minimum and low-lying isomers of B_{36}^- and B_{36}. **a** Relative energies for the isomers of B_{36}^- are given in kcal/mol at the PBE0/6-311G* and CCSD/6-31//PBE0/6-311G* (in parenthesis) levels of theory. **b** Relative energies for the isomers of B_{36} are given in kcal/mol at the PBE0/6-311G* level of theory. This figure was reproduced with kind permission of the Nature Publishing Group [132]

the VDE values 3.15 and 4.03 eV, determined respectively for the X and A bands of the B_{36}^- cluster, are respectively in excellent agreement with the measured values 3.30 and 4.08 eV. We have already learned that the spectral pattern of the experimental and simulated spectra of clusters are in fact their electronic fingerprint. In the case of the B_{36}^-–C_{2v} anion, the experimental spectrum of this global minimum, presented in Fig. 4.131a, is in a good agreement with the simulated pattern in Fig. 4.131b.

The Chemical Bonding of B_{36}^-: Piazza et al. [132] determined the chemical bonding of neutral B_{36} using the AdNDP method [111]. This cluster possesses a perfect hexagonal symmetry C_{6v} with a central hexagonal hole exhibiting bowl-like structure with an out-of-plane distortion, as shown in Fig. 4.132. However, the valence molecular orbitals of the B_{36}–C_{6v} neutral cluster are similar to those of the planar B_{36}–D_{6h} structure, which is about ~9 kcal/mol above the global minimum of the C_{6v} structure. The latter value was obtained at the PBE0/6-311G* level of theory. Therefore, the authors determined the AdNDP analysis for the D_{6h} structure, in order to avoid the mixing of the σ-, and π-bonds in the C_{6v} structure. Thus, the AdNDP analysis indicates three sets of σ bonding. The first set is composed of twelve 2c-2e σ-bonds localized pairwise between the peripheral atoms. The next set contains also

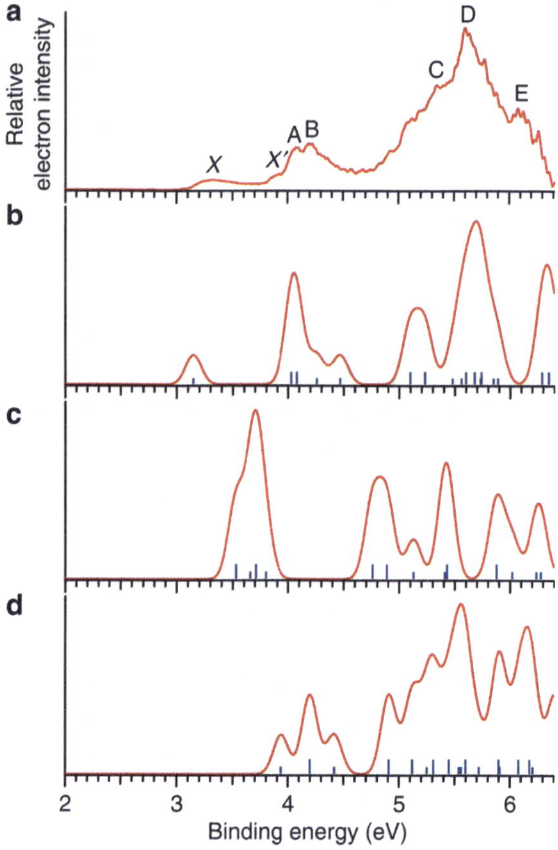

Fig. 4.131 Photoelectron spectroscopy of B_{36}^- and comparison with simulation. **a** The experimental spectrum at 193 nm. **b** The simulated spectrum for the global minimum quasi-planar structure with a hexagonal hole (I in Fig. 4.130a). **c** The simulated spectrum for the tubular isomer (II in Fig. 2a). **d** The simulated spectrum for the quasi-planar structure with two holes (III in Fig. 2a). The vertical bars in the simulated spectra represent the calculated VDEs. The simulated spectra were obtained by fitting the VDEs with a unit area Gaussian function of 0.05 eV width. This figure was reproduced with kind permission of the Nature Publishing Group [132]

twelve 3c-2e σ-bonds localized between the three rings. The third set is consisted of eighteen 4c-2e σ-bonds localized between the rings. The π bondings are composed of six 4c-2e π-bonds localized at the different four centers, while the other six 36c-2e π-bonds are delocalized over the all 36 atoms of the cluster. The B_{36} cluster can be considered as pattern composed of B_7 units (B©B_6) connected by twelve 3c-2e σ-bonds and six 36c-2e π-bonds.

Jahn–Teller Effect and B_{36} (C_{6v}): Piazza et al. [132] reported that the neutral B_{36} cluster exhibits high stability and possesses a high symmetry point group C_{6v}. Its high stability can be related to the large energy gap between its lowest unoccupied molecular orbital (LUMO) and its highest occupied molecular orbital (HOMO). High symmetry point group means the existence of degenerate molecular orbitals. In the case of the high symmetric B_{36}–C_{6v} neutral cluster, its LUMO is doubly degenerate with the orbital symmetry e_2. As soon as this orbital is occupied by only one electron, the symmetry of this orbital is broken and consequently the symmetry C_{6v} of the B_{36} cluster is distorted due to the Jahn–Teller effect causing a symmetry lowering then

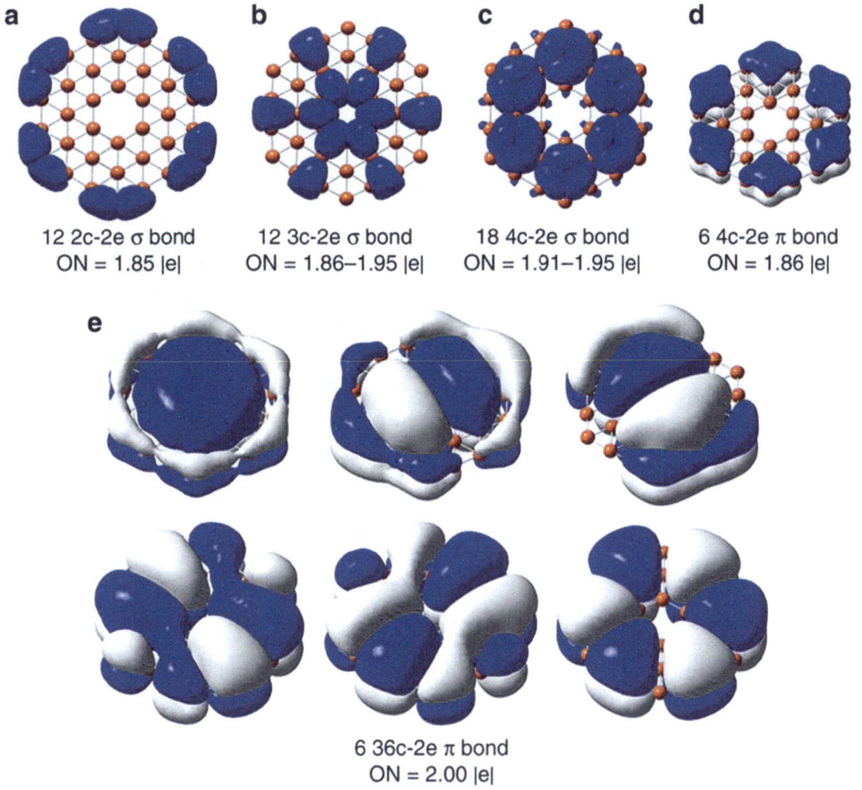

Fig. 4.132 Chemical bonding analysis for the hexagonal D_{6h} structure of B_{36}. The analysis was performed using the AdNDP method. ON stands for occupation number and is equal to 2.00 |e| (meaning a density of exactly two electrons) in the ideal case. This figure was reproduced with kind permission of the Nature Publishing Group [132]

transformation into the lower symmetry point group C_{2v} of the anionic B_{36}^{-} cluster. In the case of a dianionic boron cluster B_{36}^{2-}, which can be generated by attaching two electrons to the degenerate LUMO, then one obtains two degenerate orbitals each is occupied by one electron resulting a triplet ground state. The orbital symmetry e_2 and the cluster symmetry C_{6v} are sustained without any Jahn–Teller distortion.

The Relationship of B_{36}^{-} To 2D Boron Sheets: In 2007 Tang and Ismail-Beigi [133] investigated the triangular boron sheet proposed in 1997 by Boustani [86]. They found that creating some holes in this triangular sheet, called α-sheet, increases its stability. Thus, the hexagonal hole in the B_{36} cluster, found Piazza et al. [132], is reminiscent of the hexagonal holes in the mentioned boron α-sheets, predicted by [133]. Similar to the hexagonal C_6 unit in graphene, the B_{36} cluster can be considered as a boron unit which can form extended graphene-like boron sheets assigned by hexagonal vacancies, as shown schematically in Fig. 4.133. This figure shows that each two

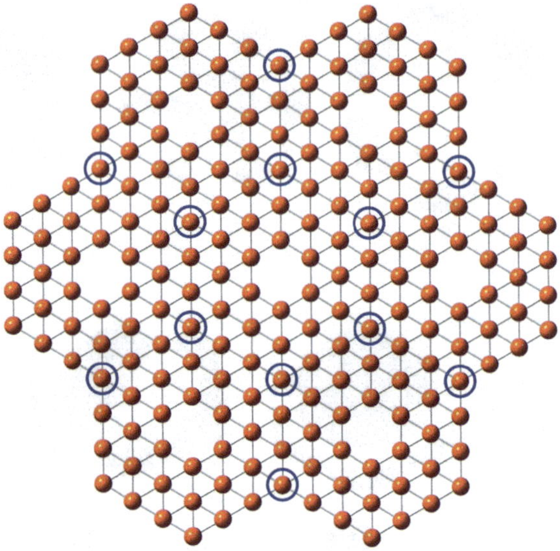

Fig. 4.133 Relationship between B_{36} and borophene. A schematic view of part of an extended one-atom-thick boron sheet, named borophene, as constructed from the hexagonal B_{36} unit. The circles represent the apex atoms in the B_{36} unit that are shared by three units. Removal of these atoms would lead to the α-sheet, proposed by Tang and Ismail-Beigi (see text) [132]

neighboring B_{36} units are shared by row of four B atoms. It also shows that every three neighboring B_{36} units are shared by apex atoms assigned by blue circles. If one counts one vacancy per 27 lattice sites in a triangular lattice, then the hexagonal hole density amounts by $\eta = 1/27$. By removing more atoms from the this unit, one receives the α-sheet with $\eta = 1/9$, as already given in [133]. Thus, it is now possible to propose stable 2D boron sheets using the building block B_{36} unit which turns to be the embryo for developing boron sheets.

Theoretical Confirmation of Quasi-planar B_{36} **and** B_{36}^-: Chen et al. [134] investigated also the isomers of the B_{36} and B_{36}^- boron clusters using the Stochastic Surface Walking (SSW) searches. This SSW method was used to search for global minimum of short-ranged Morse clusters and carbon clusters up to 100 atoms, characterized by their complex potential energy surfaces. The authors confirmed the computational evidence for the quasi-planar all-boron aromatic B_{36} (C_{6v}, $^1A_1$) and B_{36}^- (C_{2v}, $^2A_1$) clusters. Subsequently, the authors used validated density-functional method PBE0 and the basis set 6-311+G* in order to evaluate the energies of the low-lying isomeric structures. They found that the global minimum structures of B_{36} and B_{36}^- are in line with those obtained by Piazza et al. [132], as already reported in the last section. The PBE0 low-lying isomers of neutral clusters have tubular triple-ring (C_{2v}, $^1A_1$), and double-ring (D_9, $^1A_1$) structures as well as cage-like fullerene (C_{2h}, ^1A_g), as shown in Fig. 4.134a, are respectively about 1.10, 2.13 and 2.84 eV above the global minimum of the quasi-planar structure (C_{6v}, $^1A_1$). Structures 1 and 5 adopt a quasi-planar and circular shape, with a central hexagonal hole. These both structures of the B_{36} and B_{36}^- clusters can be considered as being composed of three concentric inner B_6, middle B_{12} and outer B_{18} boron ring systems. The B atoms are in-plane

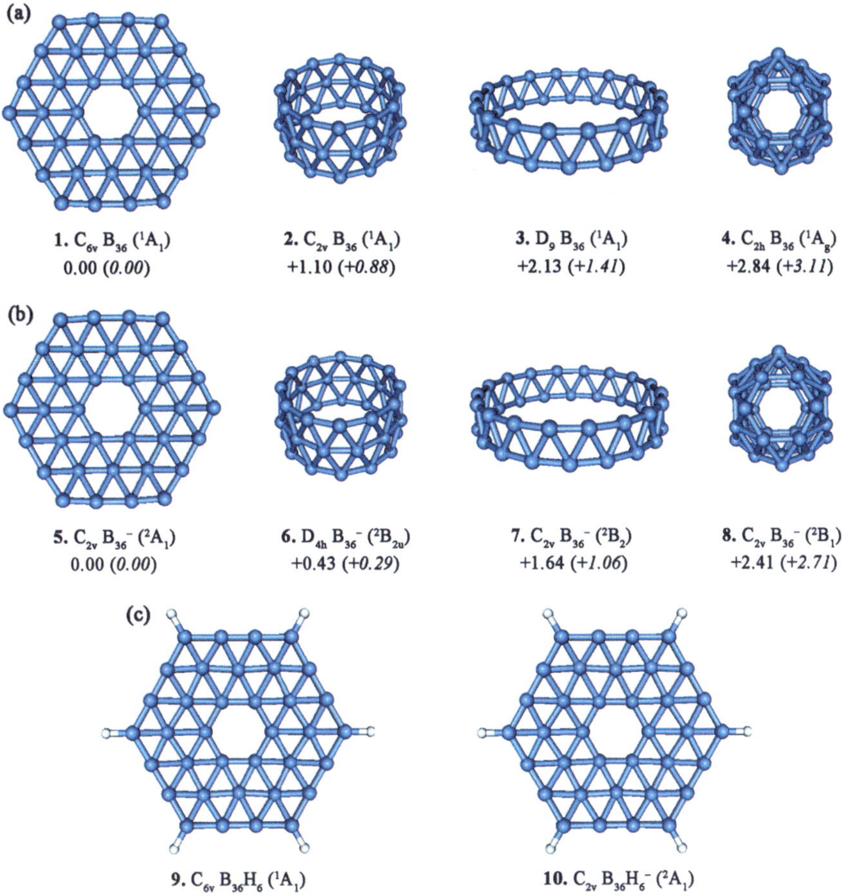

Fig. 4.134 Selected optimized structures at the PBE0/6-311+G* level for B_{36} (1-4), B_{36}^- (5-8), and their hydrogenated clusters $B_{36}H_6$ (9) and $B_{36}H_6^-$ (10). The B_{36} (1) and B_{36}^- (5) structures are established as the global minima. Relative energies are shown in eV at the PBE0 and TPSSh (in italic) levels. This figure was reproduced from the Physical Chemistry Chemical Physics with kind permission of the Royal Society of Chemistry [134]

tri-, tetra-, penta-, or hexacoordinated, leading to electron deficiency for the systems, in particular for the middle B_{12} ring with pure hexacoordinated B centers. The global minimum of the anion B_{36}^- has also a quasi-planar structure with a hexagonal hole. The low-lying isomers of the anions have tubular triple-ring (D_{4h}, $^2B_{2u}$), and double-ring (C_{2v}, $^2A_2$) structures as well as cage-like fullerene (C_{2v}, $^2B_1$), as given in Fig. 4.134b, are respectively about 0.43, 1.64 and 2.41 eV above the global minimum of the quasi-planar structure (C_{2v}, $^2A_1$). Structures 1 and 5 adopt a quasi-planar and circular shape, with a hexagonal hole positioning at the center.

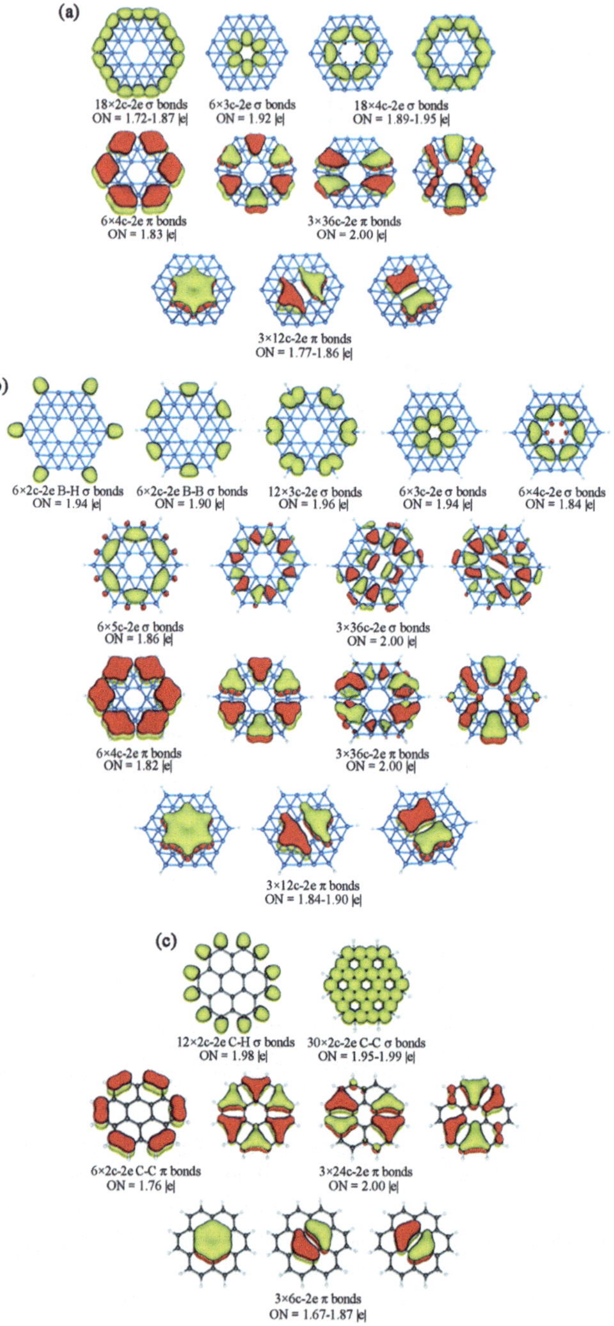

Fig. 4.135 Bonding patterns as revealed from the adaptive natural density partitioning (AdNDP) analysis of **a** B_{36} (1) and **b** $B_{36}H_6$ (9), as compared to **c** coronene ($C_{24}H_{12}$). The occupation numbers (ONs) are indicated. This figure was reproduced from the Physical Chemistry Chemical Physics with kind permission of the Royal Society of Chemistry [134]

Chemical Bonding of B_{36}: Chen et al. [134] studied the chemical bonding of the B_{36} and $B_{36}H_{36}$ clusters as well as of the hydrocarbon molecule ($C_{24}H_{12}$) using the adaptive natural density partitioning (AdNDP) method, proposed by Boldyrev et al. ([76–78] in [111]). The authors [134] found that the AdNDP analysis reveals intriguing chemical bonding pattern in the global-minimum clusters 1 and 5. The AdNDP analysis depicts the chemical bonds in n-center two-electron (nc-2e) bonds, where n is ranging between two and the total number of atoms in the clusters. The classical lone pairs and 2c-2e bonds of Lewis bonding elements and the delocalized non-classical multi-center two-electron bonds all are recovered by the AdNDP analysis. As can be seen in Fig. 4.135, the chemical bonding of the neutral B_{36} cluster consists of three sets of σ-bonds. The first one is composed of eighteen 2c-2e σ-bonds localized between the peripheral atoms. The second set contains six 3c-2e σ-bonds localized around the inner ring. The third set exhibits eighteen 4c-2e σ-bonds, six are localized between the inner and middle rings, and the remaining twelve are localized between the middle and outer rings. From the 54 pairs of valence electrons in B_{36}, the σ framework consumes 42 pairs of electrons. The delocalized chemical bondings are distributed in six 4c-2e, three 36c-2e, and thirteen 12c-2e π-bonds. Bonding analysis indicates that they possess concentric dual π-bond aromaticity, where one set of six π electrons is delocalized around the inner B_6 ring and another sextet is globally delocalized in the outer B_{30} double-chain ribbon. This π bonding pattern is in close analogy to that of coronene ($C_{24}H_{12}$), as will be explained below.

Analogy to Hydrocarbon Coronene C_{24} H_{12}: Chen et al. [134] compared the bonding pattern of the B_{36} cluster and the hydrocarbon coronene C_{24} H_{12} as revealed from the AdNDP analysis. Since a rhombic 4c-2e bond (and to a less extent, a 3c-2e or 5c-2e bond), either π or σ in the double-chain boron nanoribbon systems, is basically equivalent to a 2c-2e C-C bond in the hydrocarbons, a close connection between B_{36} and coronene $C_{24}H_{12}$ (Fig. 4.135) is readily established, making B_{36} an all-boron analog of coronene. To be specific, the three 12c-2e π-bonds in B_{36}, associated with the inner B_6 ring in (Fig. 4.135; the third row), show exact one-to-one correspondence to the three 6c-2e π-bonds in the hydrocarbon coronene (Fig. 4.135; the third row). The three 36-2e π bonds in B_{36} in (Fig. 4.135; the second row) are identical to the three 24c-2e π bonds in coronene, whereas the six 4c-2e island π-bonds in B_{36} in (Fig. 4.135; the second row) are equivalent to the six 2c-2e C-C π-bonds in coronene. This analogy can also be revealed from the π canonical molecular orbitals (CMOs).

4.4.3 Boron Clusters B_n (37 $\leq$ n $\leq$ 40)

Neutral Boron Clusters B_{38}: Lv et al. [135] studied the structures of B_{38} boron cluster using first-principles swarm structure searching calculations. The approach involves beside the ab initio energy calculations a global minimization of energy surfaces. The underlying structural relaxations and molecular dynamics simulations were performed in the frame work of density functional theory within the PBE functionals. Further computations were executed at the MP2/6-311G* and PBE/6-311-G* level of theory. They found that the vast majority of the low-lying isomers of B_{38} clusters are cage-like. The global minimum was found for a highly symmetric B_{38} (D_{2h}, ^1A_g) cluster fullerene analogue, presented in Fig. 4.136 as isomer I. The first two low-lying isomers, an irregular cage and quasi-planar structure, are respectively by 0.17 and 0.21 eV less stable than the ground-state structure D_{2h}. However, Tai et al. [136] reinvestigated the obtained structures of Lv et al. [135] carrying out higher level of theory CCSD(T) and using the 6-311G(d) basis set. The calculations show that the both isomers, the cage fullerene-like structure I and the bowl-like quasi-planar structure II indicated by two hexagonal holes, are energetically degenerate. However, the table in Fig. 4.136 shows, the CCSD level favors the 2D isomer II, while the 3D isomer I is favored by the MP2 and PBE.

Fullerene-like Anion B^-_39: Chen et al. [137] studied the structures of B_{39}^- boron cluster using the minimum hopping (MH) algorithm based on the density functional theory (DFT). They also used the Basin Hopping (BH) method for the global minimum searches. They applied the DFT-PBE0 and CAM-B3LYP approaches and the 6-311+G* basis set to optimize the low-lying structures and to evaluate their relative energies. For the 12 lowest-lying isomers of the B_{39}^- anion, they executed single-point CCSD(T) calculations of geometries, obtained at the PBE0/6-211G* level. The structures of these isomers are presented in Fig. 4.137 and placed energetically relative to the energy of the global minimum, which was placed by zero. The authors found that the cage-like chiral structure of B_{39}^- is the most stable isomer, which seems to be the global minimum. The energy of the 3D cage-like chiral structure of the isomer **1**, obtained at the CCSD(T)/PBE0/6-311G* level of theory, is about 0.28 eV below the energy of the quasi-planar structure of the isomer **7**. However, the calculations, undertaken at the CAM-B3LYP level using the Coulomb attenuating method and considering long-range corrections, confirm that the cage-like structure is only by 0.01 eV lower in energy than the quasi-planar isomer **7**. However, the cage-like B_{39}^- cluster will be completely discussed in section boron fullerenes.

The Larg 2D Quasi-planar $B^-{}_{40}$: Using photoelectron spectroscopy combined with ab initio quantum chemical and density functional methods, Zhai et al. [138] continued their investigations on boron clusters and studied the structures and chemical properties of the anion B-40. They also used the stochastic surface walking (SSW) and basin hopping (BH) algorithms to perform the unbiased global minimum searches. Furthermore, the authors applied the three PBE32, PBE033 and TPSSh DFT methods to optimize the Low-lying structures of the B_{40}^- isomers and to evaluate their relative energies. According to all DFT calculations, the quasi-planar structure of isomer 1 (C_s, $^2A'$) seems to be the global minimum for B_{40}^-. The next low-lying isomer is cage-like fullerene structure 2 (D_{2d}, $^2B_2$) seems energetically to be degenerate with the isomer 1 by an energy difference only of ~2 kcal/mol, obtained at the PBE0 level and shown in Fig. 4.138.

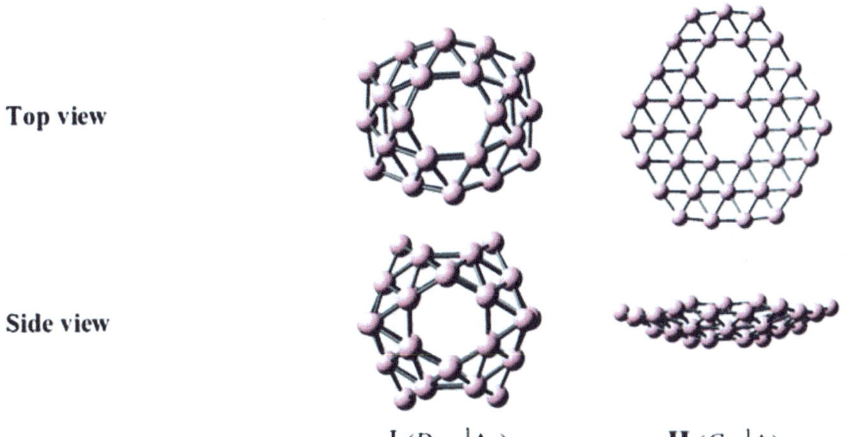

	I (D_{2h}, ^1A_g)	II (C_1, ^1A)
PBE0/6-311G(d)[a]	0.00 eV	0.21 eV
PBE/6-311G(d)[a]	0.00	0.02
MP2/6-311G(d)//PBE0[a]	0.00	3.25
PBE0/6-311+G(d)	0.00	0.38
PBE/6-311+G(d)	0.00	0.18
TPSSh/6-311+G(d)	0.34	0.00
MP2/6-311G(d)//PBE0	0.00	4.26
MP2/6-311G(d)//TPSSh	0.00	3.20
CCSD/6-311G(d)//TPSSh	0.41	0.00
CCSD/6-311G(d)//PBE0	0.50	0.00
CCSD(T)/6-311G(d)//TPSSh	0.00	0.12
CCSD(T)/6-311G(d)//PBE0	0.02	0.00

Fig. 4.136 Shapes and relative energies (eV) of isomer I and II obtained at various computational methods. Figure and table were reproduced from the Nanoscale with kind permission of the Royal Society of Chemistry [135]

The Low-lying Isomers of B_{40}^- and B_{40}: At the PBE0/6-311+G* level of theory, Zhai et al. [138] determined the structures of different B_{40}^- and B_{40} isomers. They presented various configurations of these isomers, as shown in Fig. 4.139. These representative 2D and 3D isomers, in form of quasi-planar, double-ring and triple-ring, and cage-like structures, are placed energetically relative to the energy of the global minimum. The relative energies of the ten low-lying structures of the isomers are located within ~1 eV above the global minimum of the B_{40}^- and B_{40} clusters. The tubular double-ring structure is by 1.38 eV less stable than the B_{40}^- global minimum.

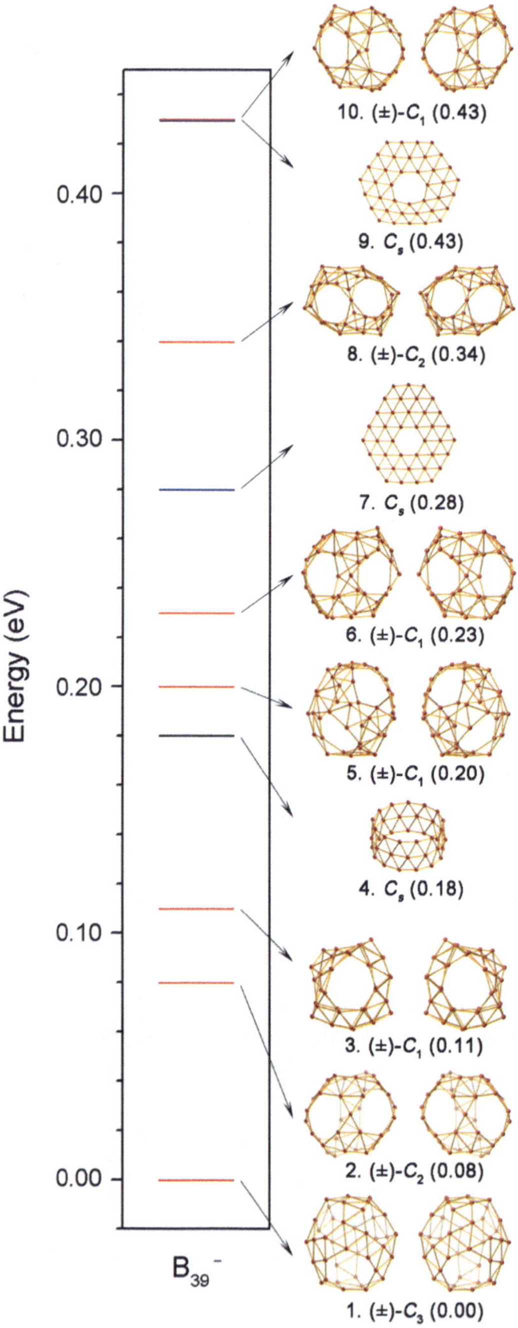

Fig. 4.137 Configurational energy spectrum of B_{39}^{-} at the single-point CCSD(T)//PBE0/6-311G* level. The energy of the global minimum is taken to be zero, and the relative energies are in eV. The red, black, and blue bars denote fullerene-like cages, triple-ring tubes, and quasi-planar structures, respectively. This figure was reproduced from the Nanoscale with kind permission of the Royal Society of Chemistry [137]

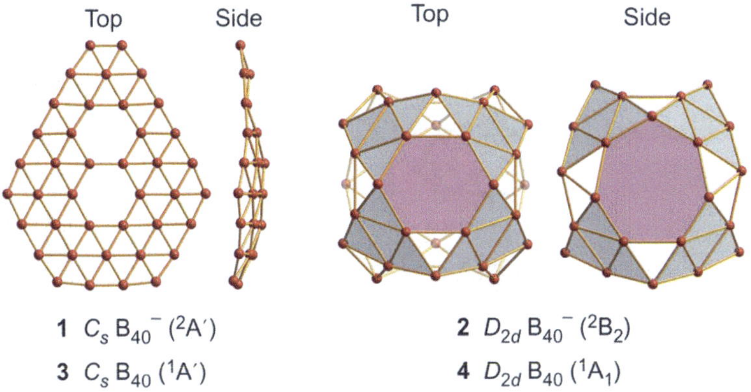

Fig. 4.138 Top and side views of the global minimum and low-lying isomers of B_{40}^- and B_{40} at the PBE0/6-311+G* level. This figure was reproduced with kind permission of the Springer Nature Publishing AG [138]

Observed and **Simulated PES of** B_{40}^-: Zhai et al. [138] simulated the photoelectron Spectra of both B_{40}^- isomers, the quasi-planar and cage-like structures, in comparison with the observed spectra, as shown in Fig. 4.140. The leading band in the spectrum, assigned by X′ and located at ~2.6 eV, is quite very weak and it must be magnified eight time in order to be visible. This band has extremely low vertical detachment energy (VDE) of about 2.62 ± 0.05 eV. The overall spectral pattern is unusually simple for such a large cluster. The three bands X′, X (VDE, 3.63 ± 0.05 eV) and A (VDE, 4.24 ± 0.03 eV) are well separated, which suggests a remarkably stable neutral B_{40} cluster with a sizable energy gap between the lowest unoccupied molecular orbital (LUMO) and the highest occupied molecular orbital (HOMO). The observed B, C and D bands are located respectively in the higher binding energy regions namely at 5.10, 5.55, and 6.08 eV. As we learned in previous sections, the observed photoelectron spectra, characterized by well-resolved spectral features, operate as genuine electronic fingerprints enabling the determination of the B_{40}^- structure in comparison with the computational simulation of the spectrum. However, the experimentally observed spectrum of the B_{40}^- cluster, as presented in Fig. 4.140, does not correlate with any of the theoretically simulated spectra, neither with those of the 2D structure **1**, nor with those of the 3D structure **2**. In other words, the observed spectrum is not reproduced by the simulated spectra of both 2D and 3D structures. Nevertheless, a mixure of the simulated spectra could accomplish a good satisfaction with the experimental data.

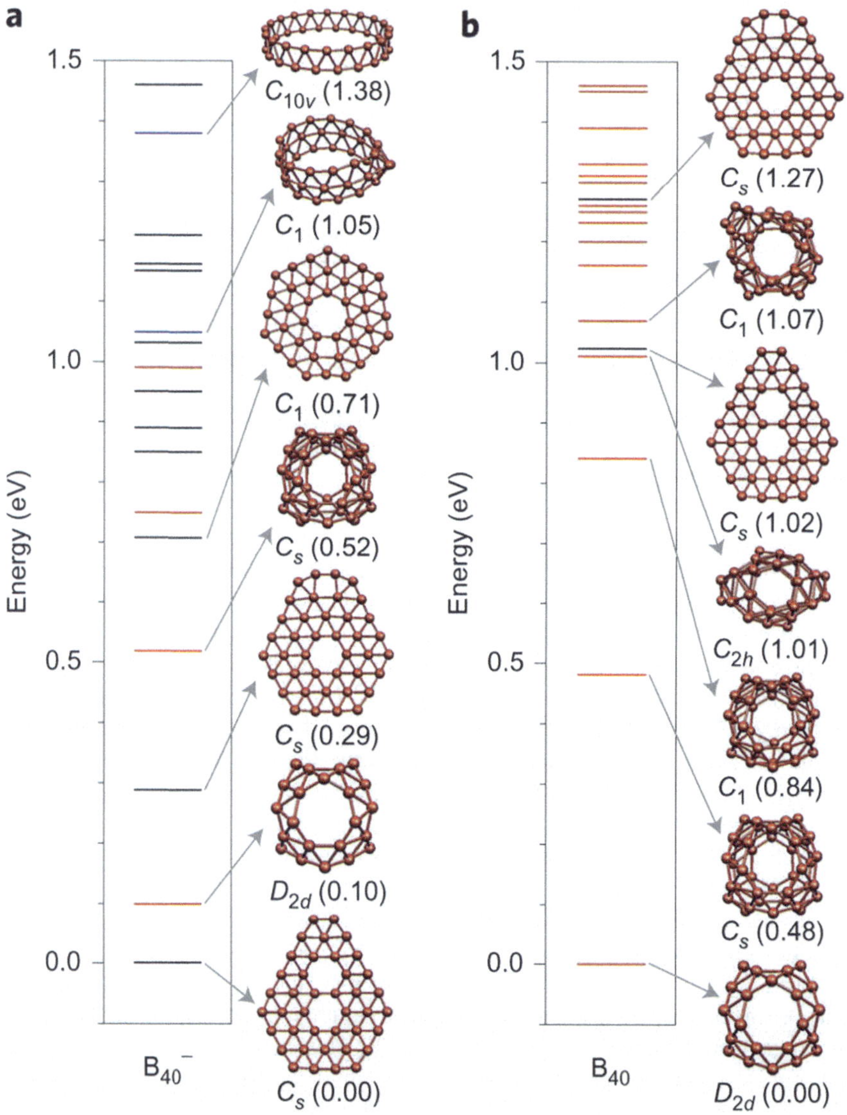

Fig. 4.139 Configurational energy spectra at the PBE0/6-311+G* level. (a) B_{40}^-, (b) B_{40}. The energies of the global minima are taken to be zero. Black, quasi-planar structures; red, fullerene-like cages; violet, double-ring tubular structures; blue, triple-ring tubular structure. This figure was reproduced with kind permission of the Springer Nature Publishing AG [138]

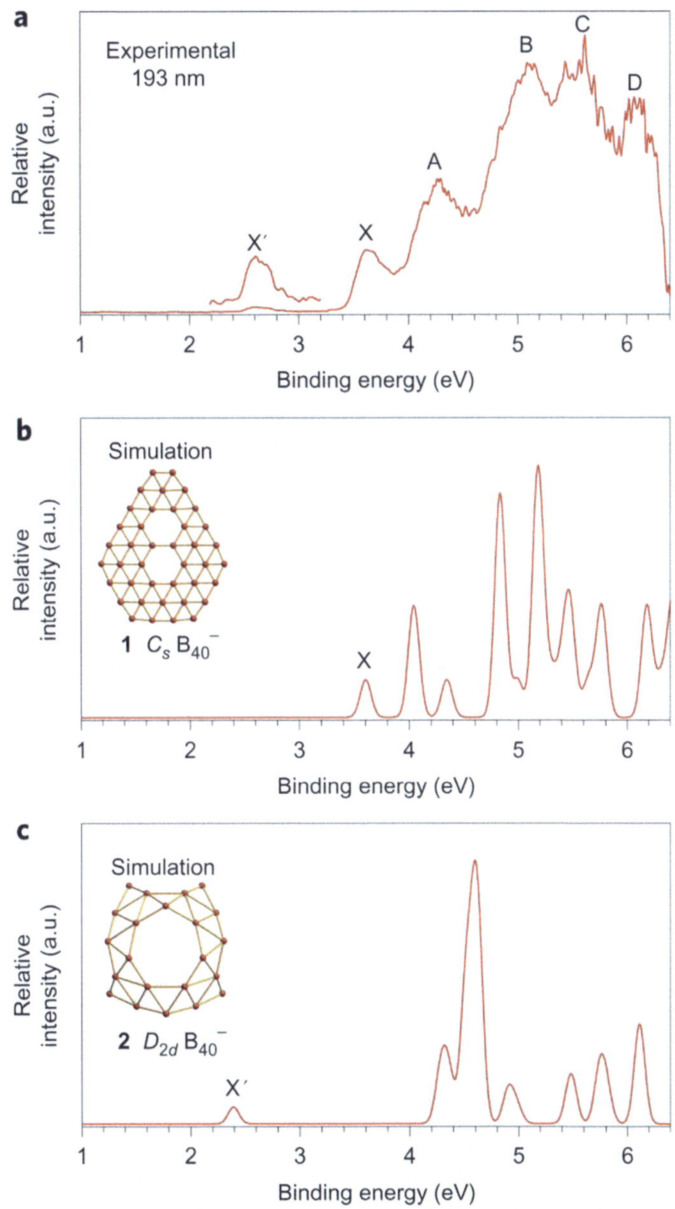

Fig. 4.140 **a** Photoelectron spectrum of the B_{40}^- cluster in comparison with simulated spectra. The experimental spectrum is observed at 193 nm (6.424 eV). **b** and **c**The simulated spectrum of the quasi-planar structure **1** (C_s, $^2A'$), and of the cage-like fullerene **2** (D_{2d}, $^2B_2$), obtained at the PBE0 level. The simulations were done by fitting the distributions of calculated vertical detachment energies with unit-area Gaussian functions of 0.1 eV half-width. a.u., arbitrary units. This figure was reproduced with kind permission of the Springer Nature Publishing AG [138]

4.5 Largest Predicted Quasi-planar Cluster B_{84}

The Largest 2D Quasi-planar Cluster B_{84}: Based on ab initio density functional calculations, Rahane and Kumar [139] investigated different isomers of the B_{84} cluster. For the exchange-correlation energy they used the generalized gradient approximation (GGA) of Perdew-Burke-Ernzerhof (PBE) and applied the projector augmented wave pseudopotential method to treat the electron-ion interaction and the cut-off energy was set to 318.6 eV. The total energy convergence was set to about 10^5 eV without any symmetry constraint and the value of the force on each ion was less than 0.005 eV Å^{-1}. Rahane and Kumar [139] investigated different isomers of the B_{84} cluster and reported that the most stable isomer exhibits a bowl-like quasi-planar structure assigned by four symmetric hexagonal holes, as shown in Fig. 4.141. Some of the 22 low-lying isomers, as presented in Fig. 4.142, possess bowl-like quasi-planar structures decorated with two, three, or four hexagonal holes. Some others have tubular structures, composed of different numbers of rings and few of them are also decorated with different number of hexagonal holes. The remaining isomers are cage-like disordered structures, either empty or filled cages composed respectively of one shell or two shells. The next low-lying isomer II with a tubular structure is by 0.45 eV higher in energy than the global minimum isomer I. All other isomers are laying in a significantly higher energy band up to 8.98 eV. However, the quasi-planar structure of isomer I seems to be dynamically stable according to the calculations of the infrared and Raman spectra. The high stability of the isomer I predicts that quasi-planar structures may belong to the low energy ones in larger clusters. Larger cluster sizes of such quasi-planar structures could lead to a higher binding energy approaching the value of the α-boron sheet, as will be discussed below. Concerning the cage-like spherical isomer with amorphous structure, the authors performed molecular dynamics simulation of the outer capped stuffed B_{84} isomer V by heating the system for 2 ps up to 600 K. They used temperature annealing in steps of 100 K approaching to 0 K where the structure was quenched and fully optimized. The optimized amorphous structure of the isomer V was $\sim$ 1.3 eV lower in energy than the initial amorphous structure, but about 1.75 eV higher in energy than isomer I.

4.5.1 Frequencies, IR and Raman Spectra of B_{84}

The authors [139] carried out further calculations to determine the vibrational frequencies for the quasi-planar structure of the lowest energy B_{84} isomer I, in order to check the corresponding dynamical stability. They applied the hybrid functional B3PW91 for the exchange-correlation energy and used the 6–21G basis set. The results show that all vibrational frequencies were found to be positive indicating that the structure of isomer I is dynamically stable. However, the IR spectrum in Fig. 4.143 exhibits three sharp peaks with high intensity located at 766.89 cm^{-1}, 992.32 cm^{-1}, and 1026.43 cm^{-1}. The latter peak at 1026.43 cm^{-1}, indicated by the highest inten-

Fig. 4.141 The most stable quasi-planar bowl-like structure of B_{84} cluster, containing four hexagonal holes [139]

sity, is belonging to the stretching and bending vibrational modes. Concerning the Raman spectra, the Fig. 4.143 shows at 1153.09 cm^{-1} a high intensity peak within a broad peak in the range of 1150–1175 cm^{-1}. Further frequencies with high intensities can be found at 1222.20 cm^{-1} and 1350.43 cm^{-1}. Thus, the Raman active modes are coupled together by stretching, bending, and breathing vibrational modes.

4.5.2 Clusters Stability in α-Sheet

Furthermore, Rahane and Kumar [139] determined the binding energy of various planar clusters with different numbers of hexagonal holes cut out from α-sheet. The calculated binding energies are 5.76, 5.82, 5.84, 5.87, 5.88, 5.89, and 5.90 eV/atom for B_{36}, B_{56}, B_{70}, B_{84}, B_{98}, B_{112} and B_{120}, respectively. They plotted these values in Fig. 4.144 and showed that by increasing the cluster size the binding energy of the clusters increases also towards the binding energy of α-sheet, which will be discussed in the chapter two-dimensional sheets.

4.5.3 Electron Distribution in B_{84}

Rahane and Kumar [139] studied the nature of the chemical bonding of the quasi-planar structures and most stable B_{84} isomer I. Therefore, they determined electron localization function (ELF) iso-surfaces and the total charge density. They plotted

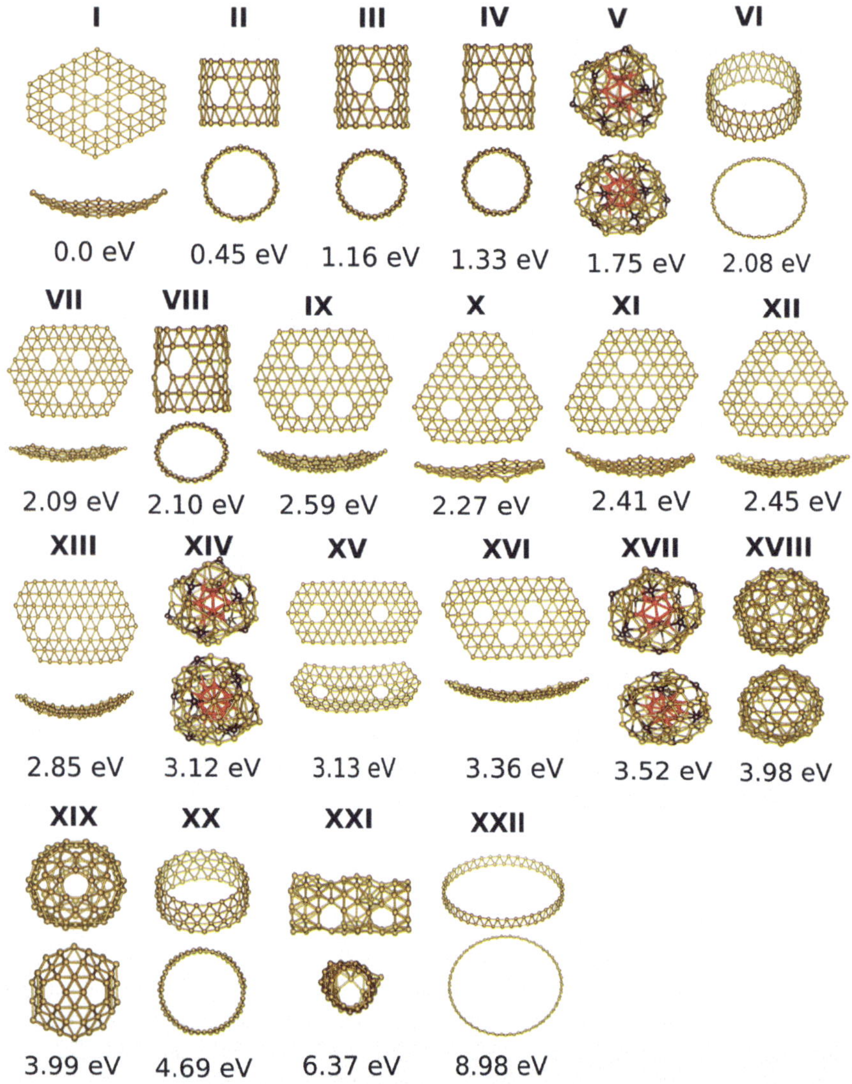

Fig. 4.142 The lowest energy and some low-lying isomers of B_{84}. For the core-shell isomers the inner B_{12} icosahedron and B atoms capping pentagons of the B60 shell in the initial structure are marked with different colors. The energy below each isomer is given w.r.t. the lowest energy isomer. This figure was reproduced from the Nanoscale with kind permission of the Royal Society of Chemistry [139]

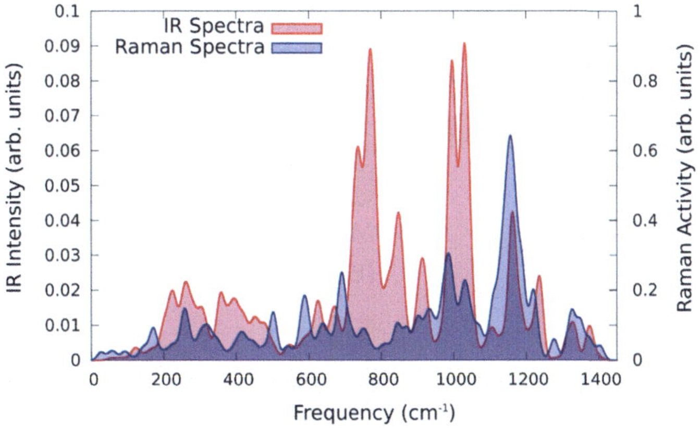

Fig. 4.143 The calculated IR and Raman spectra for the lowest energy quasi-planar isomer of B_{84}. This figure was reproduced from the Nanoscale with kind permission of the Royal Society of Chemistry [139]

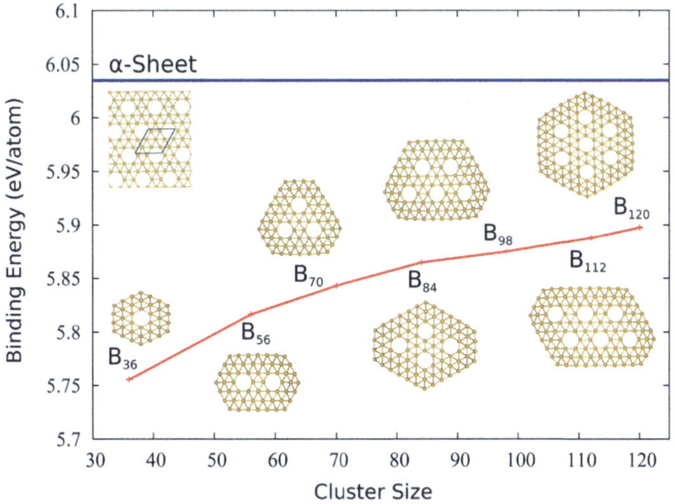

Fig. 4.144 Binding energy (eV per atom) for quasi-planar structures with different number of holes. The horizontal line (blue) shows the cohesive energy (eV/atom) for an infinite α-sheet. This figure was reproduced from the Nanoscale with kind permission of the Royal Society of Chemistry [139]

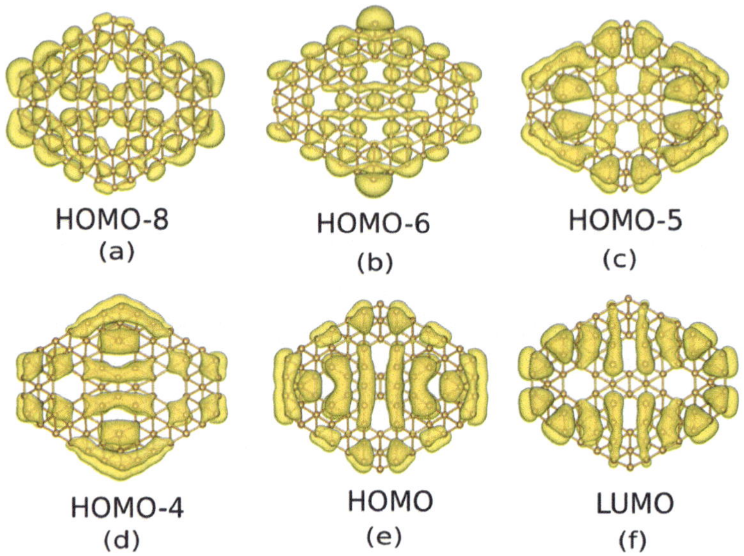

Fig. 4.145 Some of the molecular orbitals for the lowest energy isomer of B_{84}. This figure was reproduced from the Nanoscale with kind permission of the Royal Society of Chemistry [139]

some iso-surfaces for HOMO and LUMO and presented in Fig. 4.145. The molecular orbitals HOMO-8 and HOMO-6 as shown in Fig. 4.145a, b have σ-bonded molecular orbitals, while the HOMO-5 and HOMO-4 orbitals in Fig. 4.145c, d are assigned by π-bonding character. Also in Fig. 4.145e, f the both HOMO and LUMO orbitals possess also π-bonds. The authors asserted that the localized 2c-3c bonding is the characteristic feature of these quasi-planar structures as displayed by the molecular orbitals. Herewith, the authors confirmed that neutral quasi-planar boron structures can be be expected to have larger cluster sizes like the B_{84} cluster which is the largest aggregate with out-of-plane bending bowl-shaped convex structure that ever been predicted up date.

4.6 Boron Nitride and Carbide Clusters

In this section we want to introduce just two examples of small clusters, in which boron atoms are bonded either with nitrogen or with carbon atoms.

Boron Nitride Clusters: In the previous Sect. 4.1.3 we have learned that the cubic boron nitride c-BN (Fig. 4.11) has a crystal structure analogous to that of diamond. The c-BN solid is well-known with its high thermal stability, impact toughness, chemical stability and being the hardest material after diamond. This c-BN material is of great interest in industrial applications such as cutting, drilling, boring and

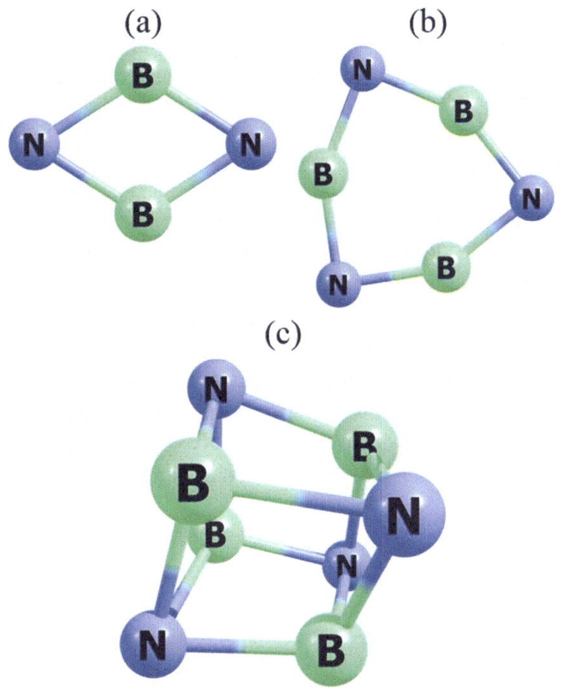

Fig. 4.146 The structures of **a** B_2N_2, **b** B_3N_3, and **c** B_4N_4 clusters. The obtained B-N bond lengths, calculated at the coupled-cluster level of theory CCSD(T), are 1.409, 1.363, and 1.511 Å for the B_2N_2, B_3N_3, and B_4N_4 clusters, respectively [143]

grading tool. Therefore, the scientific research on the c-BN and h-BN poly crystalline as well as BN sheets was much larger to that on boron nitride clusters. Nevertheless, here we will present some studies on boron nitride clusters, carried out by three groups. Based on density functional theory (DFT) as well as ab initio coupled-cluster CCSD(T) techniques, Novikov et al. [140] studied the electronic and geometrical structures of three boron nitride clusters B_2N_2, B_3N_3, and B_4N_4. They found that the B_2N_2 cluster has a rhombus form, the B_3N_3 cluster has slightly deformed triangular form, while the B_4N_4 cluster has a deformed cubic (see Fig. 4.146). The boron and nitrogen atoms are connected together in alternating manner and characterized by strong covalent bonds. Furthermore, Zahidi et al. [141] studied $(BN)_n$ clusters for n=3–5 using density functional methods in order to study the correlation between the CO and NO molecules with the boron nitride clusters. The total energies of the boron nitride clusters were determined using the hybrid functional B3LYP and the basis set 6-31G*. The optimized structures of the B_3N_3, B_4N_4, and B_5N_5 clusters have slightly deformed cyclic forms, connected together by alternating boron nitrogen atoms. Furthermore, Liu et al. [142] studied systematically larger boron nitride clusters sizes in order to understand the nucleation process in the growth of the h-BN crystalline on transition metal substrates by chemical vapor deposition (CVD), the energy of formation and stability of h-BN clusters of different geometries on the pristine Cu(111) surface using calculations based on the density functional theory. Liu et al. [142] found that the nucleation of BN clusters on the Cu(111) substrate occurs in different formations; as slightly bended chains, as cyclic rings, and as combination of square, pentagonal, hexagonal, heptagonal rings, mostly in alternating boron nitrogen atoms though avoiding B-B and N-N bonds.

Boron Carbide Clusters: Popov et al. [144] investigated a series of negatively charged boron-carbon mixed clusters $C_xB^-_{10-x}$ and studied the structural changes upon substitution of boron by carbon. They optimized the cluster at B3LYP/6-311+G(d) level and used accurate calculations based on UCCSD(T)/aug-cc-pVTZ and RCCSD(T)/aug-cc-pVTZ methods to determine the single-point energies. They found that the mixed clusters exhibit just two formations, either as molecular wheels in form of cyclic rings or as linear species. The two $C_3B^-_7$ and $C_4B^-_6$ clusters were found to be similar to the CB^-_9 and $C_2B^-_8$ clusters having a planar distorted cyclic ring with a central atom. This phenomenon of cyclic and linear formations are found by carbon clusters, which will be discussed in the next chapter. However, we can conclude that boron clusters are the only species characterized by planar or quasi-planar structures which can be extended towards planar sheets.

References

1. R.K.F. Lee, B.J. Cox, J.M. Hill, Ideal polyhedral models for single-walled nanotubes. The 2011 International Conference on Physics Science and Technology (ICPST 2011). Phys. Proc. **22**, 144–149 (2011)
2. https://nano.tu-dresden.de/pubs/slides_others/2009_04_30_Ihsan_Boustani.pdf. Accessed July 2015
3. W.G. Woods, An Introduction to boron: history, sources, uses, and chemistry. Env. Health Perspect. **102**(7), 5–11 (1994)
4. https://upload.wikimedia.org/wikipedia/commons/c/ca/Borax_-_Kramer_Borate_deposit,_Boron,_Kern_Co,_California,_USA.jpg. Accessed July 2015
5. http://www.azom.com/work/z0KxX0N7v0qNdbitBc45_files/image004.jpg. Accessed July 2015
6. http://www.liquisearch.com/boron/history_and_etymology. Accessed July 2015
7. D.H. Wang, H. Yang, Y. Zhang, X.G. Xue, Preparation and characterization of alb 12 powder. J. Chin. Ceramic Soc. **36**, 1443–1449 (2008)
8. E. Weintraub, The preparation and properties of pure boron. Trans. Am. Electrochem. Soc. **16**, 165 (1909)
9. http://www.oocities.org/ferritec_eng/chapter_1.htm. Accessed July 2015
10. http://www.jmcampbell.com/tip-of-the-month/wp-content/uploads/2013/09/Fig-31.png. Accessed July 2015
11. https://upload.wikimedia.org/wikipedia/commons/thumb/a/aa/Fe2B_structure.png/200px-Fe2B_structure.png. Accessed July 2015
12. P. Ernst, *Effect of boron on the mechanical properties of modified 12% chromium steels*; Ph.D. Thesis, ETH Zürich, *No. 8596* (1988)
13. http://www.chem.ucla.edu/dept/Faculty/hawthorne/lecture/lecture2.htm. Accessed July 2015
14. https://en.wikipedia.org/wiki/Boranes. Accessed July 2015
15. H.C. Stark, GmbH, Boron - from eye drops to rocket propellants. http://www-eng.lbl.gov/~dw/projects/DW4118_Oven_88/Refractory%20Mat%27ls/boron.pdf. Accessed Oct 2014
16. http://www.islandone.org/Launch/boron-sharp-article.htm. Accessed July 2015

17. https://books.google.de/books?id=DOEDAAAAMBAJ&pg=PA108&lpg=PA108&dq=borane+propellant+combustion& source=bl&ots=QFVU2h4UAv&sig=Am6jSRgZ4LLr7JBLzU8-iUyvtPI&hl=de&sa=X&ved=0CGAQ6AEwCGoVChMI5Y fopfG3xwIVSAssCh1MKwCS#v=onepage&q=borane%20propellant%20combustion&f=false. Accessed Oct 2014
18. http://www.russia-ic.com/img/education_science/neutron2.jpg. Accessed July 2015
19. V.I. Matkovich (ed.), *Boron and Refractory Borides* (Springer, Berlin, 1977)
20. http://www.ceramicindustry.com/CI/2002/01/Files/Images/65005.jpg. Accessed July 2015
21. http://www.ceramicindustry.com/CI/2001/01/Files/Images/8699.jpg. Accessed July 2015
22. B. Cicek, (Ed.) *Development of glass-ceramics from combination of industrial wastes with boron mining waste*. University of Bologna, Dissertation (2013)
23. http://www.wikiwand.com/en/Boric_acid. Accessed July 2015
24. https://en.wikipedia.org/wiki/Sassolite. Accessed July 2015
25. O.T. Johnson (ed.), Improvement on the mechanical properties of boron suboxide (B_6O) based composites using transition metal compounds as second phase (University of Witwatersrand, Johannesburg, 2008)
26. https://en.wikipedia.org/wiki/Boron. Accessed July 2015
27. https://crystallography365.files.wordpress.com/2014/08/cbn.png?w=887&h=413. Accessed July 2015
28. C.P. Poole Jr (ed.), *Encyclopedic Dictionary of Condensed Matter Physics*, vol. 1 (ELSEVIER Academic, Amsterdam, Boston, 2004)
29. H. Hubert, B. Devouard, L.A.J. Garvie, M. O'Keeffe, P.R. Buseck, W.T. Petuskey, P.F. McMillan, Icosahedral packing of B_{12} icosahedra in boron suboxide (B_6O). Nature **391**, 376–378 (1998)
30. http://www.nsf.gov/news/mmg/mmg_disp.jsp?med_id=51913. Accessed July 2015
31. D. He, Y. Zhao, L. Daemen, J. Qian, T.D. Shen, T.W. Zerda, Boron suboxide: as hard as cubic boron nitride. Appl. Phys. Lett. **81**, 643–645 (2002)
32. http://eathealthylivefit.com/wp-content/uploads/2014/05/Foods-Minerals-Boron-EatHealthyLiveFit_com-300x206.png. Accessed July 2015
33. http://www.chem.ucla.edu/dept/Faculty/hawthorne/lecture/lecture.htm. Accessed July 2015
34. http://physicsphunhouse.com/chemistry/unit2/Unit2StudyGuide. Accessed July 2015
35. http://mysite.du.edu/~jcalvert/phys/boron.htm. Accessed July 2015
36. http://nautilus.fis.uc.pt/st2.5/scenes-e/elem/e00530.html. Accessed July 2015
37. JEE(MAIN)/AIEEE Chemistry, Editorial Board (V & S Publishers, New Delhi, 2013), p. 173
38. http://www.webelements.com/boron/isotopes.html. Accessed July 2015
39. http://www.ustudy.in/sites/default/files/reactor.gif. Accessed July 2015
40. https://en.wikiversity.org/wiki/The_periodic_table/Boron. Accessed July 2015
41. http://www.advance-health.com/boron.html. Accessed July 2015
42. https://hairevo.com/shop/content/16-what-is-fruitex-b-calcium-fructoborate. Accessed July 2015
43. https://en.wikipedia.org/wiki/Sodium_azide. Accessed July 2015
44. https://encrypted-tbn0.gstatic.com/images?q=tbn:ANd9GcSxSPE-IE_dgmbHDmS2ZDAaO-9IDevcSf_ZjoKW3d4obcxPVm8A. Accessed July 2015
45. A.W. Laubengayer, D.T. Hurd, A.E. Newkirk, J.L. Hoard, I. Boron, Preparation and properties of pure crystalline boron. J. Am. Chem. Soc. **65**, 1924 (1943)
46. http://televaw.altervista.org/HTML/icosahedron.html.LyXconv/Icosahedron0_6.png. Accessed July 2015
47. https://en.wikipedia.org/wiki/Regular_icosahedron. Accessed July 2015
48. K. Shirai, Central and noncentral forces on the lattice dynamics of boron-rich solids. J. Solid State Chem. **133**, 215–223 (1997)
49. L.V. McCarty, J.S. Kasper, F.H. Horn, B.F. Decker, A.E. Newkirk, A new crystalline modification of boron. J. Am. Chem. Soc. **80**, 2592 (1988)
50. M. Widom, M. Mihalkovic, Symmetry-broken crystal structure of elemental boron at low temperature. Phys. Rev. B **77**, 064113 (2008)

51. C.F. Powell, C.J. Ish, J.M. Blocher Jr., The preparation of high-purity boron by hot-wire technique, in *Boron, Synthesis, Structure and Properties*, ed. by J.A. Kohn, W.F. Nye, G.K. Guale (Plenum Press, New York, 1960), pp. 7–14
52. M. Fujimori, T. Nakata, T. Nakayama, E. Nishibori, K. Kimura, M. Takata, M. Sakata, Peculiar covalent bonds in α-rhombohedral boron. Phys. Rev. Lett. **82**, 4452 (1999)
53. A. Masago, K. Shirai, H. Katayama-Yoshida, Crystal stability of α- and β-boron. Phys. Rev. B **73**, 104102 (2006)
54. M.J. van Setten, M.A. Uijttewaal, G.A. de Wijs, R.A. de Groot, Thermodynamic stability of boron: the role of defects and zero point motion. J. Am. Chem. Soc. **129**, 2458–2465 (2007)
55. T. Ogitsu, F. Gygi, J. Reed, Y. Motome, E. Schwegler, G. Galli, Imperfect crystal and unusual semiconductor: boron, a frustrated element. J. Am. Chem. Soc. **131**(5), 1903–1909 (2009)
56. G.A. Slack, C.I. Hejna, M.F. Garbauskas, J.S. Kasper, The crystal structure and density of β-rhombohedral boron. J. Solid State Chem. **76**, 52–63 (1988)
57. E.D. Jemmis, M.M. Balakrishnarajan, Polyhedral boranes and elemental boron: direct structural relations and diverse electronic requirements. J. Am. Chem. Soc. **123**(18), 4324–30 (2001)
58. W. Hayami, S. Otani, The role of surface energy in the growth of boron crystals. J. Phys. Chem. C **111**, 688–692 (2007)
59. A.R. Oganov, J.-H. Chen, C. Gatti, Y.-Z. Ma, Y.-M. Ma, C.W. Glass, Z.-X. Liu, T. Yu, O.O. Kurakevych, V.L. Solozhenko, Ionic high-pressure form of elemental boron. Nature **457**, 863–867 (2009)
60. E.Yu. Zarechnaya, L. Dubrovinsky, N. Dubrovinskaia, N. Miyajima, Y. Filinchuk, D. Chernyshov, V. Dmitriev, Synthesis of an orthorhombic high pressure boron phase. Sci. Tech. Adv. Mater. **9**, 044209 (2008)
61. M.I. Eremets, V.V. Struzkhin, H. Mao, R.J. Hemley, Superconductivity in boron. Science **293**, 272–274 (2001)
62. E.A. Ekimov, I.P. Zibrov, High-pressure high-temperature synthesis and structure of α-tetragonal boron. Sci. Tech. Adv. Mater. **12**, 055009 (2011)
63. M. Kobayashi, I. Higashi, M. Takami, Fundamental structure of amorphous boron. J. Solid State Chem. **133**, 211–214 (1997)
64. http://3.bp.blogspot.com/_lPWZO2mczQc/TTkYe0gozlI/AAAAAAAAAD0/H3VUmXgC_a0/s1600/top_down.bmp. Accessed July 2015
65. W.D. Knight, K. Clemenger, W.A. de Heer, W.A. Saunders, M.Y. Chow, M.L. Cohen, Electronic shell structure and abundances of sodium clusters. Phys. Rev. Lett. **52**, 2141 (1984)
66. H.W. Kroto, J.R. Heath, S.C. O'Brien, R.F. Curl, R.E. Smalley, C_{60}: buckminsterfullerene. Nature **318**, 162–3 (1985)
67. http://www.vias.org/tmanalytik_germ/img/hl_fulleren.png. Accessed July 2015
68. I. Boustani, W. Pewestorf, P. Fantucci, V. Bonačić-Koutecký, J. Koutecký, Systematic ab initio configuration-interaction study of alkali-metal clusters: Relation between electronic structure and geometry of small Li clusters. Phys. Rev. B **35**, 9437–9450 (1987)
69. V. Bonačić-Koutecký, I. Boustani, M.F. Guest, J. Koutecký, Ab initio configuration interaction study of the electronic and geometric structures of small sodium cationic clusters. J. Chem. Phys. **89**, 4861–4866 (1988)
70. M.C. Payne, M.P. Teter, D.C. Allan, T.A. Arias, J.D. Joannopoulos, Iterative minimization techniques for ab initio total-energy calculations: molecular dynamics and conjugate gradients. Rev. Mod. Phys. **64**, 1045 (1992)
71. L. Hanley, S.L. Anderson, Production and collision-induced dissociation of small boron cluster ions. J. Phys. Chem. **91**, 5161–5163 (1987)
72. L. Hanley, J. Whitten, S.L. Anderson, Collision-induced dissociation and ab initio studies of boron cluster ions: determination of structures and stabilities. J. Phys. Chem. **92**, 5803–5812 (1988)
73. S.J. La Placa, P.A. Roland, J.J. Wynne, Boron clusters (B_n, n = 2–52) produced by laser ablation of hexagonal boron nitride. Chem. Phys. Lett. **190**, 163–168 (1992)

74. L. Hanley, S.L. Anderson, Oxidation of small boron cluster ions (B^+_{1-13}) by oxygen. J. Chem. Phys. **89**, 2848–2860 (1988)
75. G.V. Gadiyak, YuN Morokov, Calculation of the properties of small boron and aluminum clusters. J. Struct. Chem. **30**, 738–742 (1989)
76. R. Kawai, J.H. Weare, Instability of the B_{12} icosahedral cluster: rearrangement to a lower energy structure. J. Chem. Phys. **95**, 1151–1159 (1991)
77. A.K. Ray, I.A. Howard, K.M. Kanal, Structure and binding in small neutral and cationic boron clusters. Phys. Rev. B **45**, 14247–14255 (1992)
78. H. Kato, E. Tanaka, Stabilities of small Be_n and B_n clusters ($4 \leq n \leq 8$) by vibrational analysis. J. Comput. Chem. **12**, 1097–1109 (1991)
79. H. Kato, K. Yamashita, K. Morokuma, Ab Initio MO study of neutral and cationic boron clusters. Chem. Phys. Lett. **190**, 361–366 (1992)
80. I. Boustani, Y. Li, V. Bonaĉić-Kouteckŷ, J. Kouteckŷ, Ab Initio Quantum Chemical Investigation of Small Neutral and Charged Boron Clusters. Unpublished results (1990)
81. V. Bonaĉić-Kouteckŷ, P. Fantucci, J. Kouteckŷ, Quantum chemistry of small clusters of elements of groups Ia, Ib, and IIa: fundamental concepts, predictions, and interpretation of experiments. Chem. Rev. **91**, 1035–1108 (1991)
82. I. Boustani, Systematic LSD investigation on cationic boron clusters: B^+_n(n=2-14). Int. J. Q. Chem. **52**, 1081–1111 (1994)
83. I. Boustani, A comparative study of ab initio SCF-CI and DFT. Example of small boron clusters. Chem. Phys. Lett. **233**, 273–278 (1995)
84. I. Boustani, Structure and stability of small boron clusters - a density functional theoretical study. Chem. Phys. Lett. **240**, 135–138 (1995)
85. I. Boustani, Systematic ab initio investigation of bare boron clusters: determination of the geometry and electronic structures of B^+_n (n = 2–14). Phys. Rev. B **55**, 16426–16438 (1997)
86. I. Boustani, New quasi-planar surfaces of bare boron. Surf. Sci. **370**, 355–363 (1997)
87. A. Ricca, C.W. Bauschlicher Jr., The structure and stability of B^+_n clusters. Chem. Phys. **208**, 233 (1996)
88. H.W. Jin, Q.S. Li, Structure and stability of B_4, B^+_4, and B^-_4 clusters. Phys. Chem. Chem. Phys. **5**, 1110–1115 (2003)
89. Q.S. Li, H.W. Jin, Structure and stability of B_5, B^+_5, and B^-_5 clusters. J. Phys. Chem. A **106**, 7042–7047 (2002)
90. Q.S. Li, Q. Jin, Q. Luo, A.C. Tang, J.K. Yu, H.X. Zhang, Structure and stability of B_6, B^+_6, and B^-_6 clusters. Int. J. Q. Chem. **94**, 269–278 (2003)
91. Q.S. Li, L.-F. Hong, Z.-M. Gao, Structures and stabilities of B_7, B^+_7, and B^-_7 clusters. Chem. Phys. Lett. **390**, 220–227 (2004)
92. P.-L. Cao, W. Zhao, B.-X. Li, B. Song, X.-Y. Zhou, A full-potential linear-muffin-tin-orbital molecular-dynamics study of B_7, B_10, and B_13 clusters. J. Phys. Condens. Matter **13**, 5065–5076 (2001)
93. M. Atiş, C. Özdoğan, Z.B. Güvenç, Structure and energetic of B_n (n= 2–12) clusters: electronic structure calculations. Int. J. Q. Chem. **107**, 729–744 (2007)
94. M. Atiş, C. Özdoğan, Z.B. Güvenç, Density functional study of physical and chemical properties of nano size boron clusters: B_n (n=13-20). Chin. J. Chem. Phys. **22**, 380–388 (2009)
95. K.C. Lau, M. Deshpande, R. Pandey, A theoretical study of vibrational properties of neutral and cationic B_{12} clusters. Int. J. Q. Chem. **102**, 656–664 (2005)
96. H.-J. Zhai, L.-S. Wang, A.N. Alexandrova, A.I. Boldyrev, V.G. Zakrzewski, Photoelectron spectroscopy and ab initio study of B^-_3 and B^-_4 anions and their neutrals. J. Phys. Chem. A **107**, 9319–9328 (2003)
97. H.-J. Zhai, L.-S. Wang, A.N. Alexandrova, A.I. Boldyrev, Electronic structure and chemical bonding of B^-_5 and B_5 by photoelectron spectroscopy and ab initio calculations. J. Chem. Phys. **117**, 7917–7924 (2002)
98. A.N. Alexandrova, A.I. Boldyrev, H.-J. Zhai, L.-S. Wang, E. Steiner, P.W. Fowler, Structure and bonding in B^-_6 and B_6: planarity and antiaromaticity. J. Chem. Phys. **107**, 1359–1369 (2003)

99. A.N. Alexandrova, A.I. Boldyrev, H.-J. Zhai, L.-S. Wang, Electronic structure, isomerism, and chemical bonding in B_7^- and B_7. J. Phys. Chem. A **108**, 3509–3517 (2004)
100. H.-J. Zhai, A.N. Alexandrova, K.A. Birch, A.I. Boldyrev, L.-S. Wang, Hepta- and octacoordinate boron in molecular wheels of eight- and nine-atom boron clusters: observation and confirmation. Angew. Chem. Int. Ed. **42**, 6004–6008 (2003)
101. H.-J. Zhai, B. Kiran, J. Li, L.-S. Wang, Hydrocarbon analogues of boron clusters: planarity, aromaticity and antiaromaticity. Nat. Mater. **2**, 827–833 (2003)
102. Y. Yuan, L. Cheng, B_{14}^{2+}: a magic number double-ring cluster. J. Chem. Phys. **137**, 044308-1–044308-5 (2012)
103. D.Yu. Zubarev, A.I. Boldyrev, Developing paradigms of chemical bonding: adaptive natural density partitioning. Phys. Chem. Chem. Phys. **10**, 5207–5217 (2008)
104. J.E. Fowler, J.M. Ugalde, The curiously stable B_{13}^+ cluster and its neutral and anionic counterparts: the advantages of planarity. J. Phys. Chem. A **104**, 397–403 (2000)
105. J.-I. Aihara, H. Kannp, T. Ishada, Aromaticity of planar boron clusters confirmed. J. Am. Chem. Soc. **127**, 13324–13330 (2005)
106. B. Kiran, G.G. Kumar, M.T. Nguyen, A.K. Kandalam, P. Jena, Origin of the unusual stability of B_{12} and B_{13}^+ clusters. Inorg. Chem. **48**, 9965–9967 (2009)
107. J.-I. Aihara, B_{13}^+ is highly aromatic. J. Phys. Chem. A **105**, 5486–5489 (2001)
108. N. Akman, M. Tas, C. Özdogan, I. Boustani, Ionization energies, Coulomb explosion, fragmentation, geometric, and electronic structures of multicharged boron clusters B_n ($n = 213$). Phys. Rev. B **84**, 075463-1–075463-10 (2011)
109. A.P. Sergeeva, D.Yu. Zubarev, H.-J. Zhai, A.I. Boldyrev, L.-S. Wang, A photoelectron spectroscopic and theoretical study of B_{16}^- and B_{16}^{2-}: all-boron naphthalene. J. Am. Chem. Soc. **130**, 7244–7246 (2008)
110. C. Romanesco, D.J. Harding, A. Fielicke, L.-S. Wang, Probing the structures of neutral boron clusters using infrared/vacuum ultraviolet two color ionization: B_{11}, B_{16}, and B_{17}. J. Chem. Phys. **137**, 014317-1–014317-6 (2012)
111. A.P. Sergeeva, B.B. Averkiev, H.-J. Zhai, A.I. Boldyrev, L.-S. Wang, All-boron analogues of aromatic hydrocarbons: B_{17}^- and B_{18}^-. J. Chem. Phys. **134**, 224304-1–224304-11 (2011)
112. C.R. Hsing, C.M. Wei, N.D. Drummond, R.J. Needs, Quantum Monte Carlo studies of covalent and metallic clusters: accuracy of density functional approximations. Phys. Rev. B **79**, 245401-1–245401-05 (2009)
113. W. Huang, A.P. Sergeeva, H.-J. Zhai, B.B. Averkiev, L.-S. Wang, A.I. Boldyrev, A concentric planar doubly π-aromatic B_{19}^- cluster. Nat. Chem. **2**, 202–206 (2010)
114. I. Boustani, Z. Zhu, D. Tomának, Search for the largest two-dimensional aggregates of boron: an ab initio study. Phys. Rev. B **83**, 193405-1–193405-4 (2011)
115. W. Kiran, S. Bulusu, H.-J. Zhai, S.-H. Yoo, S.C. Zeng, L.-S. Wang, Planar-to-tubular structural transition in boron clusters: B_{20} as the embryo of single-walled boron nanotubes. Proc. Natl. Acad. Sci. **102**, 961–964 (2005)
116. M.A.L. Marques, S. Botti, The planar-to-tubular structural transition in boron clusters from optical absorption. J. Chem. Phys. **123**, 014310-1–014310-5 (2005)
117. W. An, S. Bulus, Y. Gao, X.C. Zeng, Relative stability of planar versus double-ring tubular isomers of neutral and anionic boron cluster B_{20} and B_{20}^-. J. Chem. Phys. **124**, 154310-1–154310-6 (2006)
118. T.B. Tai, N.M. Tam, M.T. Nguyen, Structure of boron clusters revisited, B_n with n = 14–20. Chem. Phys. Lett. **530**, 71–76 (2012)
119. Z.A. Piazza, W.-L. LI, C. Romanescu, A.P. Sergeeva, L.-S. Wang, A.I. Boldyrev, A photoelectron spectroscopy and *ab initio* study of B_{21}^-: negatively charged boron clusters continue to be planar at 21. J. Chem. Phys. **136**, 104310-1–104310-9 (2012)
120. A.P. Sergeeva, Z.A. Piazza, C. Romanescu, W.-L. LI, A.I. Boldyrev, L.-S. Wang, B_{22}^- and B_{23}^-: all-boron analogues of anthracene and phenanthrene. J. Am. Chem. Soc. **134**, 18065–18073 (2012)
121. I.A. Popov, Z.A. Piazza, W.-L. Li, L-W. Wang, A.I. Boldyrev, A combined photoelectron spectroscopy and ab initio study of the quasi-planar B_{24}^- cluster. J. Chem. Phys. **139**, 144307-1–144307-8 (2013)

122. H.T. Pham, L.V. Duong, B.Q. Pham, M.T. Nguyen, The 2D-to-3D geometry hopping in small boron clusters: the charge effect. Chem. Phys. Lett. **577**, 32–37 (2013)
123. A.P. Sergeeva, I.A. Popov, Z.A. Piazza, W.-L. LI, C. Romanescu, L.-S. Wang, A.I. Boldyrev, Understanding Boron through size-selected clusters: structure, chemical bonding, and fluxionality. Acc. Chem. Res. **47**, 1349–1358 (2014)
124. Z.A. Piazza, I.A. Popov, W.-L. Li, R. Pal, X.C. Zeng, A.I. Boldyrev, L-W. Wang, A photoelectron spectroscopy and *ab initio* study of the structures and chemical bonding of the B_{25}^- cluster. J. Chem. Phys. **141**, 034303-1–034303-10 (2014)
125. T.B. Tai, M.T. Nguyen, Electronic structure and photoelectron spectra of B_n with n = 26–29: an overview of structural characteristics and growth mechanism of boron clusters. Phys. Chem. Chem. Phys. **17**, 13672–13679 (2015)
126. W.-L. Li, R. Pal, Z.A. Piazza, X.C. Zeng, L-W. Wang, B_{27}^-: appearance of the smallest planar boron cluster containing a hexagonal vacancy. J. Chem. Phys. **142**, 204305-1–204305-7 (2015)
127. J. Zhao, X. Huang, R. Shi, H. Liu, Y. Su, R.B. King, B_{28}: the smallest all-boron cage from an ab initio global search. Nanoscale **7**, 15086–15090 (2015)
128. W.-L. Li, Y.-F. Zhao, H.-S. Hu, J. Li, L.-W. Wang, B_{30}^-: a quasiplanar chiral boron cluster. Angew. Chem. **126**, 5646–5651 (2014)
129. T.B. Tai, L.V. Duong, H.T. Pham, D.T.T. Mai, M.T. Nguyen, A disk-aromatic bowl cluster B_{30}: toward formation of boron buckyballs. Chem. Commun. **50**, 1558–1560 (2014)
130. I. Boustani, A. Rubio, J.A. Alonso, Ab Initio study of B_{32}: competition between spherical, quasiplanar and tubular isomers. Chem. Phys. Lett. **311**, 21–28 (1999)
131. W.-L. Li, Q. Chen, W.-J. Tian, H. Bai, Y.-F. Zhao, H.-S. Hu, J. Li, H.-J. Zhai, S.-D. Li, L.-S. Wang, The B_{35} cluster with a double-hexagonal vacancy: a new and more flexible structural motif for borophene. J. Am. Chem. Soc. **136**, 12257–12260 (2014)
132. Z.A. Piazza, H.-S. Hu, W.-L. Li, Y.-F. Zhao, J. Li, L.-S. Wang, Planar hexagonal B_{36} as potential basis for extended single-atom layer boron sheets. Nat. Commun. **5**, 3113–3118 (2014)
133. H. Tang, S. Ismail-Beigi, Novel precursors for boron nanotubes: the competition of two-center and three-center bonding in boron sheets. layer boron sheets. Phys. Rev. Lett. **99**, 115501–115507 (2007)
134. Q. Chen, G.-F. Wei, W.-J. Tian, H. Bai, Z.-P. Liu, H.-J. Zhai, S.-D. Li, Quasi-planar aromatic B_{36} and B_{36}^- clusters: all-boron analogues of coronene. Phys. Chem. Chem. Phys. **16**, 18282–18287 (2014)
135. J. Lv, Y. Wang, L. Zhu, Y. Ma, B_{38}: an all-boron fullerene analogue. Nanoscale (2014). (https://doi.org/10.1039/C4NR01846J)
136. T.B. Tai, M.T. Nguyen, Comment on "B_{38}: an all-boron fullerene analogue. Nanoscale **00**, 1–3 (2014)
137. Q. Chen, W.-L. Li, Y.-F. Zhao, S.-Y. Zhang, H.-S. Hu, H. Bai, H.-R. Li, W.-J. Tian, H.-G. Lu, H.-J. Zhai, S.-D. Li, J. Li, L.-S. Wang, Experimental and theoretical evidence of an axially chiral borospherene. ASC Nano **9**, 754–760 (2015)
138. H.-J. Zhai, Y.-F. Zhao, W.-L. Li, Q. Chen, H. Bai, H.-S. Hu, Z.A. Piazza, W.-J. Tian, H.-G. Lu, Y.-B. Wu, Y.-W. Mu, G.-F. Wei, Z.-P. Liu, J. Li, S.-D. Li, L.-S. Wang, Observation of an all-boron fullerene. Nat. Chem. **6**, 727–731 (2014)
139. A.B. Rahane, V. Kumar, B_{84} : a quasi-planar boron cluster stabilized with hexagonal holes. Nanoscale **7**, 4055–4062 (2015)
140. N.V. Novikov, I.Y. Dolinskiy, M.A. Gimaldinova, K.P. Katin, M.M. Maslov, Benchmark study of the exchange-corrected density functionals: application to strained boron nitride clusters. Turkish Comp. Theo. Chem. **1**, 27–34 (2017)
141. E. Zahedi, A.H. Pangh, H. Ghorbanpour, DFT Study of CO and NO adsorption on boron nitride $(BN)_{n=3-5}$ nanoclusters. Surf. Rev. Lett. **22**, 1–14 (2015)
142. Z.Q. Liu, J. Dong, F. Ding, The geometry of hexagonal boron nitride clusters in the initial stages of chemical vapor deposition growth on a Cu(111) surface. Nanoscale **12**, 13366–13376 (2019)

143. K. Katin, *Department of Condensed Matter Physics, National Research Nuclear University MEPhI, Moscow, Russia*. Private communication
144. I.A. Popov, V.R. Popov, K.V. Bozhenko, I. Černušák, A.I. Boldyrev, Structural changes in the series of boron-carbon mixed clusters $C_xB^-_{10-x}$ (x = 3-10) upon substitution of boron by carbon. J. Chem. Phys. **139**, 114307-1–16 (2013)

Chapter 5
Carbon and Inorganic Binary Clusters

Abstract Beyond the conventional allotrope of carbon, graphite, diamond and related compounds, the structures of small carbon clusters were investigated in order to understand their transition to nanoclusters and nanostructures, like fullerenes and graphenes. The alternation between even and odd carbon clusters was also studied. In addition, the transition from 1D to 2D and to 3D structures is explained and at at which cluster size can occur.

Introduction: Carbon is the next element to boron and one of the 118 elements listed in the periodic table and very widely distributed in nature. Carbon is known to be abundant in our sun system and in the atmospheres of many planets, but also in comets and starts. Carbon occurs as a variety of allotropes. The well-known two crystalline forms are diamond and graphite. As kids we have known the pencil as a slender of wood containing a black core or strip, which is nothing else as graphite. But carbon builds also the most beautiful and expensive materials diamonds, as presented in Fig. 5.1. important biological molecules in our life, from plants, animals, minerals, to air and earth and up to humans. Carbon is with 18.6% of the human body weight after 65.0% of Oxygen. One of carbon characteristics is its unique ability to build complex structures in form of chains, rings and cages. Carbon is very important material in many industrial applications like in aerospace. Carbon also provides common fuels–coal, coke, oil, gasoline, and natural gas, as shown in Fig. 5.2 [2].

5.1 History of Carbon

Carbon Origin: The element carbon, (Latin: carbo), is the most well-known specie since the prehistoric time. It was familiar to the people, who manufactured it to produce charcoal by burning organic materials in insufficient Oxygen. The charcoal material was a precondition in order to enable the development and preparation of manufacturing and metallurgy of metals.

Carbon Discovery: First in the seventeenth century and as proposed by Robert Boyle, carbon was recognized as an element, because it was not possible to decom-

I. Boustani, *Molecular Modelling and Synthesis of Nanomaterials*, Springer Series in Materials Science 290, https://doi.org/10.1007/978-3-030-32726-2_5

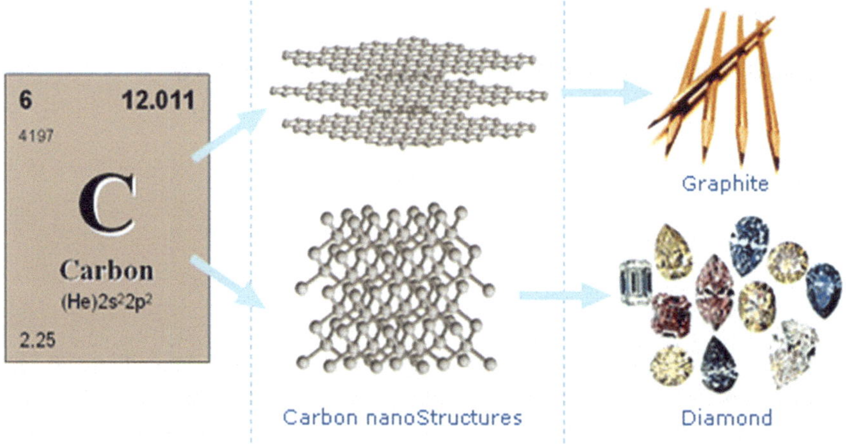

Fig. 5.1 The well-known crystalline forms of carbon and the corresponding graphite and diamond structures [1]

Fig. 5.2 RAS TANURA Raffinerie, Saudi-Arabian, by W. Robert Moore, National Geographic, [2]

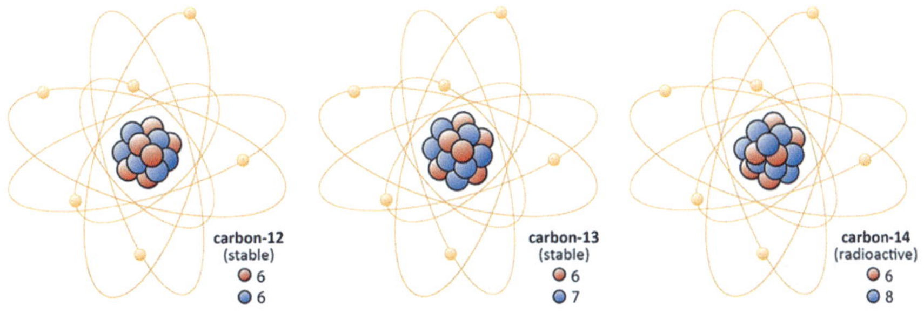

Fig. 5.3 The carbon isotopes ^{12}C, ^{13}C are stable, while ^{14}C is radioactive [4].

pose the substance of the element into much simpler substances. However, at the beginning of the 18th century started the proficiency about the carbon as a chemical element to develop [3]. Joseph Black in Scotland showed in the middle of this century that a gas composed of carbon dioxide is produced when charcoal substance is burned. Some years later, in 1772, Antoine Lavoisier confirmed that carbon is the only element which forms diamonds. At the beginning of the 19th century, Guyton de Morveau succeeded to produce steel from iron, by fusing soft iron in a small crucible and adding a diamond with a weight of one kg.

Carbon Isotopes: Carbon has around 15 known isotopes, from ^{8}C to ^{22}C. Three of them, ^{12}C, ^{13}C, and ^{14}C (Fig. 5.3) are natural and stable, as shown in Fig. 5.3 [4]. Both atoms ^{12}C and ^{13}C are occurring in a natural proportion of approximately 99:1. Thus, ^{12}C was adopted by the (IUPAC)[1] as the basis for atomic weights [5]. ^{14}C is a radioactive isotope assigned by a half-life of 5730 years. Carbon is produced by thermal neutrons from cosmic radiations in the upper atmosphere and is transported down to earth to be absorbed by living materials. The amount of ^{14}C is one of the methods used to determine the age of materials like wood and archaeological species. However, the artificial ^{11}C, with a half-life of around 20 min, is the most stable isotope [6].

Why Carbon: What are the unique properties of the element carbon that allows it to produce millions of numbers of organic compounds? First of all, carbon has the ability to form covalent bonds with itself, but it is whether shy nor picky to bond with any different element. Carbon makes in fact arguably the best bonding than any element else. Carbon can assemble covalent bonds making chains, rings and cages, and can also form single, double and triple bonds, as shown in Fig. 5.4. In addition, it is a fact that carbon-based molecules are very stable. Carbon can react to make new molecules with amazing geometrical structures. Carbon forms not only metallic but also organic components like proteins, chromosomes, muscles and fats. Therefore, it is really a true question, why our life is carbon-based but not based off of one of other 118 elements. In simple terms, it is impossible to imagine the life on earth without carbon.

---

[1]The international Union of Pure and Applied Chemistry.

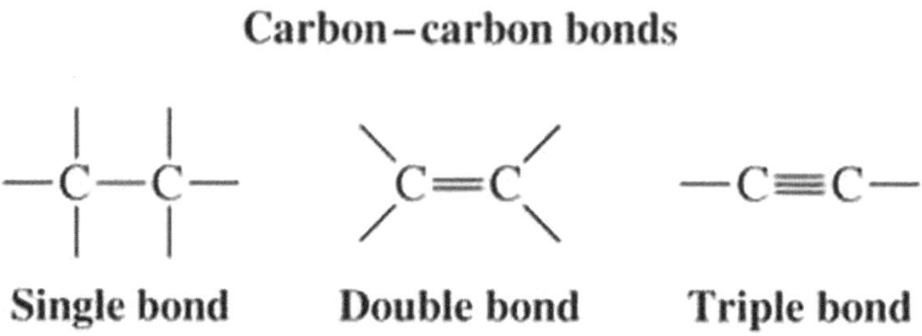

Fig. 5.4 Carbon can make single, double, and triple bonds. The number of bonds it makes, determines the structure [7]

5.1.1 Chemistry of Carbon

Atomic Carbon: Carbon is one of the most common elements on earth. It belongs to the Group 14 (IVa) of the periodic table and defined as a nonmetallic chemical element. It has the highest melting point and pertains to p-block elements. Carbon has six electrons, two of them are core electrons being in the $1s$ orbital close to the nucleus, and the other four are valence electron reside in the $2s$ and $2p$ subshells. According to the Hund's rule, the two valence electrons in the $2p$ subshell are unpaired and separated in two different orbitals. In fact, the orbitals have no physical reality of their own, because they are simply mathematical functions that portray particular standing-wave patterns, which can be presented in form of graphics. Due to the wave-like nature of the orbitals, commonly represented by Ψ functions, are usually combined both in-phase and out-of-phase to obtain a pair of orbitals. The squares of these Ψ functions describes the electron distributions in the individual atoms, molecule, or solids [8].

Valence Electrons: We have discussed above that carbon has six electrons, composed ot two core electrons, and of four valence electrons. The core electrons do not participate in any chemical bonding or reaction, while the four valence electrons are active in the chemical bonding and chemical reactions. Thus, the four valence electrons, residing in the $2s$ and $2p$ orbitals, need further four more electrons in order to complete its octet. Completing the octet can occur by sharing the valence electrons of carbon atoms with those of other atoms forming the covalent bonds, known as the tetra-valency of carbon. The orbitals of these four valence electrons are oriented towards the vertices of a tetrahedron (Fig. 5.6), and forming angles of 109° 28′, similar to diamond. Therefore, due to its different electronic configurations, carbon can form single, double and triple C-C bonds in compounds, but also quadruple bond in diatomic carbon C_2 [9]. Due to the small size of carbon, it is able to form multiple bonds with itself and with other small atoms. Due to its capability to form multiple bondings, there are millions of carbon compounds counting the majority under all chemical compounds.

Electron Configuration and Hybridization: The electron configuration of carbon atom cab be written as $1s^2 2s^2 2p^2$. The fact that carbon is able to build up various

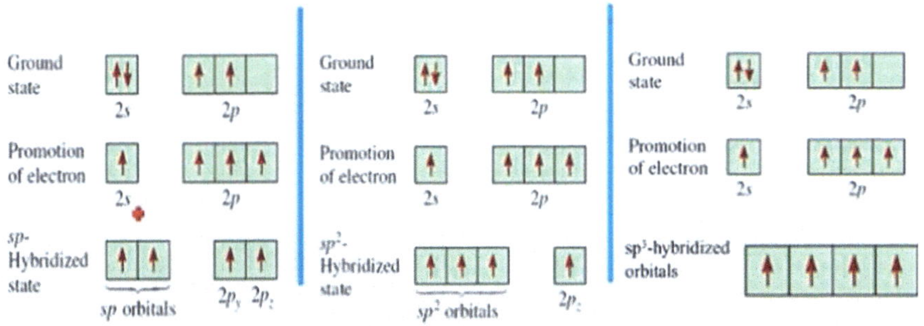

Fig. 5.5 Electron configurations of carbon atom at the ground state (1st row), after electron promotion from $2s$ into $2p$ (2nd row), then after hybridization to sp, sp^2, and sp^3 orbitals, [10]

covalent bonds, due to its ability to form hybrid orbitals. Hybridization in general is a mixing of s, p, d and f atomic orbitals in different atomic components. The ground state configuration of carbon valence electrons is $2s^2 2p^2$, as given at the upper row of Fig. 5.5. One or both electrons in the $2s$ orbital can be energetically promoted into the third nearly empty orbital of $2p$, then one gets the excited state configuration $2s^2 2\text{p}_x^1 2\text{p}_y^1 2\text{p}_z^1$, as to be seen at the middle row of Fig. 5.5 [10]. Thus mixing via excitation of singly occupied $2s$ orbital with one, two or three of the 2p_x^1, 2p_y^1 and $2p_z^1$ orbitals leads to the three hybridization forms sp, sp^2 and sp^3, respectively, as given at the lower row of Fig. 5.5. These hybrid orbitals form linear, planar and tetrahedral geometries, as can be seen in Fig. 5.6.

Sigma (σ-) & Pi (π-) bonds in Carbon: The chemical bonding in chemistry appears in two kinds of directed bonds: the sigma (σ), and pi (π) bonds. They are the strongest covalent chemical bonds. They are formed by overlapping between the atomic orbitals. Sigma bonds are formed when orbitals from two separate atoms overlap along the same axis. By Carbon, sigma bonding occurs mostly between the orbitals s, p, and d, them-self, or between each other like $s + s$, $s + p$, or $p + p$, where p is the axial component of the bond. The pi bonds are formed when p-orbitals from two separate atoms are parallel to each other, but still overlap because of the closeness of the two atoms. Pi bonds are only formed when a sigma bond brings atoms close enough to each other that the parallel orbitals overlap. When two atoms approach each other, they share one axis and two parallel axes. Therefore, σ-, and π-bonds could occur between sp, sp^2, and sp^3 hybrids forming single-, double-, and triple-bonds between carbon-carbon atoms [12], as illustrated in Fig. 5.7.

Carbon Compounds: The first compound of carbon we experience every day and produced by cars and power stations is the well known molecule carbon dioxide (CO_2), which is a colorless gas dissolved in all natural waters and is about 0.04% of the earth's atmosphere, while it is 96.2% of the atmosphere of Mars. Other compounds are known as carbonate salts like calcium carbonate (limestone) and hydrocarbons like coal and petroleum. Carbon can form chemical bonds with a large number of atoms, like hydrogen, oxygen, nitrogen, and many other elements building a large

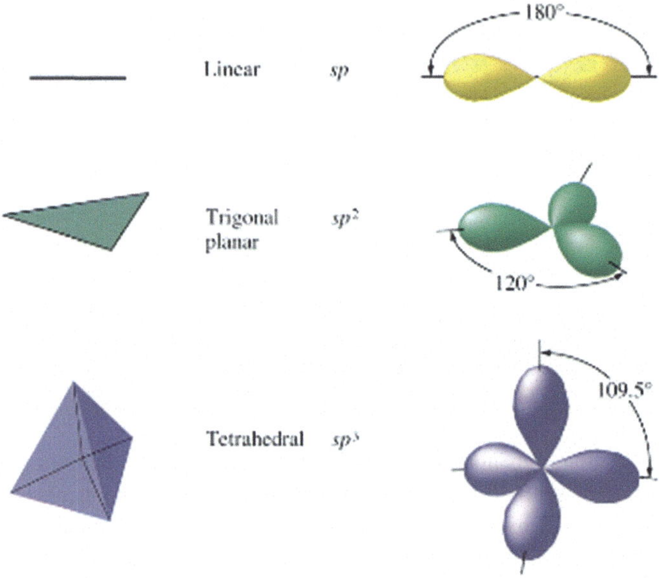

Fig. 5.6 The hybrid orbitals sp, sp^2, and sp^3 correspond to linear, trigonal planar and tetrahedral binding forms of carbon atoms [11]

variety of compounds. These carbon compounds can be counted to ten millions, few thousands of them are essential in organic molecules and life procedures. Obviously, it seems to be impossible to imagine the life without carbon. One thought to replace carbon by silicon due to its hosting of similar compounds, but however it is not possible to build very long chains of silicon atoms in stable compounds. Other well known carbon compounds are acetylene (C_2H_2), benzene (C_6H_6), methane (CH_4), ethylene (C_3H_4), chloroform ($CHCl_3$), glucose ($C_6O_6H_{12}$), carbon monoxide (CO), carbon disulfide (CS_2), carbon tetrachloride (CCl_4), and their derivatives [13]. The melting and boiling points of carbon occur at very high temperatures 3652 and 4827° C, respectively. Figure 5.8 [14] shows some of these compounds.

5.1.2 Carbon Allotrope

Introduction: Due to its valency, the capability of carbon to form various allotropes is enormous. These allotropes appear in different formations, like diamond, graphite, and fullerenes as the cage C_{60} and as nanotubes. Carbon occurs also in amorphous form, which does not have a definite crystalline structure. The white carbon is another formation, which seems to exist [13]. Another type of crystalline carbon named ceraphite, known to be one of the softest materials, is for iron an excellent hardener. Modern constructions benefit a lot from the variety of these yielded steel alloys. The

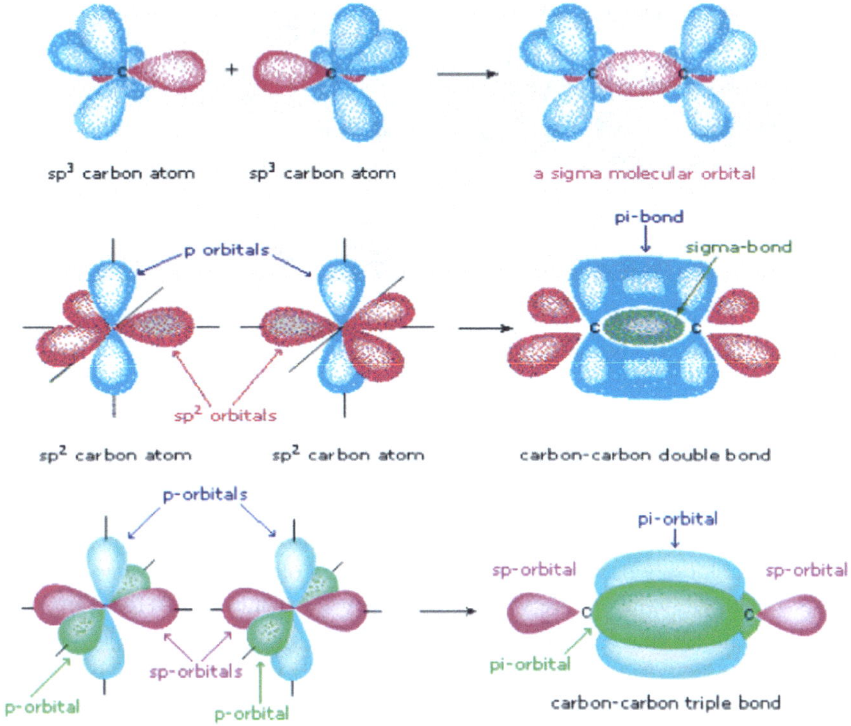

Fig. 5.7 The hybrid orbitals sp, sp^2, and sp^3 correspond to linear, trigonal planar and tetrahedral binding forms of carbon atoms [12]

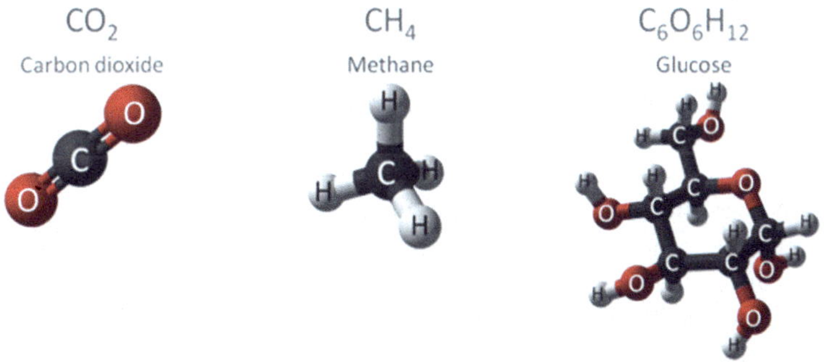

Fig. 5.8 Some examples of carbon compounds: carbon dioxide (CO_2), methane (CH_4), and glucose ($C_6O_6H_{12}$) [14]

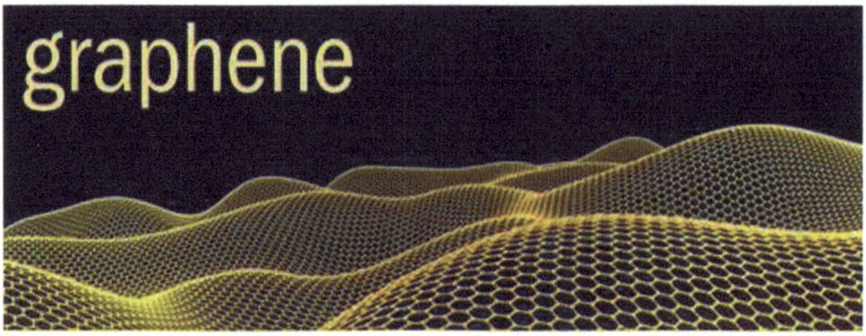

Fig. 5.9 Graphene is single-layer hexagonal honeycomb sheet, assigned by σ-bonds between the in-plane sp^2 hybrids of carbon atoms [16]

crystal structure of diamond, which is considered in nature to be the hardest naturally occurring substance, can be related to the tetrahedral sp^3 hybrid orbitals of the carbon atoms, which undergo very strong covalent bonds between each C-C pair. The next allotrope of carbon is the well known graphite. It is known as the softest substance. It is arranged in loosely parallel planes attracted to each other by weak van der Waals forces. Each plan exhibits a hexagonal honeycomb structure. Therefore, graphite is used as a lubricant. As mentioned above, the new non-crystalline and recently discovered formations of carbon are called fullerenes. They present the new class of carbon allotropes, in which carbon appears in single-wall or multi-wall hollow spheres or nanotubes. The family of fullerenes are called after the American architect Richard Buckminster Fuller. The spherical structures of carbon like C_{60} are named Buckyballs. These new formations of carbon present a new class of carbon materials, which have also new application in industry.

Graphene Sheet: Novoselov et al. [15] synthesized a single Layer of graphite for the first time and called it graphene. Therefore graphene constitutes a new allotrope of carbon. The bond length of C-C pair in graphene is about 1.42 Å. For their successful isolation, Novoselov and Geim were awarded in 2010 the Nobel prize for physics. Graphene is only one atom thick. The bonding between the carbon atoms in plan are σ bonds, whereas the remaining electrons of the P_z lie above and below the graphene sheet forming the π-bonds. The high mobility of these π-electrons, when they put into motion, means conductivity or simply electricity. This material of single carbon layer displays besides electrical, extraordinary mechanical, optical and thermal properties. The carbon atoms of this single planar one-atom thick carbon sheet are bonded by sp^2 hybrid orbitals composed of carbon hexagons in a honeycomb formation, as presented in Fig. 5.9 [16]. By creating holes within the sheet or doping with foreign atoms or impurities, it is possible to made semiconductors with desired band gaps.

Graphetic Carbon: Graphete is nothing else as multi-layered graphene sheets, arranged and shifted on top of each other. The layers of graphene are separated by 3.41 Åand held together by a weak dispersion as van der Waals forces. Figure 5.10 [17, 18] shows the graphite structure and the σ-bonds within the layers and the π-bonds between the layers. Each carbon atom in the plane provides three valence

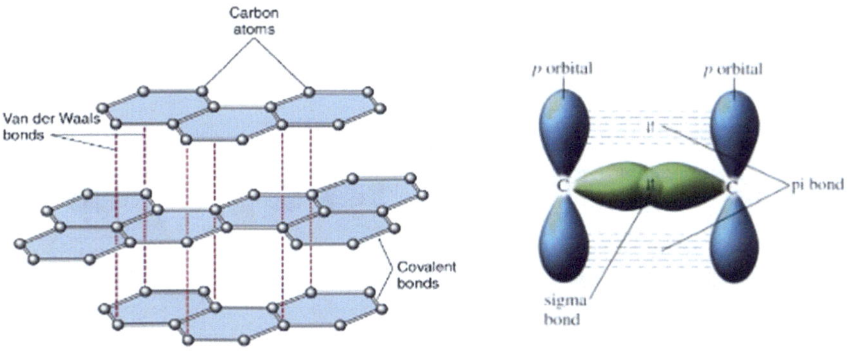

Fig. 5.10 Graphite is multi-layered hexagonal honeycomb sheets, assigned by the covalent σ-bonds between the in-plane sp^2 hybrids of carbon atoms and by the weak Van der Waals π-bonds, between the layers [17, 18]

electrons to the σ-bonds, and one electron to the π-bonds in order to be covalently bonded to the three neighboring carbon atoms. The σ-bonds are localized between the C-C atoms, while the π-bonds in form of delocalized electron density are distributed above and below the planes of the carbon atoms. The delocalization of the π-bond electrons to move free throughout the plane leads to an electrical conductivity. This means that the electricity can conduct along the plane of the layers, but it conducts weakly in the vertical direction to the plane. Graphite is known to be used in pencils and lubricants. However, it has other applications related to its heat and conductive properties. For example, due to its resistance to high temperatures up to 3000°C, it is used in heat-resistant materials.

Diamond Structure: The structure of the diamond is a tetrahedral configuration in which each carbon atom is bonded to four other carbon atoms located at the four apexes of a tetrahedron. The carbon atoms are connected together by strong covalent σ-bonds through sp^3 hybrid orbitals. Each p-component of the sp^3 hybrid orbital overlaps with the neighboring p-component of the other forming σ-bond, as can be seen in Fig. 5.11. The diamond structure can be interpreted as being combined of two interweaving face-centered cubic (FCC) sub lattices, shifted along the body diagonal of the cubic cell. However, the centered tetrahedrons build together high stable 3D network, in which six carbon atoms form an armchair, is assigned by incredibly strong covalent bonds. Due to the fact that the four σ-bonds of the four valence electrons are localized between the central carbon atom in the tetrahedron and the its four carbon apexes. The electron localization means that there is no electron movement and the residence of the electron density is restricted inside the tetrahedron and thus there is no any possibility to induce an electrical current or conductivity. Therefore, diamond is an insulator with a very wide band gap of about 5.5 eV. Chemically, diamond is insoluble in any organic solvents and needs a high temperature of around 4000° C in order to be melted.

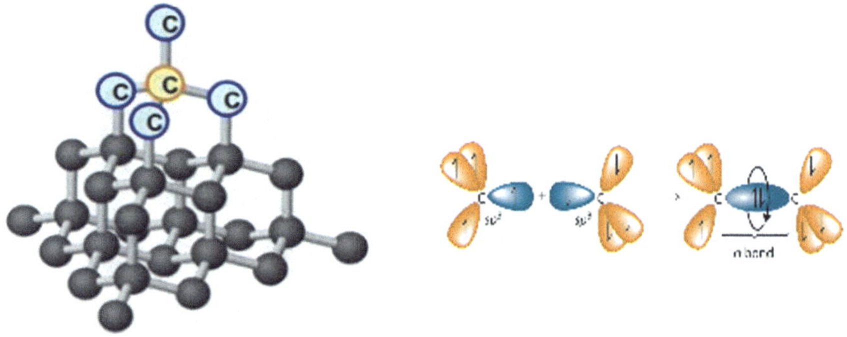

Fig. 5.11 Diamond has a tetrahedral structure of two nested FCC lattices, assigned by covalent σ-bonds between the sp^3 hybrid orbitals of carbon atoms, so that each atom has four covalent σ-bonds [19, 20]

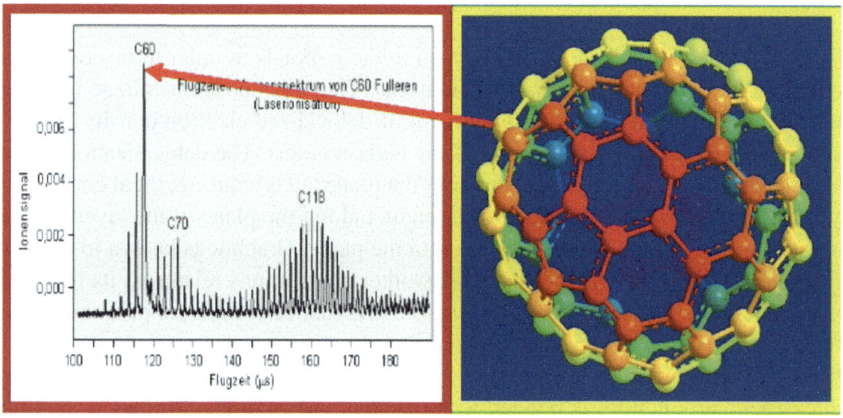

Fig. 5.12 The famous Buckminster fullerene C_{60} and the corresponding peak of the carbon mass spectra determined by Kroto et al. [21]

Buckminster Fullerenes: are a new allotrope of the element carbon, in which the atoms are arranged in closed hollow shells or cages. The most famous specie was C_{60}, a cluster of 60 carbon atoms. Its icosahedral structure is composed of 12 pentagonal and 20 hexagonal connected rings or faces, very similar to a European football. The mass spectrum of carbon clusters, determined via time-of-flight experiments, was published by Rohlfing et al. [22] in 1984. They found a strong signal corresponding to C_{60}, as to be seen in Fig. 5.12. However, the authors were not interested in the peak of C_{60} in their mass spectra but in studying small carbon clusters. Due to the successful interpretation of the C_{60} signal and the recognition of its geometrical structure, Kroto et al. [21] were awarded in 1996 with the Nobel prize in chemistry for the discovery of C_{60}. A critical review about carbon nanostructures is given by Shenderova et al. [23].

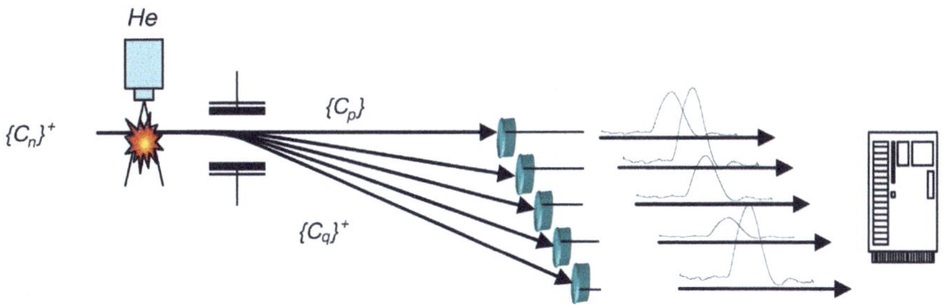

Fig. 5.13 Schematic view of the experimental set-up [24]

5.2 Carbon Clusters

5.2.1 Mass Spectra of Carbon Clusters

Experimentally: Charged carbon clusters were the most investigated particles, due to the fact that the ionized fragments are experimentally easier to identify when electric or magnetic fields are used [24]. After the collision of a swift C_n^+ projectile with atoms in a gaseous stream, various neutral C_p and cationic C_q^+ carbon clusters are generated, and the fragments are then deflected according to their charge over mass ratio via an intense electric field. The separated fragments impinge on solid-state surface barrier detectors providing transient currents. Thus, the number and the mass of each fragment impinging on the detector are resolved, as soon as the current shape is analyzed. A schematic view of the spectrometer is shown in Fig. 5.13.
The Mass Spectra: of cationic carbon clusters C_n^+ (for n $\leq$ 15) were first observed in 1943 by Mattauch et al. [25] using high-frequency sparking between graphitic electrodes. Further work on the mass spectra of cationic carbon clusters up to a cluster size of C_{28}^+ was carried out in 1959 by Dörnerburg and Hintenberger [26]. They found in their experiments that the cationic carbon clusters have alternate odd-even character, were odd clusters having higher intensity than those of even n for n $\leq$ 10. Few years later Berkowitz and Chupka [27], see Fig. 5.14, employed a magnetic mass spectrometer to analyze the composition of ejected vapor of graphite. They found that their mass spectra are comparable with the results obtained by Dörnenburg et al. [28]. The spectra of the carbon ions have also odd-even pattern. However, the odd clusters have higher intensity than the even ones.

5.2.2 Small Carbon Clusters C_n ($2 \leq n \leq 10$)

In the subsequent years theoretical studies on carbon clusters were investigated. Based on quantum chemical methods and related classical interatomic potential,

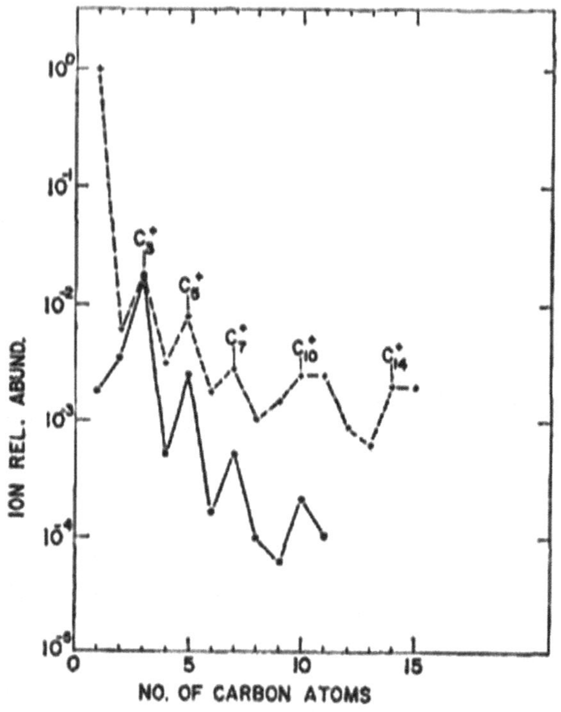

Fig. 5.14 Relative Carbon ion intensities as given by Berkowitz and Chupka (Lower solid line) [27] and (Upper dashed line) as given by Dörnenburg et al. [28]

Kosimov et al. [29] studied small carbon clusters. This potential was proposed in 1990 by Brenner [30], as a very accurate empirical potential, based on bond-order expression which was fitted on diamond, graphite crystal and small hydrocarbons and developed by Tersoff [31, 32]. The Brenner potential seems to be reliable also for small carbon systems like carbon clusters and chains beside the accurate description of different bulk condensed phases of diamond and graphite and being in relative magnitude faster than the ab initio schemes. However, for the carbon-carbon interaction pair potential the recently modified version of the Brenner potential was used [29]. The so called Brenner second-generation reactive empirical bond order (REBO) potential is one of various bond order potentials, which were parameterized particularly for carbon systems. The advantage of this Brenner potential, which is based on the empirical bond order formalism, is to enable the covalent bonds to break or to bind according to the hybridization changes in the atoms. Therefore, such flexible potential is appropriate to describe the various complex chemical bonds in large systems composed of many atoms. In order to investigate the carbon clusters, Kosimov et al. [29] used the modified Brenner potential to carry out the energy and structural minimization of the neutral carbon clusters C_n (n = 2 − 10). The most stable and energy low-lying isomers of these clusters are shown in Fig. 5.15. The structures of the most stable carbon clusters C_n for n $\leq$ 5 are exhibiting linear forms, while for n $\geq$ 6 monocyclic forms, which are in good agreement with several previous theoretical

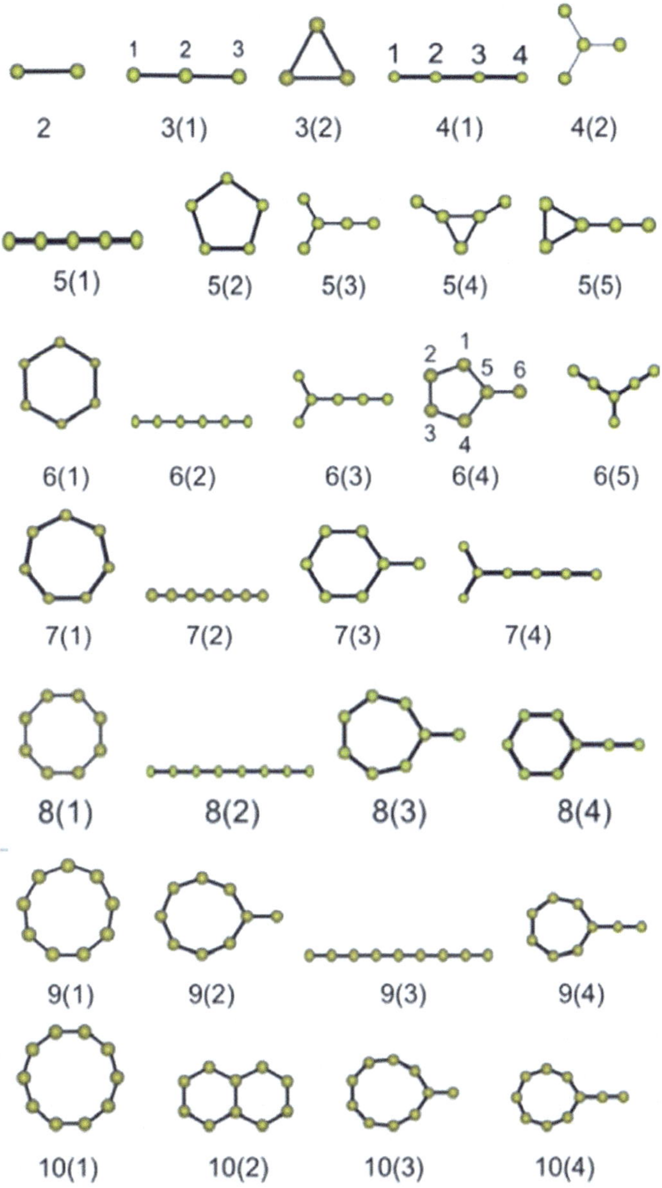

Fig. 5.15 The stable configurations for the isomers of C_n (n = 2 − 10) [29]

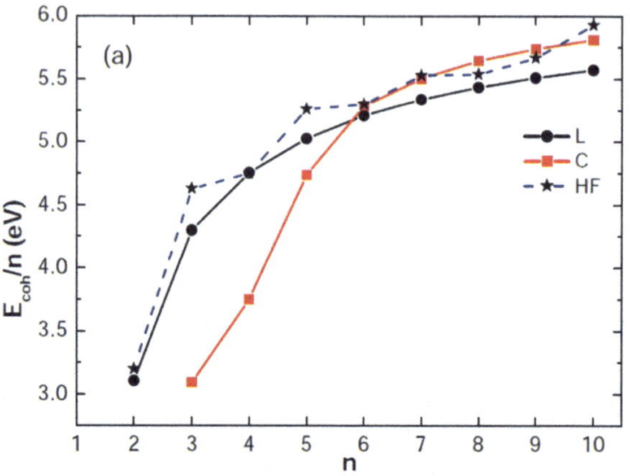

Fig. 5.16 Binding energy Ecoh of carbon clusters with linear (L) and cyclic (C) structures calculated in present study and by HF as a function of cluster size n [29]

calculations. All the monocyclic structures of the studied carbon clusters are regular with rotational symmetry. However, earlier ab initio calculations on carbon clusters reveal irregular monocyclic formations. The most stable carbon cluster favor planar structures, while those with non-planar structures having higher energies. Furthermore, the hexagon C_8 and the decagon C_{10} of carbon clusters seem to be the seeds for better comprehension in studying sub-monolayer of planar graphene. The authors determined the stability of the carbon clusters in form of binding energy per atom (E_{koh}). They found that the E_{koh} of the linear and cyclic clusters is increasing with increasing the cluster size, as presented in Fig. 5.16. Their results are in good agreement with earlier experimental and calculated results for $n = 2 - 5$. However, no any n-even-odd oscillations in structures were found for the investigated carbon clusters.

Ab Initio Calculations of Carbon Clusters: Raghavachari and Binkley [33] performed the first accurate post Hartree–Fock ab initio calculations. They studied carbon clusters C_n (for n = 2 to 10) using polarization functions and including electron correlation. They found that there is an odd-even alternation and confirmed that the linear structures having the odd numbered clusters, while the cyclic structures are preferred by the even numbered clusters, as shown in Fig. 5.17. They also found that the cyclic rings assigned by the even numbered clusters up to C_7 are less stable than the linear structures with the odd numbered clusters. The results of Raghavachari and Binkley [33] were confirmed by Menon et al. [34]. They found that the more stable even clusters also prefer to be chains, while the even cluster have mostly cyclic configurations. However, the results of Menon et al. [34] show that carbon clusters exhibit important differences from the element silicon. The consistency of both structures with the ring and linear forms can be related to the strain energy provoked by the smaller covalent radius of carbon, which in turn is responsible to prevent any closed-pack formation which found in atomic clusters with larger covalent radii like silicon. However, any symmetry distortion of the cyclic rings leads to relieve the strain energy enabling convenient overlapping between the π orbitals and yielding unique 2D carbon structures. Due to the fact that carbon can build structures

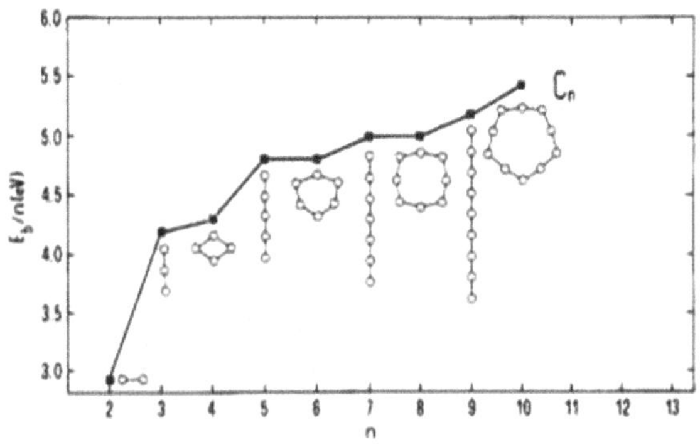

Fig. 5.17 The binding energy of carbon clusters as a function of the cluster size, assigned by its odd-even alternation corresponding respectively to linear and cyclic cluster forms, [33]

with different bond lengths and directions, it is generally considered to be a difficult element to deal with.

Confirmation of Odd-Even Alternation via Theory: Based on Anderson's formalism, Jo and Lee [35] used parameterized semi-empirical tight binding method of Porezag et al. [36], in order to determine the ionization potentials and cohesive energies of small carbon clusters C_n for (n = 2 to 13). They also used the universal parameters of Harrison in order to obtain the matrix elements from the Slater-Koster terms. The structures of carbon clusters, obtained at the Hartree Fock level, were initially adopted from the semi-empirical total energy scheme, at which the bond lengths of the structures were adjusted. The obtained cohesive energies are in very good agreement with other results, while the ionization potentials are in reasonably good agreement with others. The authors adopted for the carbon clusters C_n for $3 \leq n \leq 10$ the HF structures, the odd numbered clusters for linear chains and the even numbered clusters for cyclic rings. According to the calculated total energies, the authors compared the relative stability of the chains and the cyclic forms and found that the even numbered cyclic rings are more stable than the linear chains, in agreement with the HF and TB methods, as shown in Fig. 5.18. Thus, they confirm the results obtained by Raghavachari and Binkley [33] as well as by Menon et al. [34]. However, up to that date, it was still unclear what is behind the structural alternation between the chain and ring.

5.2.3 *Small Carbon Clusters C_n ($12 \leq n \leq 32$)*

Cyclic & Cage Carbon Clusters: Jones [37] performed density functional calculations using both local spin density (LSD) and gradient corrected corrected (Becke–Perdew) approximations to the exchange correlation energy to study neutral carbon clusters C_n ($4 \leq n \leq 32,\ n$ even). The structures of the most stable isomers of carbon clusters exhibit 2D and 3D configurations like chains and cyclic rings, bowls and cages, as well as graphitic formations. Also stable carbon clusters C_n for $n \geqslant 8$

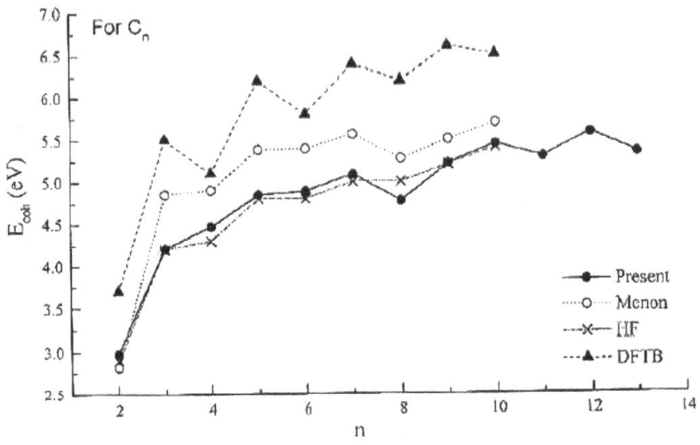

Fig. 5.18 Cohesive energies of small carbon clusters: HF, Menon, and the DFTB are from [33, 34, 36], respectively [35]

possess spherical structures, while those monocyclic rings are more stable than the double ring structures. Jones [37] studied first small carbon clusters with even numbers. He found that most stable C_4 exhibits a chain structure, while C_6, C_8, and C_{10} clusters exhibit cycle like structures, as already found in previous studies. He investigated different isomers of C_{12} and found that the cyclic form is the mot stable structure, as shown in Fig. 5.19. The energies of different isomers are given relative to that of the monocyclic ring structures, and the energies are for the clusters, not per atom. The lines joining the LSD and BP values represent the structural type: Full lines indicate cages, dotted lines with solid triangles represent double rings, and graphitic structures squares include both planar plate and bent bowl forms. Similarly, he studied many isomers of C_{14}, C_{16} and C_{18} clusters. He found that the cyclic ring of C_{14} is the most stable structure, while the cage shape was the third in stability. The most stable isomer of C_{16} clusters exhibits also cyclic ring, while the bowl shape is the next stable structure. In contrast, the most stable isomer of C_{18} clusters has a cage form, while the next stable one exhibits a cyclic ring, and therewith the C_{18} cluster announces the transition from 2D cyclic rings into 3D cage forms for larger carbon cluster sizes. This trend proceeds up to C_{32} clusters, in which the cyclic rings are displaced and degraded to the third or fourth rank of stability.

The Smallest Carbon Fullerene: R. O. Jones [37] investigated the stability of different isomers of C_{20}. He confirmed that C_{20} is smallest fullerene structure, if the item "fullerenes" can be defined as n threefold coordinated carbon atoms building polyhedral cages with $(n/2 - 10)$ hexagonal and 12 pentagonal faces, as shown in Fig. 5.20. The isomer (1) with the cage structure of the C_{20} fullerene is about ~4 eV lower in energy than the cyclic ring structure (6), obtained at the LDA level, while at the PB level it is only 0.1 eV. It is interesting to note that both cages of the (3) and (5) structures in Fig. 5.20 can be transformed in each other by changing the positions of pairs of pentagons and hexagons according to the Stone–Wales transformation.

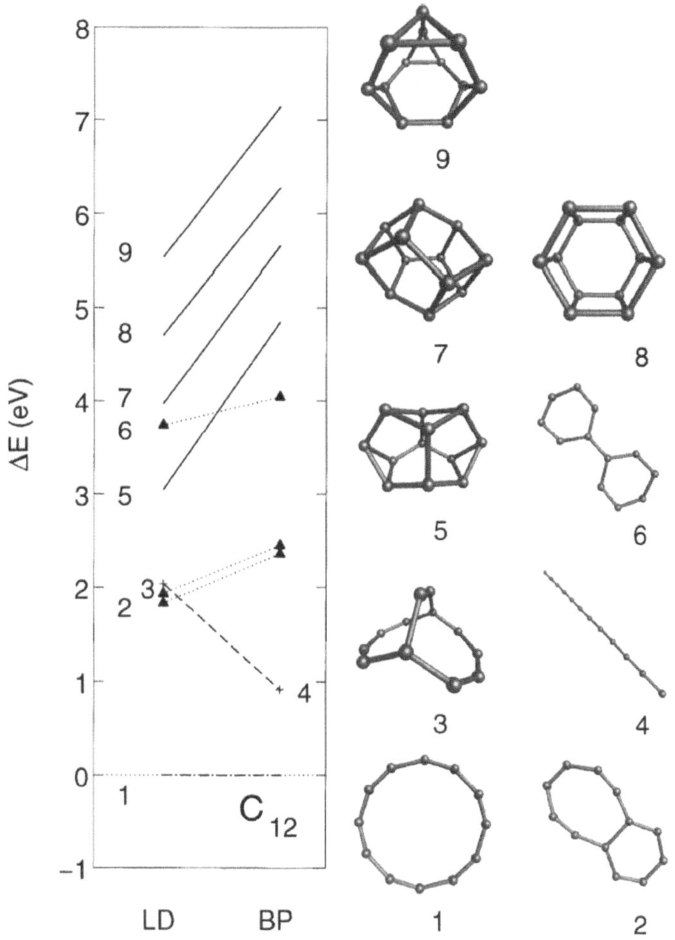

Fig. 5.19 Structures and energies (relative to the monocyclic ring) of isomers of C_{12} (eV). LSD and GC values are connected by lines representing the structural type full: Cages, dotted: Double rings, crosses: Chains, squares: Graphitic, dash–dot: Monocyclic rings. The structures are ordered in pairs from below according to the LSD energies [37]

The cage structure (3) is by 1.15 eV more stable that the cage one (5). However, the appearance of four membered rings in these structures implies that they are not fullerenes as explained above. However, in comparison with the smallest fullerene C_{28}, both cages (3) and (5) of C_{20} remain rather small.

The Relative Stability of Carbon Clusters: As mentioned above, R. O. Jones [37] determined the stability of different isomers of C_n clusters, which is given as binding energies per atom, known also as cohesive energies. The most stable isomers of each type of structures, determined at the local spin density (LSD) and the Becker–Perdew gradient-correction (PB) methods, are compared together and presented in Fig. 5.21.

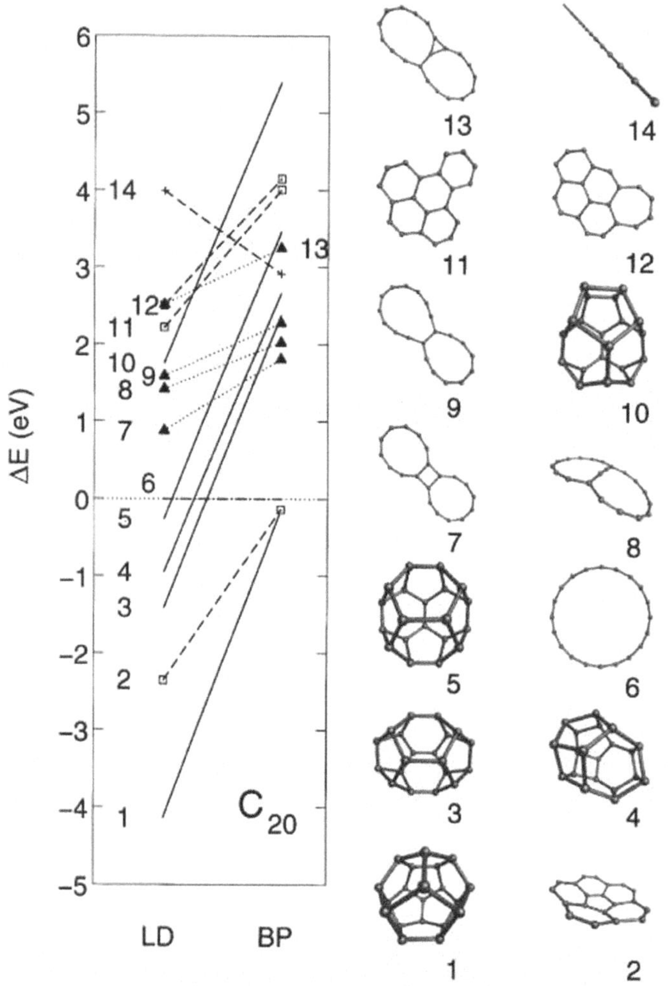

Fig. 5.20 Structures and energies (relative to the monocyclic ring) of isomers of C_{20} (eV) [37]

It shows that the values obtained at the BP gradient corrections are about ~1 eV lower than the LSD values. Similar difference was found in other atomic clusters of the main groups. The Fig. 5.21 displays that the stability of the structures is increasing with increasing the cluster size. The increasing stability of the cage structures is also observed without any alternation, in opposite to the stability of the ring structures, which exhibits zigzag behavior with increasing cluster size. Therefore, the stability of the monocyclic rings can be expressed as periodic and the curves of both methods LSD and BP are almost parallel. The distinctive value of C_{20} at the both LSD and BP levels can be associated that C_{20} being the first fullerene structure. However, the

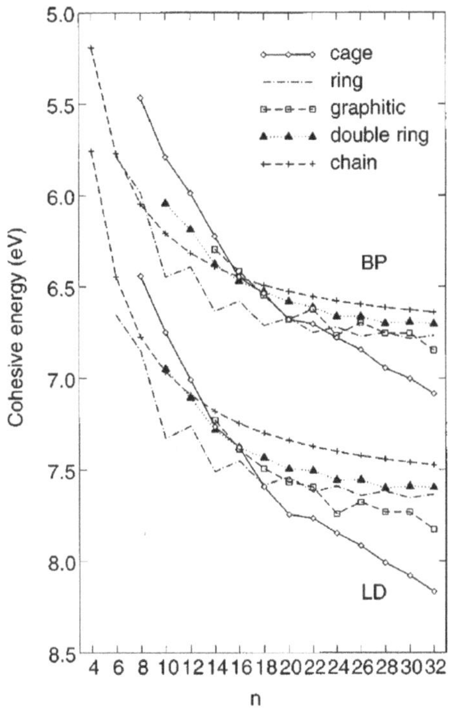

Fig. 5.21 Cohesive energies (binding energies per atom eV) of the most stable of different isomer classes in C_4–C_{32}. [37]

stability of some other non-fullerene structures like C_{18} and C_{22} can be estimated by values extrapolation of other larger cages of fullerene structures.

Graphitic Sheet & Cage Carbon Clusters: Based on quantum mechanical tight-binding methods [38], density functional theory (DFT) approximations, and highly accurate diffusion quantum Monte Carlo methods (DMC) [39], Kent et al. [40] investigated different isomers of even numbered carbon clusters C_n for (n = 24, 26, 28, and 32), and determined their electronic and structural properties. The authors were not only interested in the structural stability and the ordering of these even numbered clusters but also in the question: "which is the smallest stable fullerene?". Though, the answer is rather difficult due to the sensitivity of cluster formation in experimental conditions and of course a challenge to theoretical approximations. However, they found that the C_{26} (C_{2v}) and C_{28} (T_d) carbon clusters are the smallest stable fullerenes, and the C_{24} (C_s) isomer is the smallest stable graphitic fragment, exhibiting a combination of one-pentagon and six-hexa-gon bowl structure. However, an accurate treatment of electron correlation was given by the DMC method, which was able to reproduce the experimental binding energies of small hydrocarbon molecules to within 1%. The accuracy of the DMC technique can be demonstrated by the cohesive energy calculations of the bulk of diamond. The experimental cohesive energy value of the diamond is about 7.37 eV, and the calculated value, obtained with the

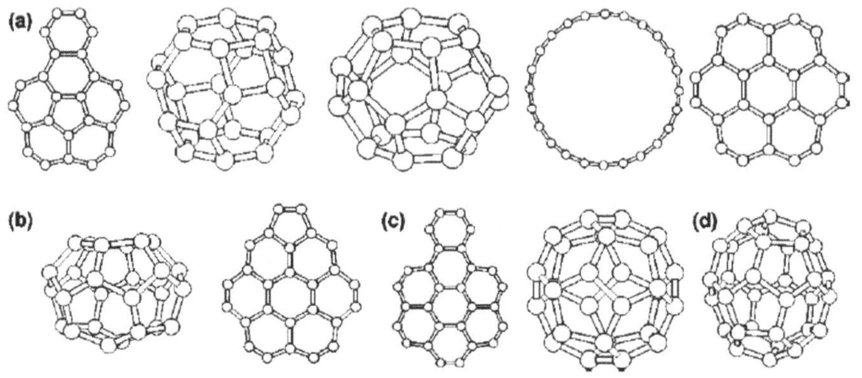

Fig. 5.22 The structures of the C_n clusters. **a** C_{24} one-pentagon bowl, O_h cage, D_6 fullerene, ring, and sheet (most stable), **b** C_{26} fullerene and one-pentagon sheet, **c** C_{26} sheet and fullerene, **d** C_{28} fullerene. The $n = 26$, 28, and 32 rings are not shown [40]

DMC method, is 7.46(1) eV per atom. Thus, the experimental and the theoretical values are indeed in a very good agreement.

The Isomers of C_{24} Carbon Clusters: The low-lying C_{24} isomers are presented in Fig. 5.22. The structures of these isomers are exhibiting either cage-like or graphitic formations. One cage is composed of tetragonal and hexagonal faces with O_h symmetry, while the other is fullerene composed of pentagonal and hexagonal faces with D_6 symmetry. One of the graphitic like structure is the bowl-shaped one consisted of a central pentagon surrounded and edged by six hexagons, while the other is a true flat graphitic sheet. The remaining C_{24} isomer is a polyacetylenic monocyclic ring. Other configurations with three-pentagon bowl or bicyclic ring structures were excluded from the DFT calculations due to their high energies. The predicted most stable C_{24} isomer is the graphitic sheet being 1.3(2) eV lower in energy than the monocyclic ring, which was almost isoenergetic with the D_6 fullerene. The latter $C_{24} - D_6$ fullerene is by 0.7(2) eV lower in energy than the O_h cage and by 0.5(2) eV lower than the bowl-like structure. The high stability of the C_{24} graphitic sheet can be explained by the strain-less compact structure composed of seven hexagonal honeycomb pattern. However, the carbon sheet of C_{24} seems to be the smallest stable graphitic fragment, because smaller graphitic sheets have higher energy.

The Carbon Clusters C_{26}, C_{28}, and C_{32} : The most stable isomers of the C_{26} and C_{28} clusters are presented in Fig. 5.22b,c, respectively. Only one isomer of the C_{32} clusters is shown in Fig. 5.22d. The cage like structure seems to be the most stable isomer of C_{26}, which is by 2.2(4) eV lower in energy than the graphitic like isomer. Therefore, the cage like structure of the C_{26} isomer is considered as the smallest stable fullerene. Due to the DMC calculations, the authors [40] predicted that the fullerene of the C_{28} isomers is the most stable structure indicating the possibility of its production. The C_{28} fullerene is by 4.2(3) eV lower in energy than the cyclic ring and by 3.2(3) eV lower in energy than the graphitic sheet. However, the DMC binding energies per atom of all low-lying graphitic like, bowl, cage like and fullerene structures are

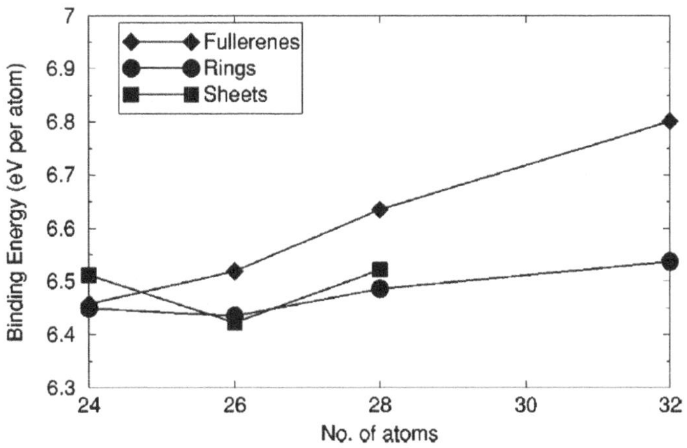

Fig. 5.23 The DMC binding energies per atom of the lowest energy ring, sheet/bowl, and fullerene C_n structures. The lines drawn are for guidance only. The statistical error bars are smaller than the symbols [40]

compared and presented in Fig. 5.23. Due to the reduction in strain energy, the binding energy per atom for the cyclic rings and fullerenes increases gradually. The Fig. 5.23 indicates also that the 3D structures is favored over the 2D structures and the binding energy per atom of fullerenes rises steeply than the corresponding binding energy of the rings when cluster size is increased. The faster Íncreasing binding energy can be related to the strain in smaller fullerenes, which declines by larger fullerenes. As can be seen in Fig. 5.23 the curves of the rings and fullerenes are crossing nearly at C_{24}. Below this critical point the rings are more stable than the fullerenes, while above this point the fullerenes are more stable than the rings. The DMC calculated binding energy per atom of the C_{32} fullerene 6.8 eV is in comparison with the experimental binding energy of the C_{60} fullerene by 1 eV less stable. However, by larger cluster sizes the function of the binding energy of fullerenes flattens off essentially.

5.3 Inorganic CdSe-Like Binary Clusters

Introduction: Besides the alkali-metal clusters of lithium atoms (Sect. 3.2.6) and of sodium atoms (Sect. 3.4.3), we have discussed in Sect. 4.3.1, small boron clusters, in Sect. 4.6, small boron nitride clusters, and in Sect. 5.2.2, small carbon clusters, composed of atoms belonging to the p-block non-metal elements, such as boron, carbon, and nitrogen. We reported the experimental findings of the boron, carbon, and boron nitride clusters starting from the mass spectra up to infrared and Raman spectroscopy as well as to vibrational frequencies, combined with theoretical calculations confirming and predicting the experimental data and related geometrical and molecular structures. However, in this section we are reporting clusters of the inorganic (carbon-less) binary nanoclusters.

5.3.1 *Clusters-Structures Versus Clusters-Sizes*

We have already seen that the structural configurations of lithium and sodium alkali-metal clusters, as mentioned in Sects. (3.2.6, and (3.4.5, respectively, are depending on the cluster size. Lithium clusters by sizes less that 7 atoms exhibit 2D structures (see Fig. 3.14, while by larger sizes the clusters possess 3D structures. In the case of boron clusters B_n, we have seen that they possess 2D planar or quasi-planar structures until cluster size of $n \leq 19$ atoms. When the cluster size $n \geq 20$ then the clusters possess 3D structures. For example, when $n = 20$ we have the smallest boron nanotube, composed of two staggered rings (see Fig. 4.76. By larger boron cluster sizes there is a competition between 2D and 3D structures. In the case of carbon clusters, the structures depend not only on the cluster size but also on whether the size number is odd or even. For example, the C_n clusters for $3 \leq n \leq 10$ exhibit linear form when n is odd, and cyclic forms when n is even (see Fig. 5.17. For larger clusters carbon sizes up to $n \leq 19$, there is a competition between cyclic rings and cages strcutures. But when $n = 20$, then the C_{20} cluster exhibits 3D structure and presents the smallest carbon fullerene, as can be seen in Fig. 5.20. Therefore, we can conclude that the structural configuration of clusters depends strongly of the cluster size. In the next section will learn whether the structure development is a function of nanosized clusters.

5.3.2 *Small Inorganic Binary Clusters*

By means of Monte Carlo method based global optimization routines via evolutionary and stochastic quenching algorithms within a new software package, namely the Knowledge Led Master Code (KLMC) [42], Farrow et al. [41] investigated the inorganic binary of nanoclusters and searched for the atomic structures of the low-lying minima on the energy hypersurfaces of selected compounds like zinc oxide $(ZnO)_n$, magnesium oxide $(MgO)_n$, cadmium selenide $(CdSe)_n$, and potassium fluoride $(KF)_n$, where $n = 1$–12 formula units. The search was carried out directly and after pre-screening on the DFT energy landscape. Based on the interatomic potentials, structural relaxation via pre-screening procedure was performed to minimize the energy function. This energy minimization leads to the reduction of the DFT calculations steps in order to find the local minima on the energy landscape. The optimized structures with the lowest DFT energy of the ZnO, MgO, KF and CdSe compounds, obtained with the software package KLMC, are presented in the Figs. 5.24, 5.25, and 5.26. The most stable structure with the lowest energy is assigned by the global minimum (GM). The next local minimum corresponds to the second lowest energy

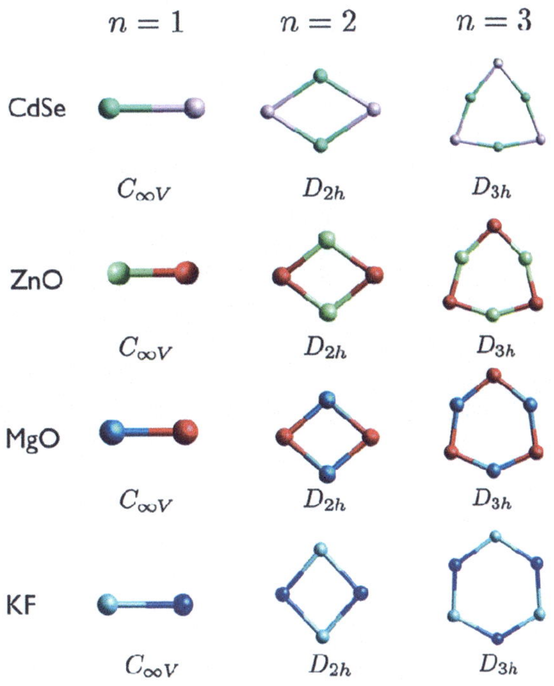

Fig. 5.24 $(CdSe)_n$, $(ZnO)_n$, $(MgO)_n$, and $(KF)_n$ GM nanoclusters, as determined on the PBEsol energy landscapes, for $n = 1$–3, arranged by size and degree of polarization (with point symmetry group labels). Colors: turquoise is Cd, pink is Se, sea green is Zn, red is O, blue is Mg, light blue is K, and dark blue is F. This figure was reproduced from the Physical Chemistry Chemical Physics with kind permission of the Royal Society of Chemistry, [41]

structure assigned by (LM2), and the third local minimum corresponds to the third lowest energy structure assigned by (LM3). For $n = 1$, as shown in Fig. 5.24, the bond lengths of the dimers CdSe, ZnO, MgO, and KF are 2.35, 1.68, 1.73, and 2.17 Å, respectively. The corresponding binding energies of CdSe, ZnO, MgO, and KF are 2.04, 3.80, 4.39, and 6.37 eV, respectively. Thus, the CdSe dimer has the weakest bond, while the KF has the strongest bond. For $n = 2$, all clusters exhibit rhombus structure with D_{2h} symmetry and a bond length of 2.46, 1.87, 1.90, and 2.36 Å, respectively, for the $(CdSe)_2$, $(ZnO)_2$, $(MgO)_2$, and $(KF)_2$ clusters. The largest rhombus and the weakest bond is the $(CdSe)_2$ compound, while the smallest rhombus and the strongest bond is the $(ZnO)_2$. For $n = 3$, the $(KF)_3$ has hexagonal structure, while the remaining compounds have deformed hexagonal structures adopting rather triangular forms. For $n = 4$, the global minimum (GM) structures of $(CdSe)_4$ and $(ZnO)_4$ are nearly quadratic rings, while the next low-lying isomers have a cuboid local minimum $(LM)_2$ structures, as can be seen in Fig. 5.25. In contrast, the GM for the $(MgO)_4$, and $(KF)_4$ compounds is the cuboid motif, while the local minimum $(LM)_2$ structures have planar shapes. The Fig. 5.25 shows beside n = 4 also the lowest energy structures for n = 5, and the dependence of the structural motifs on the cluster sizes. Thus, for $n = 5$, the $(MgO)_5$, and $(KF)_5$ compounds correspond to the lowest energy nanoclusters having similar armchair like structural motifs. The GM of $(CdSe)_5$ exhibits 3D distorted structure, while the GM of the $(ZnO)_5$ compound is a pentagonal structure with D_{5h} symmetry. The structural transition from 2D to

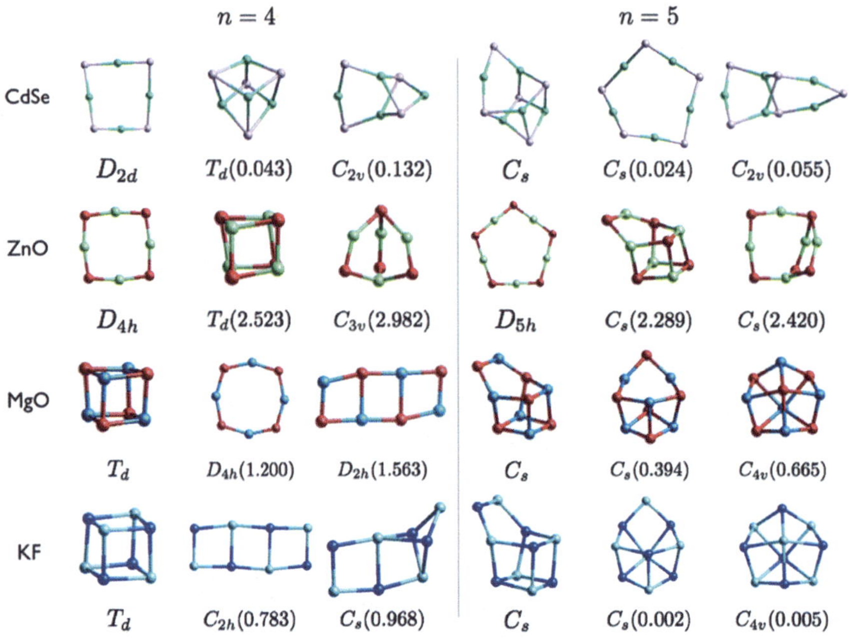

Fig. 5.25 $(CdSe)_n$, $(ZnO)_n$, $(MgO)_n$, and $(KF)_n$ lowest PBEsol energy nanoclusters, for $n = 4$–5. Energy differences from the GM (in eV) to the local minimum 2 (LM2), and to local minimum 3 (LM2), are given in brackets, and notations and colors are as in Fig. 5.24. This figure was reproduced from the Physical Chemistry Chemical Physics with kind permission of the Royal Society of Chemistry, [41]

3D motifs occurs for the MgO and KF compound between $n = 3$ and 4, while the transition for the ZnO and CdSe compounds occurs by larger cluster sizes. The trend of nanoclusters stability is expected to be improved when the cluster size or the average coordination are increased. However, the $(MgO)_n$ and $(KF)_n$ compounds have similar motifs by smaller cluster sizes, and the motifs of larger $(KF)_n$ nanoclusters are expected to form cuboids similar to the rock-salt phase. Due to the structure similarity of the compounds $n = 10$, 11 to those for $n = 12$, we present the lowest structures of $n = 12$ in Fig. 5.26. It shows that $(MgO)_{12}$ has a tubular form, the $(KF)_{12}$ compound is similar to rock-salt cuts, and the $(ZnO)_{12}$ exhibits a sodalite cage. Not only structural optimization was carried out but also electronic properties like infra red spectra, differences between the HOMO and LUMO energies ΔE, and electron affinity for the GM of the $(KF)_n$ and $(CdSe)_n$ compounds ($n = 1$ to 12). They asserted that ΔE lays at around 2–3 eV, within the visible part of the UV spectrum region. They found that electron affinity of the ZnO and CdSe compounds increases by increasing the cluster size, in opposite to the MgO where the affinity decreases by increasing the cluster size. The authors reported that ΔE does not confirm the conventionally assumed trend of increasing ΔE with decreasing cluster size. Finally,

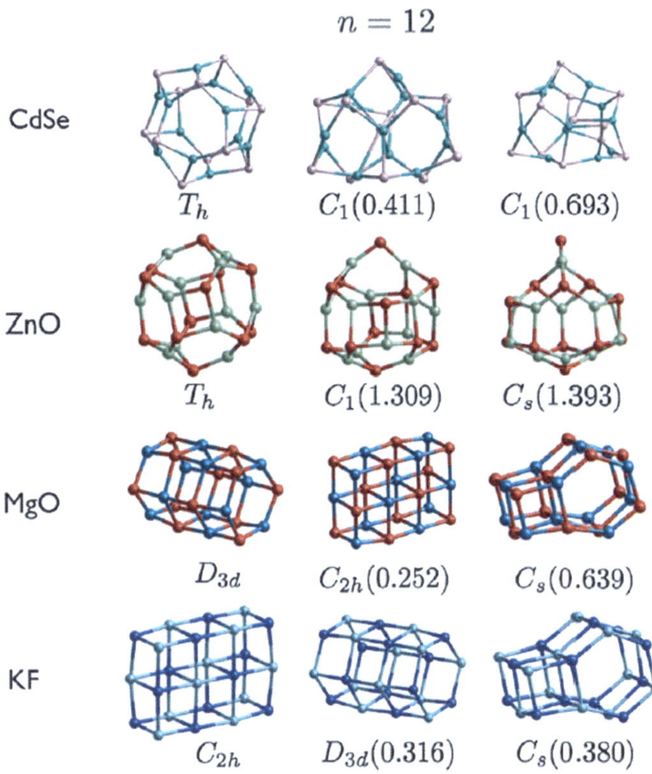

Fig. 5.26 $(CdSe)_n$, $(ZnO)_n$, $(MgO)_n$, and $(KF)_n$ lowest PBEsol energy nanoclusters, for $n = 12$. Energy differences from the GM (in eV) to the local minimum 2 (LM2), and to local minimum 3 (LM2), are given in brackets, and notations and colors are as in Fig. 5.24. This figure was reproduced from the Physical Chemistry Chemical Physics with kind permission of the Royal Society of Chemistry, [41]

the authors have shown that a careful application of interatomic potential leads to successful results and offers an efficient algorithm search at the energy potential surface.

References

1. https://nano.tu-dresden.de/pubs/slides/others_2009_04_30_IhsanBoustani.pdf. Accessed on July 2015
2. Ras Tanura-Raffinerie, Saudi-Arabian, http://www.posterlounge.de/rastanuraraffineriesaudi arabienpr156200.html. Accessed on October 2014
3. The Discovery of Carbon, http://en.wikipedia.org/wiki/Carbon. Accessed on October 2014
4. http://pixgood.com/carbon-isotopes.html. Accessed on October 2014

5. P. Enghag, *Encyclopedia of the Elements* (Wiley-VCH Verlag GmbH & Co, KGaA, Weinheim, 2004)
6. G. Audi, O. Bersillon, J. Blachot, A.H. Wapstra, The NUBASE evaluation of nuclear and decay properties. Nuclear Phys. A **729**, 3–128 (2003)
7. https://www.saddlespace.org/whittakerm/science/cms_page/view/7795271. Accessed on October 2014
8. http://www.chem1.com/acad/webtext/chembond/cb06.html. Accessed on October 2014
9. S. Shaik, D. Danovich, W. Wu, P. Su, H.S. Rzepa, P.C. Hiberty, Quadruple bonding in C_2 and analogous eight-valence electron species. Nat. Chem. **4**, 195–200 (2012)
10. http://chemistryistorture.blogspot.de/. Accessed on October 2014
11. http://depts.washington.edu/chemcrs/course_index.cgi?bulkdisk/chem162Awin07. Accessed on October 2014
12. https://www2.chemistry.msu.edu/faculty/reusch/VirtTxtJmlImagessporbs.gif. Accessed on October 2014
13. D.R. Lide, *CRC Handbook of Chemistry and Physics*. (CRC Press, Boca Raton, FL, 2005). http://www.hbcpnetbase.com. Accessed on October 2014
14. http://m7science.wikispaces.com/file/view/CarbonCycleCompounds.png/205835566/CarbonCycleCompounds.png. Accessed on October 2014
15. K.S. Novoselov, A.K. Geim, S.V. Morozov, D. Jiang, Y. Zhang, S.V. Dubonos, I.V. Grigorieva, A.A. Firsov, Electric field effect in atomically thin carbon films. Science **306**, 666–669 (2004)
16. http://www.trendtopics.com/wp-content/uploads/2014/01/graphene-header.jpg. Accessed on October 2014
17. http://www.geo.arizona.edu/xtal/geos306/9_8.jpg. Accessed on October 2014
18. http://www.meritnation.com/img/shared/discuss_editlive/2309551/2011_12_23_00_30_49/7.png. Accessed on October 2014
19. http://www.weldons.ie/wp-content/uploads/2011/03/tetrahedral-carbon-1.gif. Accessed on October 2014
20. http://2012books.lardbucket.org/books/principles-of-general-chemistry-v1.0/section_28/258be47d749353d519d7e578b564e9d5.jpg. Accessed on October 2014
21. H.W. Kroto, J.R. Heath, S.C. O'Brien, R.F. Curl, R.E. Smalley, C_{60}: Buckminsterfullerene. Nature **318**, 162–3 (1985)
22. E.A. Rohlfing, D.M. Cox, A. Kaldor, Production and characterization of supersonic carbon cluster beams. J. Chem. Phys. **81**, 3322 (1984)
23. O.A. Shenderova, V.V. Zhirmov, D.W. Brenner, Carbon nanostructures. Crit. Rev. Solid State Mater. Sci. **27**, 227–356 (2002)
24. K. Béroff, M. Chabot, F. Mezdari, G. Martinet, T. Tuna, P. Désesquelles, A. LePadellec, M. Barat, Fragmentation of small carbon clusters, a review. Nucl. Instr. Meth. Phys. Res. B **267**, 866–871 (2009)
25. J. Mattauch, H. Ewald, O. Hahn, F. Strassmann, Hat ein Caesium-Isotop langer Halbwertszeit existiert? Ein Beitrag zur Deutung ungewöhnlicher Linien in der Massenspektrographie. Zeitschrift für Physik **120**, 598–617 (1943)
26. E. Dörnbenburg, H. Hintenberger, Das Auftreten vielatomiger Kohlenstoffmoleküle im Hochfrequenzfunken zwischen Graphitelektroden. Z. Naturforschg. **14a**, 765–767 (1959)
27. J. Berkowitz, W.A. Chupka, Mass spectrometric study of vapor ejected from graphite and other solids ba focused laser beam. J. Chem. Phys. **40**, 2735–2736 (1964)
28. E. Dörnenburg, H. Hintenberger, J. Franzen, Über die Struktur der im Hochfrequenzfunken entshenden vielatomigen Kohlenstoffmoleküle. Z. Naturforschg. **16 a**, 532–534 (1961)
29. D.P. Kosimov, A.A. Dzhurakhalov, F.M. Peeters, Theoretical study of the stable states of small carbon clusters C_n (n = 2–10). Phys. Rev. B **78**, 235433 (2008)
30. D.W. Brenner, Empirical potential for hydrocarbons for use in simulating the chemical vapor deposition of diamond films. Phys. Rev. B, **42**, 9458 (1990). Erratum Phys. Rev. B **46**, 1948 (1992)
31. J. Tersoff, Empirical interatomic potential for silicon with improved elastic properties. Phys. Rev. B **38**, 9902 (1988)

32. J. Tersoff, Modeling solid-state chemistry: Interatomic potentials for multicomponent systems. Phys. Rev. B **39**, 5566(R) (1989)
33. K. Raghavachari, J.S. Binkley, Structure, stability, and fragmentation of small carbon clusters. J. Chem. Phys. **87**, 2191 (1987)
34. M. Menon, K.R. Subbaswamy, M. Sawtarie, First-principles molecular dynamics study of carbon clusters. Phys. Rev. B **48**, 8398 (1993)
35. C. Jo, K.-Y. Lee, Ionization potentials and cohesive energies of small carbon clusters from the semiempirical tight binding method. J. Korean Phys. Soc. **41**, 200–204 (2002)
36. D. Porezag, Th Frauenheim, Th Köler, G. Seifert, R. Kaschner, Construction of tight-binding-like potentials on the basis of density-functional theory: Application to carbon. Phys. Rev. B **51**, 12947 (1995)
37. R.O. Jones, Density functional study of carbon clusters C_{2n} $(2 \leq n \leq 16)$. I. Structure and bonding in the neutral clusters. J. Chem. Phys. *110*, 5189–5200 (1999)
38. J. Cioslowski, *Electronic Structure Calculations on Fullerenes and Their Derivatives* (Oxford University Press, London, 1995)
39. J.C. Grossman, L. Mitas, K. Raghavachari, Structure and stability of molecular carbon: Importance of electron correlation. Phys. Rev. Lett. **75**, 3870. Erratum: Phys. Rev. Lett. **76**, 1006 (1995)
40. P.R.C. Kent, M.D. Towler, R.J. Needs, G. Rajagopal, Carbon clusters near the crossover to fullerene stability. Phys. Rev. B **62**, 15394–15397 (2000)
41. M.R. Farrow, Y. Chow, S.M. Woodley, Structure prediction of nanoclusters; a direct or a pre-screened search on the DFT energy landscape? Phys. Chem. Chem. Phys. **16**, 21119 (2014)
42. S.M. Woodley, Knowledge led master code search for atomic and electronic structures of LaF_3 nanoclusters on hybrid rigid Ion-Shell model-DFT landscapes. J. Phys. Chem. C **117**, 24003–24014 (2013)

Part III
Modelling of Nanostructures

Chapter 6
Two-Dimensional Sheets

Abstract Geim and Novoselov synthesized in 2004 single-atom thick carbon sheet graphene and opened the door for research of 2D carbon materials. Graphene and related derivatives called graphynes as well as nanoribbons have unique electronic, optical, and mechanical properties. However, with the prediction of quasi-planar boron clusters and sheets in 1997, I. Boustani also opened the door for the 2D boron sheets research known nowadays as borophene. The synthesis of borophene encouraged researchers to explore the mechanical, optical, magnetic and electronic properties as well as their potential applications in nanotechnology.

6.1 Boron-Based Nanosheets (BNSs)

Before starting with the nanosheets, one needs to know the definition of the term "nano" and the relationship of nano to angstrom and to nanostructures.

6.1.1 Nano (Concept)?

Since we are dealing with atomic domain, we first define the term angstrom as a measure in this domain. Angstrom, as the atomic measure, is named after Anders Jonas Ångström, a Swedish scientist (1814–1874), is assigned by (Å) and considered to be a unit length of 10^{-10} m. The angstrom is used to measure such quantities like the wavelength of light, molecular diameters and thickness of films on liquids. The counterpart, the nano as a unit of a nanometer is defined as $1\,\text{nm} = 1\,\text{nm} = 10^{-9}\,\text{m}$ (one billionth of meter).

In order to estimate the nanometer unit illustratively, we consider ten hydrogen atoms. The diameter of a single hydrogen atom in its ground state is around 1 Å in diameter, while atoms like carbon, sulfur or chlorine are about 2 Å in diameter. Thus, one nanometer is nothing else as the width of ten hydrogen atoms lied side by side, as schematically presented in Fig. 6.1. Hereupon, the nanometer is equal to 10 angstrom, or simply written: $1\,\text{nm} = 10$ Å.

I. Boustani, *Molecular Modelling and Synthesis of Nanomaterials*, Springer Series in Materials Science 290, https://doi.org/10.1007/978-3-030-32726-2_6

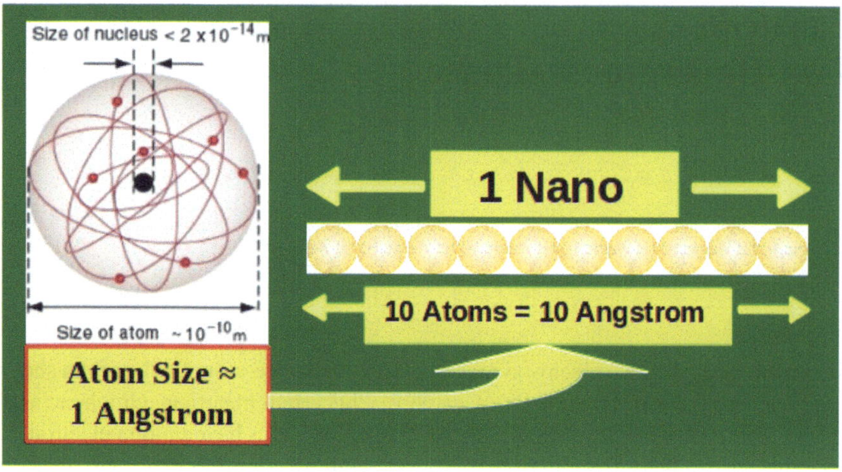

Fig. 6.1 Nanometer is simply the length of ten hydrogen atoms aligned side by side

6.1.2 Nanostructure (Concept)?

The small boron clusters, investigated in Sect. 4.3.1, or carbon clusters, investigated in Sect. 5.2, all have different length and width, depends on whether they are circular or elongated. For example, after considering 1.65 Å as an average distance of B–B atomic bond for all boron clusters, the elongated B_{16} cluster of Fig. 4.63 (Sect. 4.4.1), has a length of around 8 Å and a width of around 2.8 Å. The average length and width of the elongated B_{24} cluster, presented in Fig. 4.94 (Sect. 4.4.3.), is about 12.4 Å and 9.6 Å, respectively. The larger circular cluster B_{36} in Fig. 4.130 (Sect. 4.4.1) has an average diameter of around 9.6 Å. The average length and width of the planar cluster B_{40} in Fig. 4.138 (Sect. 4.4.3) is 11.8 Å and 8.4 Å, which is respectively 1.18 and 0.84 nm. The proposed structure of borophene in Fig. 4.133 (Sect. 4.4.2), as extended shape of B_{36} simulating boron sheet, has an average diameter of around 26 Å or say 2.6 nm. Herewith one can assert that as soon as a length of a structure accomplishes the nanometer domain, then it is classified as a nanostructure.

6.1.3 The "Aufbau Principle" and Nanosheets

Boustani found in 1994 for the first time [1] that small boron clusters B_n for cluster sizes between 10 and 14 exhibit nearly planar or convex shapes. He predicted that the nearly planar or quasi-planar clusters can be considered as fragments of a one-atom-thick boron sheet, and the convex boron clusters as fragments of nanotubes or hollow boron sphere. Boustani proposed the so-called "Aufbau Principle" [2] (see Fig. 4.38, Sect. 4.3.2.), which says that there are two basic units B_6 and B_7, from which highly stable boron can be constructed, like cluster, sheets, spheres and nanotubes. These B_6 and B_7 units have pentagonal and hexagonal pyramidal forms, respectively. Thus, the B_7 unit can serve as building block forming single-layered boron sheets.

6.1.4 The First Flackes of Boron Sheets

Furthermore, different sizes of small quasi-planar boron clusters, like B_{15}, B_{16}, B_{20}, B_{22}, B_{32}, B_{42}, and B_{46}, were proposed and investigated by Boustani for the first time [3]. Some of these sheets, or boron flacks, are presented in Fig. 6.2. The boron atoms are ordered in up and down manner forming a buckled surface. The high stability of the quasi-planar clusters can be related to the sp^2 hybridization of the atomic orbitals of boron atoms. These hybridized atomic orbitals form strongly directed chemical bonds. Those in-plane σ bonds are inward-pointing hybrid orbitals and those out-of-plane π bonds are outward-pointing hybrid orbitals [3], as already shown for B_{13} cluster in Fig. 4.36 (Sect. 4.3.2). Thus, according to the Aufbau Principle, boron sheets are constructed by adding the hexagonal pyramidal B_7 units consecutively. This procedure shows the growth of small clusters towards stable boron sheets. The stability as well as the quasi-planarity (chemically aromaticity) of the sheets are on the one side due to the in-plane σ bonds and on the other side due to the increasing population of the electronic π-system[1] by increasing the cluster size. Thus, the first-principles investigations on quasi-planar boron clusters opened the door to 2D boron research, as reported by Boustani [3].

6.1.5 Prediction of BNSs via Theory

Conductive Hexagonal Net-work of Boron Sheets: By means of sequential ab initio based investigations, Boustani et al. [4] studied the unique properties of the nearly planar and nanotubular boron clusters and predicted the existence of novel boron structures, up to that date unknown in nature. These proposed structures could have single layered planar or tubular sheets, or quasi-crystalline solids, all are composed of only one single unit, the B_7 hexagonal pyramid. The authors compared their boron structures with similar layered structures composed of carbon atoms or boron-nitride network. In order to compare the structures and properties of different boron, carbon, and boron–nitride species, the authors [4] applied different ab initio quantum chemical methods in the form of self-consistent field (HF-SCF) as well as spin-polarized density functional theory (DFT) [5] in form of local spin density (LSD) [6]. In addition, they used the tight-binding approaches (TB) [7] and the linearized muffin-tin orbital based band structure methods (LMTO-ASA) [8]. Furthermore, they deployed the nonlocal normconserving pseudopotential plane-wave (PW) methods

[1] The population of π system occurs by filling the π-orbitals with valence electrons.

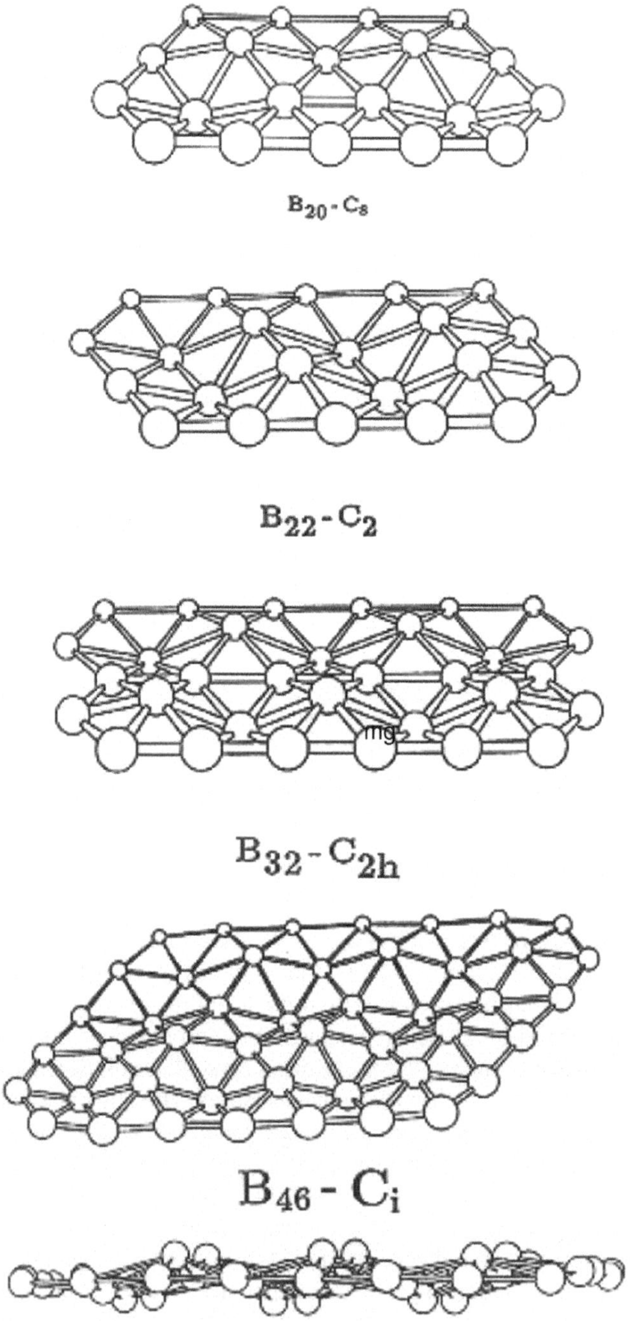

Fig. 6.2 Boron sheets with buckled surfaces, proposed by Boustani for the first time, are composed of 20, 22, 32 and 46 atoms. Reference [3]

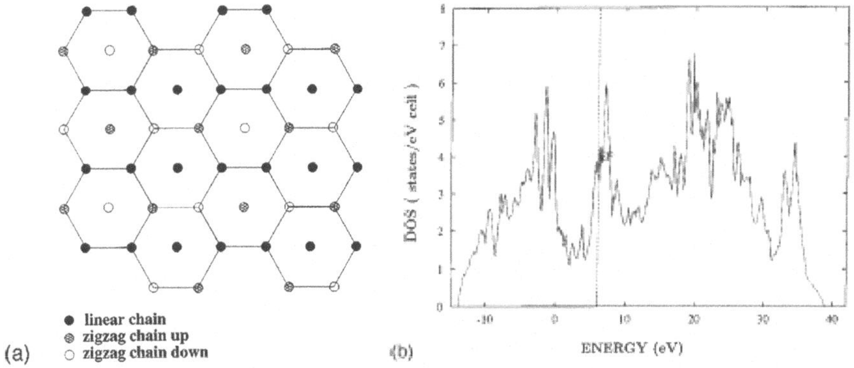

Fig. 6.3 The structure of monolayer for a model of buckled boron sheet and the corresponding total density of states. Reference [4]

[9]. The analytical gradient methods were carried out for the structural optimization and the final structures and energies were determined by the calculations at fixed configurations. They found that hexagonal network of boron sheets with buckled character is metallic in opposite to carbon or to boron-nitride sheets. The model of the quasi-planar boron sheet, as presented in Fig. 6.3a, exhibits definitely a metallic character indicated by the density of states (DOS) as given in Fig. 6.3b. Therefore, the boron sheets are expected to be a good conductors and better than graphene promising potential applications in nanotechnology, in high-temperature light materials and electronic devices, and in field emission and coating materials.

Boron Sheet and α-Rh. Boron: As already discussed in Sect. 4.2.2, the unit cell of α-rh.-boron is composed of eight boron icosahedra connected together in a rhombohedral cell, as shown at the top of Fig. 6.4, containing 96 boron atoms. In order to estimate whether the proposed nonexisting boron sheet is competitive with the real existing crystalline α-rh.-boron, Boustani et al. [10] constructed a wavy planar boron sheet of 96 atoms (Fig. 6.4) and calculated its energy and also the energy of the α-rh.-boron unit cell, using first-principles methods. The all-electron calculations were computed at the Hartree–Fock self-consistent field level to determine the total energies of the structures of theory using the standard STO-3G basis set. Due to the large number of degrees of freedom of investigated systems, the geometries of all isomers were optimized under reasonable symmetry constraints, but using analytical gradient methods. In order to compare both energies, the binding energy per atom E_b defined as $E_b = (nE_1 - E_n)/n$, where E_1 is the atomic energy and E_n is the cluster (sheet or α-cell), and n is the number of atoms in cluster. For $n = 96$ is the binding energy per atom $E_b = E_1 - E_{96}/96$. The obtained E_b for the α-rh.-boron unit cell is 4.76 eV, while the E_b for the sheet is 4.82 eV. Thus, the resulting binding energy per atom shows that the unit cells of α-rh.-boron crystals, which is composed of 96 boron atoms, is less stable than the proposed 96 atomic quasi-planar sheet. Consequently, the predicted quasi-planar sheet is a novel material with new properties.

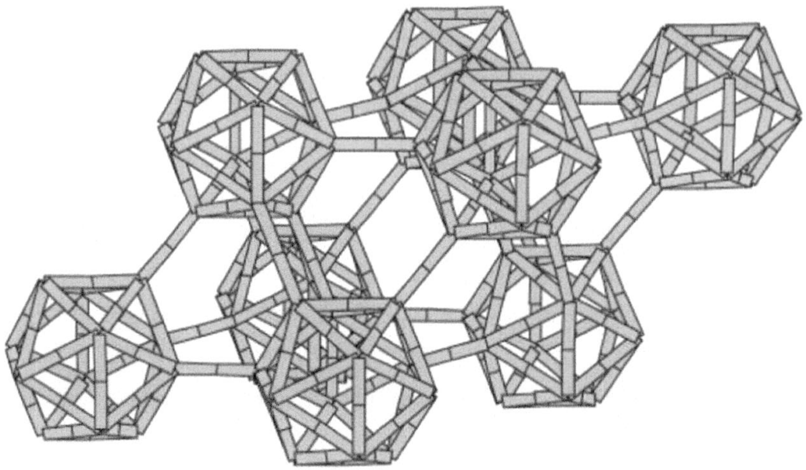

B_{96} (α-Boron) – C_{2h}

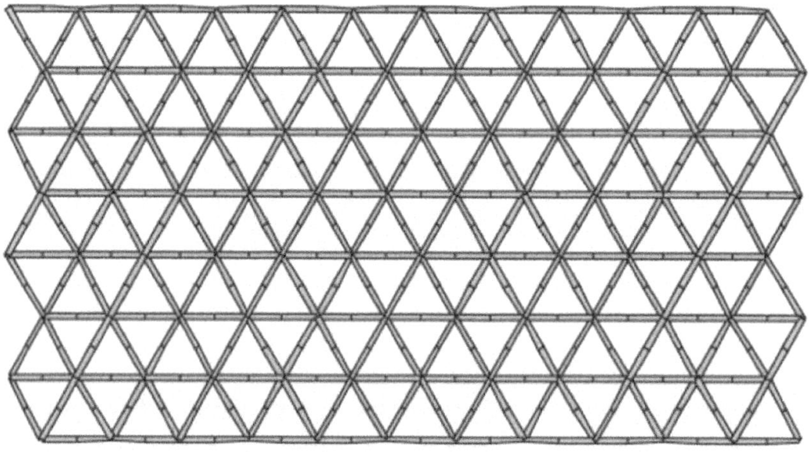

B_{96} – C_i, Top view

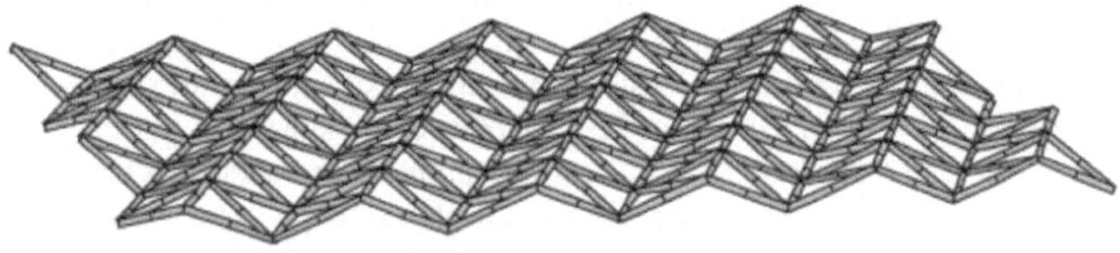

B_{96} – C_i, Side view

Fig. 6.4 The α-Rh. Boron unit cell of α-Boron crystals (top), and a fragment of quasi-planar boron sheet in wavy character, both indicated by the size and point group symmetry. Reference [10]

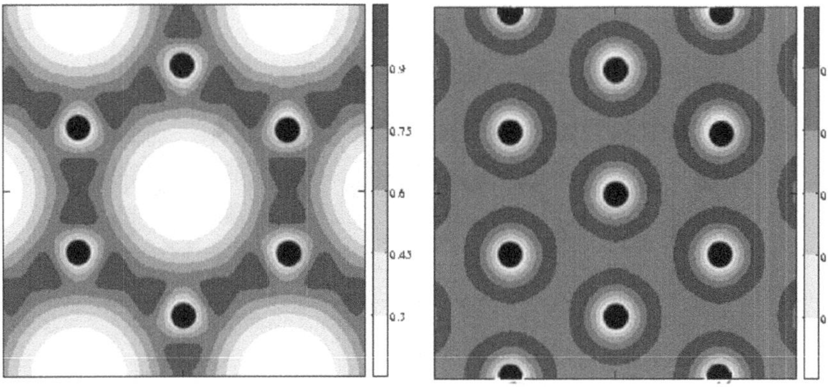

Fig. 6.5 The electron density for boron planes, in electrons/Å^3. The ionic cores are shown as black circles. Density fluctuations near the cores are characteristic of the pseudopotential. **a** The hexagonal boron plane. Note the sp^2 bonding, also found in graphite planes. **b** The triangular boron plane. Outside of the ionic cores, the density is nearly uniform. Reference [11]

Electronic Properties of Triangular Boron Sheet: Based on the density functional theory (DFT) and the local density approximation (LDA), Evans et al. [11] investigated in detail the electronic structure and relative stabilities of the planar and tubular boron structures. They used a plane-wave basis set, and pseudopotentials to represent the ionic cores. The experimental and theoretical parameters for the α-rhombohedral structure were accurately reproduced in the pseudopotentials. For Brillouin zone integration, the K-point meshes were sufficient to converge total energies to 1 mHa per atom. The in-plane electron density for the hexagonal plane in the graphene-like boron structure, as presented in Fig. 6.5a, is assigned by high charge concentrations within the bond regions and by sp^2 hybridization. In comparison with the cohesive energy (CE) of the α-rhombohedral phase 7.37 eV, the CE of the triangular boron sheet 6.53 eV and the CE of the hexagonal boron sheet 5.96 eV are considerably lower. Thus, the CE of triangular sheet is just by 0.84 eV less than that of the boron α-phase. However, the covalent hexagonal plane of boron sheet is not energetically favorable. As shown in Fig. 6.5b, the in-plane electron density for the lowest energy flat triangular plane of boron sheet is in sharp contrast to the electron density of the graphene-like boron sheet, as given in (a). This contrast is namely distinctive and can be expressed by the density, which is distributed nearly uniform between the boron atoms, connected by little covalent bonds. However, the electronic bands are far from being free-electron-like, even though the electron density is nearly homogeneous. Herewith, Evans et al. [11] conclude their results that the triangular planar boron sheet is metallic assigned by a nearly homogeneous electron density and consequently confirm the conductivity of boron sheets predicted by Boustani et al. [4] and illustrated in Fig. 6.3b.

Puckering Properties of Triangular Boron Sheet: Based on numerical ab initio density functional theory and the local-density approximation (LDA), as well

Fig. 6.6 Top view of a quasi-planar boron sheet. In a planar projection the atoms form an almost perfect triangular lattice. The basic structural unit is a hexagonal pyramidal B_7 cluster, as suggested by the Aufbau principle, described in [2]. Reference [12]

as the Perdew-Zunger-Ceperley-Alder exchange-correlation functional, Kunstmann and Quandt [12] studied a structure model for a broad boron sheet (BS) using plane-wave basis sets and a supercell approach to model solid materials, surfaces, or clusters. They used ultrasoft pseudopotential, as supplied by Kresse and Hafner, to represent the ionic cores of the system. Furthermore, the authors used the method of Methfessel and Paxton in first order to estimate the k-space integration and employed a smearing width of around 0.3 eV. During the structural optimization procedure, they considered for all interatomic forces a reduction of less than 0.04 eV/Å. For the relaxation runs they adapted the cutoff energy of around 257.1 eV in order to expand the electronic wave functions in terms of plane-waves, and for the static determination of the total energy they adapted 321.4 eV. However, the structure of the boron sheet as a quasi-planar arrangement of B_7 units, which have a hexagonal pyramidal form, follows the Aufbau principle as predicted by Boustani [2]. When this boron sheet is projected on a plane then we would always have a sheet system with a triangular lattice, as already shown in Fig. 6.6. The out of plane modulation (i.e., the buckling) will be considered after setting up a suitable supercell that allows a systematic generation of various periodic schemes as soon as the ab initio structural optimizations are carried out. The authors adapted some basic buckling schemes from the configurations of the B_{22}, B_{32}, and B_{46} boron cluster, which were provided in [3]. However, the calculations of the buckled triangular supercell, which contains 16 atoms, are repeated periodically. However, the local minimum of the flat boron sheet, as modeled in Fig. 6.7a, has a cohesive energy of about 6.76 eV. The buckled boron sheet, as also modeled in Fig. 6.7b, occupies a local minimum on the energy hypersurface and indicates a cohesive energy of about 6.94 eV. Thus, the model (b) seems to exhibit the largest cohesive energy of all calculated models of boron sheets. This value is by 0.57 eV below the calculated cohesive energy of the well known α-rh.-boron crystal, which amounts to 7.51 eV.

Mechanical Properties of Triangular Boron Sheet: Kunstmann and Quandt [12] investigated also the mechanical properties of model (a) and of the most stable boron sheet model (b), as shown in Fig. 6.7. As already mentioned above, the model (a) of the planar sheet forms a triangular lattice assigned by a single atom in the unit

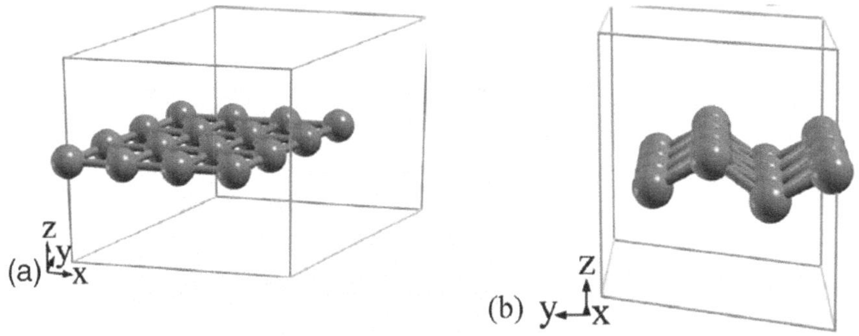

Fig. 6.7 Different models for broad boron sheets. Each supercell (thin lines) contains 16 atoms. **a** A simple flat sheet is metastable. **b** A simple up and down puckering seems to be the most stable modulation. Models **b** is periodic repetitions of structural motives taken from B_{22} clusters, described in [3]. Reference [12]

cell with a only one lattice constant A. This constant A depicts a bond length of about 1.69 Å, typical for boron-boron bonds. Considering a symmetry with a prefect triangular system, the both elastic C_x and C_y Young's modeli are equal and being C_x=C_y ~ 750 GPa, which are surprisingly relative big. The authors determined the elastic modulus of model (b) and found that their Young's modeli strongly depend on the stretching directions. The calculated the elastic Young's modulus of the BS model (b) were found to be 420 and 870 GPa for C_x and C_y, respectively, which is meaning that $C_y \approx 2\ C_x$. The question now arises how to declare the appearance of the anisotropies?. However, an analysis of electron localization function (ELF) shows that there are homogeneous multi-center bonds, which hold the sheet together. As presented in Fig. 6.8, the chemical bonds in the model (b) are lying along the armchair direction, characterized by parallel linear chains with *sp* hybridized strong σ bonds. It is obvious to recognize that the stretching of the BS along along the zigzag (=x) direction, assigned by slightly weaker multi-center bonds, is easily to perform than the stretching along the armchair (=y) direction, which is rather difficult and harder to perform due to the strong σ bonds. The results of Kunstmann and Quandt [12] are in contrast to the results of Evans et al. [11] (Sect. 6.1.5), who assumed that the σ bonds are strong but soft.

Confirmation the Buckled Triangular Boron Sheet: Cabria et al. [13] studied boron nanotubes and sheets and performed first-principles calculations to highlight the relevant structural, electronic, and bonding characteristics of these novel boron materials. They also investigated the chemisorption and physisorption of hydrogen on the surfaces of these boron sheets and nanotubes. They applied the density functional theory (DFT), the local density functional approximation (LDA), and the methods of Vosko, Wilk and Nusair (VWN) and the generalized gradient approximation (GGA) of Perdew and Wang (PW91) for the exchange and correlation functionals, and used the plane-waves and ultrasoft pseudo-potentials. In order to avoid the interactions

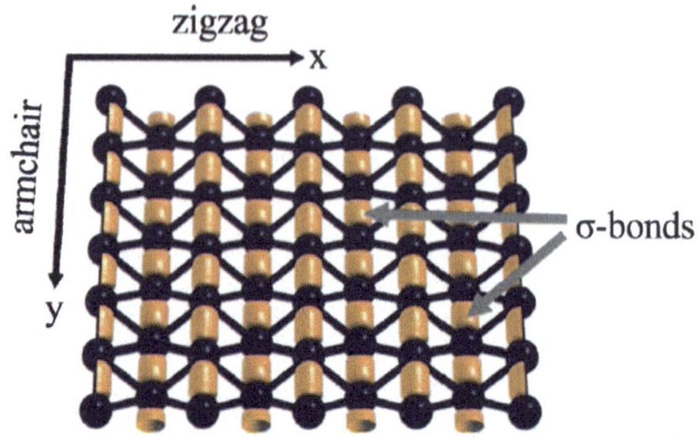

Fig. 6.8 Orange: charge density contours of the boron sheet in model (b) at 0.9 e/Å^3. One observes parallel linear chains of *sp* hybridized σ bonds lying along the armchair direction. Reference [12]

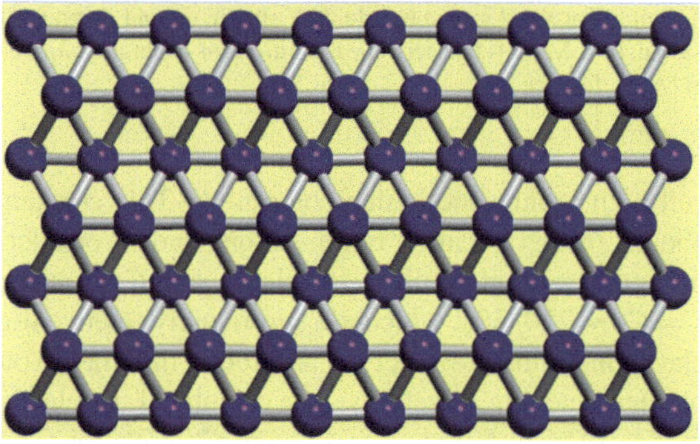

Fig. 6.9 Triangular buckled network of boron atoms. The 2nd, 4th and 6th row are out of the plane. The buckling height (h) is about 0.8 Å

between hydrogen molecules and between sheets or nanotubes of the neighboring cells, they considered the unit cells to be large enough. They used the DACAPO code [14] and considered a cut-off value of around 300 eV, in order to determine the Monkhorst–Pack K-point sampling. The energies of the boron nanotubes and sheets as well as of the adsorption binding energies were obtained by a precision of 0.01 eV/atom and of 1 meV/molecule, respectively. However, They found that the metastable flat sheets are by 0.21 eV/atom less stable than the buckled boron sheets, which have a buckling height (h) with about 0.8 Å, as indicated in Fig. 6.9. The buckled sheet is assigned by a wavy character forming parallel rows with up and down alternating behavior in the surface.

Polymorphism of Boron Sheets: Lau and Pandey [15] performed the electronic and geometric structures of various pattern of boron sheets with different morphologies. They applied the density functional theory (DFT) and the exchange correlation functionals of Perdew-Wang (PW91) using a plane-wave basis set. They also used ultrasoft pseudopotential (US-PP) to describe the valence-core interaction. They exerted the Methfessel and Paxton method in the first order for the k-space integrations employing smearing width of 0.05 eV for both the cell parameters and atomic positions optimization. For the plane-wave expansion they considered an energy cut-off of 260 eV and for the augmented charge they used a cutoff of about 443 eV. With a

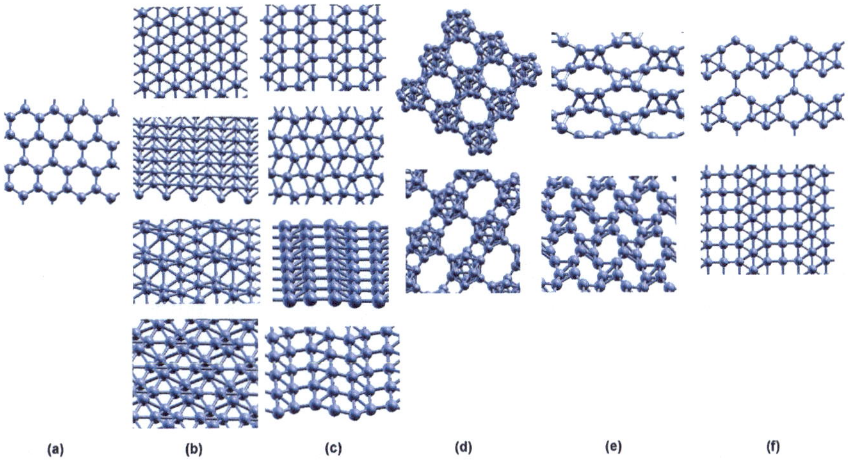

Fig. 6.10 Boron sheets in hexagonal graphene-like, idealized flat triangular, buckled triangular structures, and icosahedral sheet composed of B_{12} icosahedra. This figure was reproduced from the Journal of Physical Chemistry with kind permission of the American Chemical Society, [15]

precision of 5 meV/atom are the sizes of the k-point sampling converged individually namely for the different systems with related unique cells. The total energy of each optimized structure was determined using a cutoff energy value of around 320 eV in the plane-wave expansion, as well as the tetrahedron method including the Blöchl corrections. The authors constructed a supercell for each monolayer boron sheet as a rectangular grid putting inside a basic unit of the sheet in the xy-plane. The separation between the sheet and its image was given by ~10 Å in the z-direction, in order to avoid or to minimize the interaction between the surfaces. Depending on the proposed configuration of the sheet, the constructed basic unit was composed of 8–12 boron atoms which generates the considered sheet configuration by a periodical repetition. For the integration in the reciprocal space, they used a 8 X 8 X 8 Monkhorst–Pack grid to sample the Brillouin zone. The convergence of the calculations is considered when the interatomic forces were less than 0.01 eV/Å and the changes in the total energy were less than 10^{-5} eV. The authors determined the binding energy per atom (BE) for all boron sheets, which are presented in Fig. 6.10. The EB of the hexagonal graphene-like boron sheet is about 4.96 eV/atom, of the flat triangular sheet is 5.48 eV/atom, of the 2D icosahedral network is about 5.60 eV/atom, and of the buckled sheet is 5.70 eV/atom. Thus, the most stable structure is the buckled triangular sheet, while the second lowest isomer is the icosahedral network sheet.

Prediction of α-Sheets With Hexagonal Holes: Tang and Ismail-Beigi [16] performed first-principles calculations based on density functional theory (DFT), the local density approximation (LDA), and the generalized gradient approximation (GGA) for exchange and correlation. They used the ab initio supercell plane-wave pseudopotential total energy approach and determined the structures, energetics, electronic states of various boron sheet structures and provided a clear picture about the nature of the chemical bonding in these sheets. They presented a new class of boron sheets that are more stable than till then accepted ones. Most of the obtained results were employed and verified by the LDA and GGA methods, respectively. However, both LDA and GGA methods produce results, qualitatively similar, but quantitatively with minor variances. The authors considered the cutoff energy of 32 Ryd for the plane-wave basis, and 1 meV/atom threshold for the K-point samplings convergence of the total energies. Both cutoff radii $r_c^s = 1.7$ and $r_c^p = 2.1$ a.u. were considered for the normconserving pseudopotentials. For the boron sheets, extended in the xy-plane, a separation of around 7.4 Å was taken along the z direction between the sheets and their periodic images, in order to minimize the interaction between the surfaces and being sufficient for the convergence. The structural optimization procedure was carried out until the values of the Hellmann–Feynman forces were less than 1 meV/Å, and all in-plane stresses values approach below 5 MPa.

Three-Center Bond in α-Sheets: As shown in Fig. 6.12, Tang and Ismail-Beigi [16] proposed for the stability of the new flat triangular boron sheets a three-center bonding scheme with crucial implications. They asserted that creation of hexagonal holes in the sheets, induced by removing systemically boron atoms, improves the stability of these new boron sheets. They further explained that the stabilization holes effect occurs as soon as the Dirac point and the Fermi level are crossed. Indeed, the valence and conduction bands π and π^* are meeting at the Dirac points, as presented in Fig. 6.24c, Sect. 6.1.5. They divided the chemical bonds into σ in-plane bonds and π out-of-plane bonds. Those in-plane bonds are sp^2 hybrids contain bonding and antibonding orbitals, as shown in Fig. 6.12. However, as presented in Fig. 6.11, the in-plane PDOS of the fourth structure, assigned by α-sheet, are zero at the Fermi energy level E_F, at which the bonding and antibonding states are clearly separated. The α-sheet would be ideally the most stable structure, if the valence bands characterized by b-derived in-plane boding are fully occupied and the conduction bands characterized by a*-derived in-plane antibonding are total empty. In other words, the ideal stability can be realized if some features were fulfilled like (i) when each atom provides two electrons to fill the valence bands with b-derived in-plane bonding character, (ii) the conduction bands with a*-derived antibonding character were empty, and finally (iii) if each atom would provide an electron to occupy and fill the low-energy the p_z-derived bonding bands. That means the E_F should separate the bonding and antibonding bands of the in-plane PDOS by zero. Therefore, the highest stability of the mixture between the centered hexagons and hexagonal holes can be achieved when the Fermi level E_F is precisely placed at zero-point of the in-plane PDOS and occupying all in-plane bonding states but none of the antibonding states. The low-energy p_z-derived states should be filled by the remaining electrons in order to establish a system with metallic character.

The Density of Hexagonal Holes in α-Sheets: In addition, the authors [16] calculated the binding energy per atom of the buckled and flat triangular boron sheets as well as the α- and β-sheets. They found that the buckled triangular sheet is by 0.12 eV/atom less stable than the α-sheet and by 0.08 eV/atom less stable than the β-sheet. The α-sheet can be produced directly from the flat triangular sheet by removing the

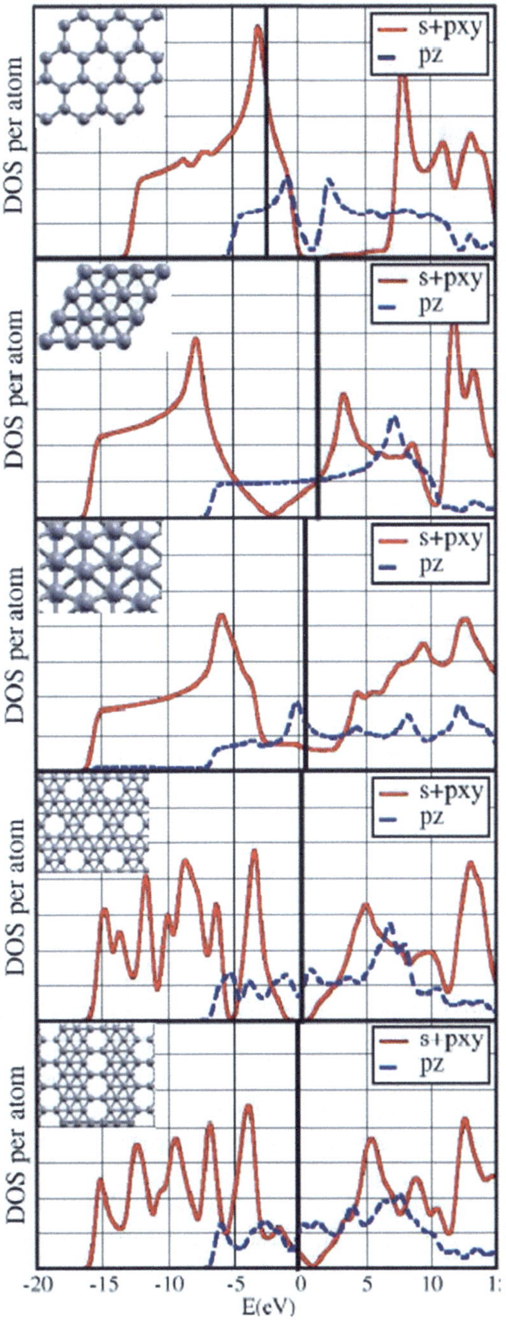

Fig. 6.11 Projected Densities of States (PDOS) for five BS: graphene-like, flat and buckled triangular, α and β sheets. Projections are onto in-plane (sum of s, p_x and p_y, solid red) and out-of-plane orbitals (p_z, dashed blue). Thick vertical solid lines show the Fermi energy E_F. This figure was reproduced with kind permission of the Physical Review Letters of the American Physical Society, [16]

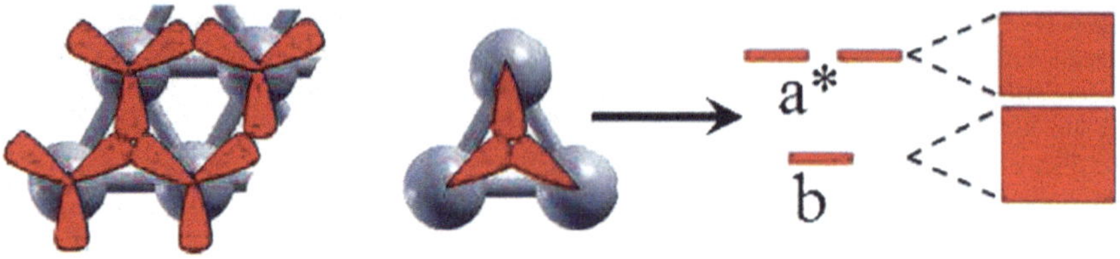

Fig. 6.12 Three-center bonding scheme in flat triangular sheets. Left: orientation of sp^2 hybrids. Center and right: overlapping hybrids within a triangle (D_3 symmetry) yield one bonding (b) and two antibonding (a*) orbitals. These then broaden into bands due to inter-triangle interactions. This figure was reproduced with kind permission of the Physical Review Letters of the American Physical Society, [16]

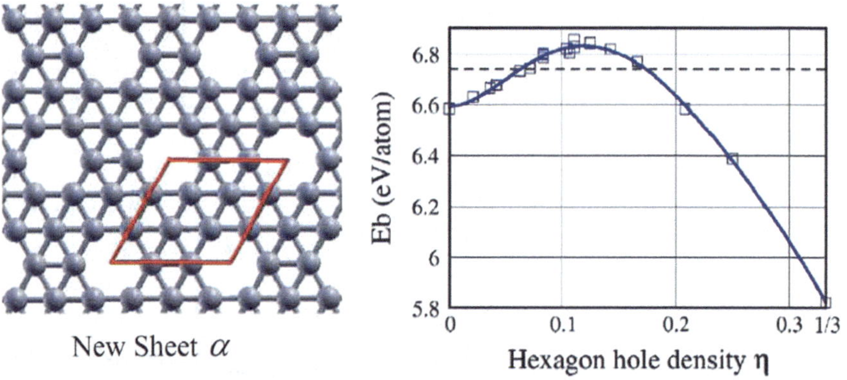

Fig. 6.13 The proposed α-sheet, and the LDA binding energy per atom (E_b) versus the hexagon hole density η for sheets with evenly distributed hexagons. The dashed line shows E_b for buckled triangular sheet. The solid blue curve is a polynomial fit. The two limiting cases $\eta = 0$ and $\eta = 1/3$ correspond to the flat triangular and hexagonal sheets, respectively. Maximum E_b occurs for α-sheet ($\eta = 1/9$). This figure was reproduced with kind permission of the Physical Review Letters of the American Physical Society, [16]

central atoms of the hexagons, creating hexagonal holes, and constituting a mixture of centered hexagons and hexagonal holes as well as triangles. Since the removal of atoms and creation of holes is arbitrary, the authors proposed the hexagonal density η to distinguish the boron sheets decorated with hexagonal holes. This η is the ratio of the number of holes over the number of atoms of the triangular sheet (η = Nr. of holes/Nr. of atom). For example, η is equal to 0 for the triangular sheet (zero holes), $\eta = 1/3$ for the hexagonal graphene-like sheet, $\eta = 1/9$ for the α-sheet, and $\eta = 1/7$ for the β-sheet. The authors calculated the binding energy per atom of different sheets as a function of the hexagonal hole density and presented in Fig. 6.13. They found that only one hexagonal hole inside nine triangular atoms leads to the most stable α-sheet.

Competition Between α- and γ-Sheets: Based on ab initio quantum chemical and density functional methods, Özdogan et al. [17] investigated different structures of boron sheets and nanotubes. For the ground state non-periodic systems, an optimization procedure was carried out for closed shell systems using the restricted Hartree–Fock (RHF) procedure, and for open shell systems using the unrestricted Hartree–Fock (UHF) one. The hybrid functionals (B3LYP) of Becke (exchange) and of Lee, Yang, and Parr (correlation) ([27, 32], Chap. 1) was used for the gradient-corrected exchange correlation energy. For the ground state periodic systems the DFT methods were used within plane-wave basis sets as well as a supercell approach to sample clusters, surfaces, and other solid materials. For the calculations of the 2D monolayered boron sheets, a large vacuum distance of about 10 Å in z-direction was considered, in order to avoid the interactions between the layers. Beside the α-rh. boron unit cell, three boron sheets were considered in the calculations. The first one, proposed by Tang and Ismail-Beigi [16], is the α-sheet, the other two, proposed by Özdogan et al. [17], are the γ-sheet, decorated with parallel hexagonal holes, and the triangular buckled sheet, as presented in Fig. 6.14. It seems that the modalities of the calculations affect the competition between these three monolayered boron systems. On the one side, as free-standing systems, the three boron sheets were calculated as clusters composed of 96 atoms, the size of α-rhombohedral boron unit cell. The free-standing B_{96}-α-rh. boron unit cell was considered for comparison. The obtained binding energies per atom were found to be 6.366, 6.328, 6.385, and 6.429 eV for α-rh. boron unit cell, α-sheet, γ-sheet, and triangular buckled sheet, respectively. On the other side, as periodic systems, the calculated binding energies per atom of the α-rh. unit cell and of the three planar sheets: α-, γ-, and triangular-sheets, were found to be 7.660, 6.300, 6.259, and 6.197 eV, respectively. As free-standing systems, the most stable structure is the triangular buckled sheet, followed by γ-sheet. But as periodic systems, α-rh. boron lattice is obviously the most stable structure, followed by α-sheet, γ-sheet, and triangular buckled sheet. To summarize, if the calculations of the boron nanosheets were carried out as periodic systems, then after the α-rh. boron lattice is α-sheet the most stable infinite single-layered planar sheet. In contrast, if the calculations of the boron nanosheets were carried out as free-standing noninteracting clusters, then the triangular buckled-sheet is more stable structure favored over the α-, and γ-sheets.

Electron Counting of B_7 Unit in α-Sheets: Galeev et al. [18] investigated the nature of the chemical bonding in the α-sheet, which was proposed few years ago by Tang and Ismail-Beigi [16]. Galeev et al. [18] tried via the AdNDP chemical bonding analysis of the model fragments, to decipher the mystery of the hexagonal holes in this α-sheet and proposed for an infinite all-boron α-sheet the following chemical bonding picture depicted in Fig. 6.15b. They found that in each filled hexagon, B_7, there are three triangles with 3c-2e σ-bonds (solid small triangles), while the other three triangles with 4c-2e σ-bonds (solid rhombi). The 3c-2e σ-bonds are bordering upon the holes, while the 4c-2e σ-bonds are the junction of two filled hexagons. In addition, there are delocalized 6c-2e π-bonds (circles) over the hexagonal holes, and 7c-2e π-bonds (circles) over the filled hexagons [19]. According to the model of fragments, each B_7 fragment provides six and rather three valence electrons arising from the three 3c-2e and 4c-2e σ-bonds, respectively. In addition, two further valence

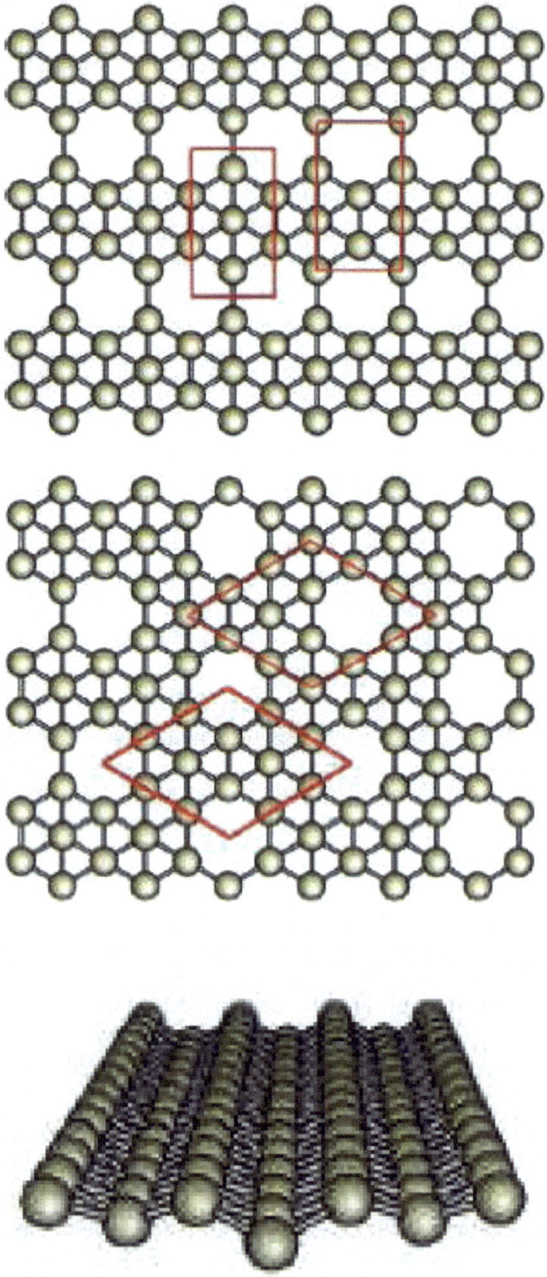

Fig. 6.14 The proposed γ-sheet (top), α-sheet, and triangular buckled sheet as free-standing B_{96} clusters. This figure was reproduced from the Journal of Physical Chemistry with kind permission of the American Chemical Society, [17]

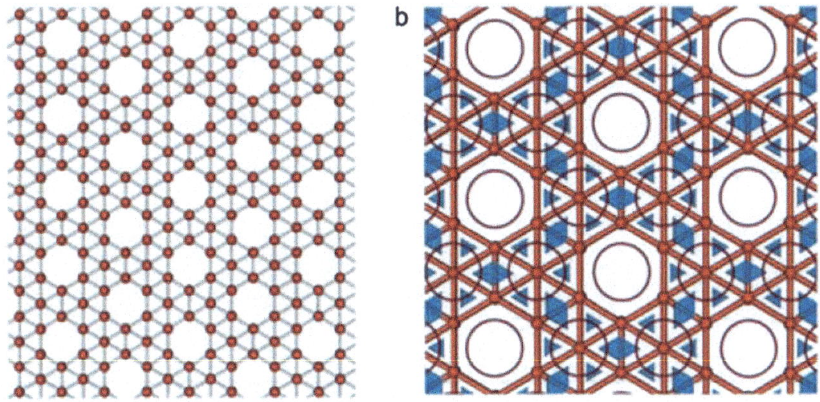

Fig. 6.15 **a** Geometric structure of the all-boron α-sheet. **b** The proposed bonding pattern for the all-boron a-sheet: 3c-2e σ-bonds (solid triangles), 4c-2e σ-bonds (solid rhombi) and delocalized π-bonds (circles). This figure was reproduced from the Physical Chemistry Chemical Physics with kind permission of the Royal Society of Chemistry, [18]

electrons are coming from the 7c-2e π-bond and herewith achieving eleven electrons, the total number of valence electrons in each fragment. However, one can calculate the total number of valence electrons for a filled hexagon being as a part of solid lattice as follows: when each of the six-membered ring-atoms provides the half of its valence electrons, then it reveals a total number of nine electrons, on the one hand. On the other hand, the boron atom at the center of the hexagon supplies its three valence electrons receiving totally twelve valence electrons from each filled hexagon.

The Electronic Donor Acceptor in α-Sheets: The electron counting in the model fragments above gives one extra valence electron in each filled hexagon, which is in fact not involved in the above presented bonding. In the lattice picture, this extra electron on the filled hexagon, which can be considered as an electronic donor, is shared by the surrounding three hexagonal holes, which in turn can be considered as three electronic acceptors. This means that each of these three acceptors (holes) becomes 1/3 electron. Thus, since each of the six filled hexagons provides the encircled hole with 1/3 electron then this hole becomes two extra valence electrons. As can be seen in Fig. 6.15b, the two extra electrons are shown as circles over hexagon acceptors or holes representing the 6c-2e π-bonds [19]. Illustratively, one can describe the electron deficient hexagonal holes as thieves or scavengers which steal the extra electrons from the neighboring centered hexagons.

Polymorphism of 2D Boron Sheets: In the latter sections we have experienced that creating hexagonal holes in planar triangular boron sheets improves their stability and produces a pattern composed of triangles and holes and the α-, and β-sheets are some examples. However, the combination of triangular boron sheets with hexagonal holes leads to a huge number of configurations and pattern. Therefore, the direct

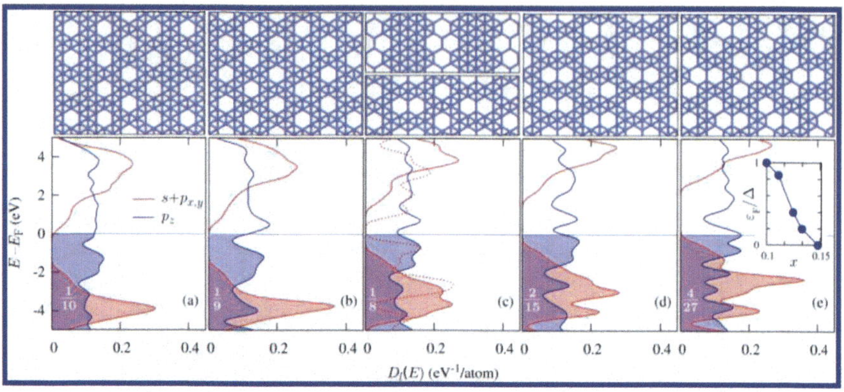

Fig. 6.16 The pattern and the projected density of states $D_l(E)$ of some most stable sheets in the range of $x \simeq 0.1 - 0.15$. The familiar and famous α-sheet with x = 1/9 is the second pattern. This figure was reproduced from the Nanoletters with kind permission of the American Chemical Society, [20]

use of the first-principles calculation methods will be rather complicated due to the huge combinatorial possibilities of decorated structures. New methods were recently developed by some researchers like Penev et al. [20], Oganov et al. [21], and Wang et al. [22] to overcome this problem, as reported by Li et al. [23]. Penev et al. [20] used the cluster expansion (CE) method in combination with the first-principles calculations and succeeded to solved this problem by exploring the configuration space of 2D boron sheets, thoroughly and systematically. The authors considered the 2D boron sheet, composed of closed-pack triangles and vacant hexagons, as pseudoalloy. They calculated different decorated 2D boron sheets and obtained some boron polymorph, as shown in Fig. 6.16, exhibiting cohesive energies of nearly identical values. These with hexagonal holes decorated sheets are holding vacancy concentrations in a narrow range of 10–15%. Therefore, the variety of decorations in the 2D boron sheets with multiple phases is unprecedented in all other known 2D materials.

Searching the Global Minimum of Boron Sheet: As mentioned above, Penev et al. [20] succeeded to predict several 2D boron sheets with different pattern and close energies and the determination of the ground state of these 2D boron sheets with their various structures was definitely a challenge [23]. Thus, beside the cluster expansion methods used by Penev et al. [20], powerful tools and an efficient structural optimization methods are needed to search for the ground states global minima of the complex 2D boron sheets. Thereupon, a new method assigned as ab initio evolutionary algorithm (USPEX) was developed by Oganov et al. [21]. Furthermore, ab initio global minimum particle-swarm optimization (PSO) method was firstly developed and used by Xiaojun Wu et al. [24] for exploring the 2D boron sheets. For the prediction of stable 2D structures under given conditions, Wang et al. [22] designed and proposed the crystal analysis particle-swarm optimization technique called (CALYPSO). However, Xiaojun Wu et al. [24] determined different structures of boron sheets with various hexagonal pattern and confirmed that the low-energy α-sheet with the hexagonal hole density $\eta = 1/9$, as proposed by Tang and Ismail-Beigi [16], is among the obtained structures. However, beside the α-sheet, they predicted three low-lying structures of boron sheets: one α-sheet with buckling feature (assigned by α'-sheet), and two boron sheets with different motives called α_1 and β_1, as presented in Fig. 6.17. According to the PBE0 functionals, the cohesive energy per atom (CE) 5.619, 5.718, and 5.678 eV, respectively, of the predicted α', α_1, and β_1 sheets are larger than that of the original α-sheet 5.582 eV. In addition, these three predicted α', α_1, and β_1 sheets are dynamically stable in contrast to the α-sheet. As shown in Fig. 6.18b, the phonon spectrum of the α-sheet has some imaginary frequencies (at the bottom below the Fermi level) and therefore is α-sheet dynamically unstable. These negative phonon frequencies appeared in the Brillouin zone, directly below the Fermi energy at point zero. Thus, the largest negative frequency is located around the M(0.5, 0.0, 0.0) k-point and amounts to about ($-72\,\text{cm}^{-1}$). Thus, the negative frequency eigenvectors are in fact vibrations with an out-of-plane bending characters. In order to remove these negative frequencies, the α-sheet underwent a structural relaxation and becomes afterward a slightly buckled structure, which is named the α'-sheet. This new α'-sheet can be described as being assigned by two adjacent boron atoms moving inward and outward from the plane with about 0.17 Å in vertical distance from the plane. This behavior of these buckled atoms is similar to atoms in the quasi-planar triangular boron sheet (Fig. 6.2, Sect. 6.1.5) and proposed by Boustani [3], which exhibits a buckled character with dovetailed hexagons with up and down centers. All of these outward and downward atoms in the α'-sheet and triangular sheets are possessing coordination number of six. Finally, according to band structures calculations, determined at the PBE0 level, the α-sheet is semiconductor, while the dynamically stable α_1-sheet is conductor or namely metallic.

Global Searches of Sheets with Twin-Hexagonal Holes: Based on density functional theory and ab initio global searches, H. Lu et al. [25] investigated 2D monolayer boron sheets (BSs) using in conjunction with the Perdew–Burke–Ernzerhof (PBE) and the hybrid PBE0 functionals the projector augmented wave (PAW) method. They also used the FINDSYS module of ISOTROPY package to determine the selection of supercells and to convert the unit cell to its primitive cell by symmetrization. The authors found that all boron sheets with low-lying energies are exhibiting unit cells decorated with 1 to 4 hexagonal holes. They asserted that the most stable boron sheets possess binary structures in nature due to their unique compositions assigned either by isolated-hexagon holes or by twin-hexagon holes. However, due to an extensive structural analysis, two types of the close-lying most stable single-layered boron sheets were indicated exposing a novel polymorphism of boron sheets. The first one is the type-I with the $\eta_{1/9}$-, and $\eta_{2/15}$-units, the second one is the type-II with the $\eta_{1/8}$-, and $\eta_{2/14}$-units. Both types I and II, as presented in Fig. 6.19, are pieces of evidence for binary nature of boron sheets. At the PBE0 level, there are two low-lying nearly degenerate boron sheets, $\eta_{4/28}$-, and $\eta_{4/33}$-sheets, possessing cohesive energies per atom of 5.987 and 5.986 eV, respectively. The first most stable sheet, indicated by the $\eta_{4/28}$-unit, possesses the highest hexagonal hole density of $\eta = 1/7$. As can be seen

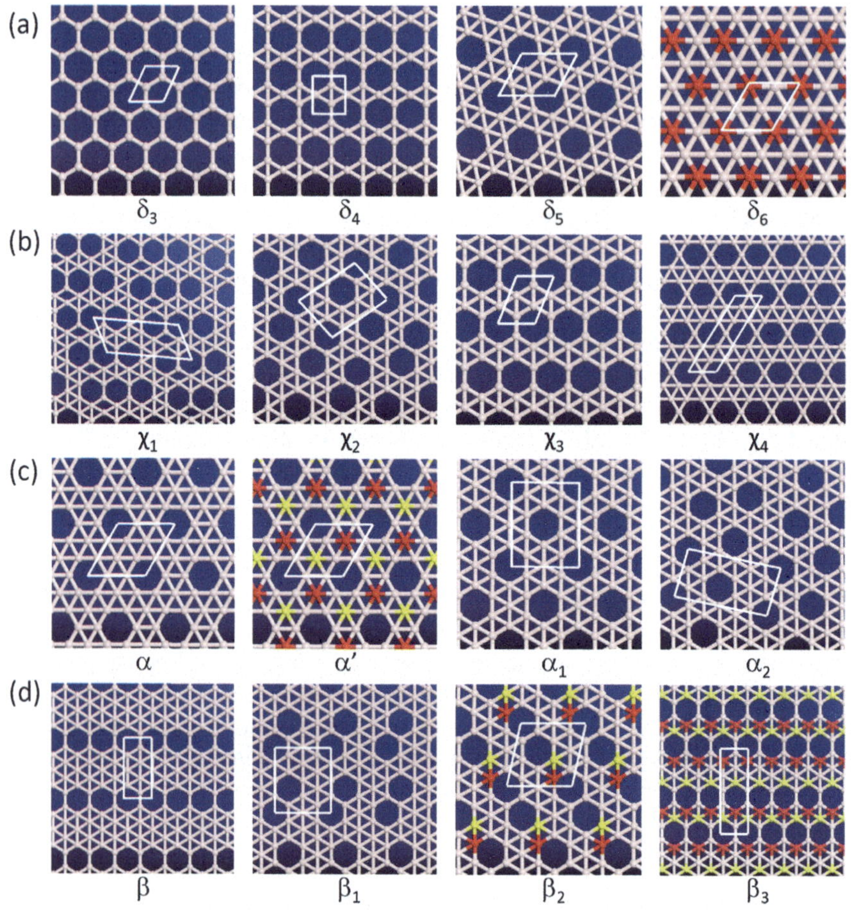

Fig. 6.17 Top view of various low-lying energy structures of **a** δ-, **b** χ-, **c** α-, and **d** β-type monolayered sheets. The cohesive energies of α, α', α_1, and β_1 sheets are respectively 5.58, 5.62, 5.71 and 6.68 eV, obtained at PBE0 functional. Red and yellow balls denote boron atoms moving outward or inward from the plane, resulting in buckled boron sheets. This figure was reproduced from the ACS Nano with kind permission of the American Chemical Society, [24]

in Fig. 6.19b structure (5), the $\eta_{4/28}$-sheet is consisted of planar D_{2h} B_{12} units (two joined filled hexagons) connected with two twin-hexagonal holes. The second most stable sheet, indicated by the η4/33-unit, as shown in Fig. 6.19a structure (4), has a hexagonal hole density of $\eta = 1/9$, and is composed of three fused filled hexagons and one twin-hexagonal holes and two isolated hexagonal holes.

Swarm Optimization Algorithm of Boron Sheet: Based on first-principles methods and particle swarm optimization algorithm (PSO), Xiao Yu et al. [26] investigated the structures of 2D boron sheets. In comparison with the α-sheet, many metastable boron sheets are exhibiting an energy difference less than 0.1 eV/atom. The predicted